The Hatfield SCT Lunar Atlas

Anthony Charles Cook

The Hatfield SCT Lunar Atlas

Photographic Atlas for Meade, Celestron, and Other SCT Telescopes

A Digitally Re-Mastered Edition

Springer

Anthony Charles Cook
Newtown, UK

ISBN 978-1-4939-3826-1 ISBN 978-1-4614-8639-8 (eBook)
DOI 10.1007/978-1-4614-8639-8
Springer New York Heidelberg Dordrecht London

Softcover reprint of the hardcover 2nd edition 2014

Printed on acid-free paper

Springer is part of Springer Science+Business Media (www.springer.com)

This book is dedicated to Sir Patrick Moore (1923–2012)
for inspiring generations of astronomers and space scientists.

About the Editor

Anthony Charles Cook, Ph.D., FBIS, FRAS, is both an amateur astronomer and a professional lunar and planetary cartographer. He has worked with many leading planetary cartography research groups in the UK, Germany, and the USA. He produced a planet-wide digital elevation model (DEM) of the Moon in the year 2000, and this was used by the European Space Agency to guide its SMART-1 spacecraft down to a precise impact. Six previously unknown impact basins larger than 300-km across were discovered using this lunar DEM. Presently, he is a lecturer at the Institute of Mathematics and Physics, Aberystwyth University, in the UK. He is Assistant Director of the British Astronomical Lunar Section and coordinates the Transient Lunar Phenomena observing program alongside a similar program for the Association of Lunar and Planetary Observers. He has also had Digital Planetary Cartography papers published in several refereed scientific journals.

Contents

Introduction

1

Why a New Version of This Atlas?

Past versions of this atlas have inspired many amateur astronomers to take up the observing and studying of the Moon, including the present editor. When Commander Henry Hatfield devised the atlas it consisted of a straightforward 4×4 chart format covering the Moon's nearside, with photos at a variety of different lighting conditions. The atlas was ideal for taking outside to use at the telescope because it was neither too simplified nor too hefty, nor over detailed for use under a flashlight. This made it a very practical tool for finding and identifying craters and other features on the Moon while working at the telescope – or indeed, later on indoors while examining sketches, photographs, digital images, or video.

After Commander Hatfield's first edition of the Hatfield Lunar Atlas, later revised editions were made by my father and included new crater names, the transition from imperial to metric units, and a change in coordinate systems from classical (east towards Grimaldi and west towards Mare Crisium) to the modern day International Astronomical Union (IAU) definitions (west towards Grimaldi, and east towards Mare Crisium).

With the increasing use of amateur telescopes having a more compact light path than the conventional Newtonian, a further edition of this lunar atlas was eventually produced because many of these compact telescopes have their usual viewing position at right angles to the telescope optical axis, allowing greater comfort when observing but resulting in an image reversal. This change in direction of the optical path has the effect for northern hemisphere observers of reversing south–north so that north now appears at the top of the image but leaving the positions of east and west unchanged. The result is that the image in the eyepiece now appears with lunar north at the top and south at the bottom, but Mare Crisium (east IAU) remains to the left with Grimaldi (west IAU) to the right, which is contrary to the convention. Note however that viewing from the rear of such a telescope along its optical axis still produces a conventional astronomical telescope view.

Many past editions of this atlas are undoubtedly now showing some wear and tear from excessive use, but at least this demonstrates how useful they have been. Now the opportunity has arisen to produce a revised edition, and this has allowed the use of digital augmentation techniques to illustrate lunar surface detail beyond which Commander Henry Hatfield would have been able to photograph with his 30 cm (12-in.) Newtonian telescope during the 1960s.

Since the first edition was published in 1968, 12 astronauts have set foot upon the Moon, and in the subsequent four and a half decades, we now have spacecraft images that cover many parts of the lunar surface with sufficient resolution to make out the Apollo lunar modules, rovers, and even astronaut footpaths. The Moon has been mapped in three dimensions and in many different colors and wavebands (including infrared and ultraviolet). Its weak surface magnetism and gravity have also been charted. Many in the lunar science community were surprised to learn from NASA's LCROSS spacecraft that ices exist in some permanently shadowed areas of the lunar poles, and India's *Chandrayan-1* mission found evidence for tiny traces of water bound to minerals at lower latitudes on the Moon. Clearly both lunar science and mapping have come a long way since the first edition of this atlas, but there still remains much we do not know about our Moon.

Now, anyone considering starting to observe the Moon may wonder, apart from enjoying and learning one's way around the Earth-facing lunar surface, what science is there left for amateurs to do? Admittedly the scope of new science for the amateur astronomer has become more limited over the decades since Apollo, but surprisingly it is still possible to discover previously unrecognized structures and markings on the Moon in telescopic images when looking near the sunrise/sunset terminator. This is because, despite having very high resolution spacecraft imagery, much of this is at local noon, when shadows are at their shortest, and so low-lying structures, such as domes and rilles, may not always show up. Even when spacecraft images are taken at more shallow Sun angles, the choice of imagery at different illuminations is more limited than what amateur astronomers can achieve.

In addition, telescopic studies of the Moon have, in the past, proven useful in finding very occasional errors in the official Lunar Aeronautical Chart series of maps produced during the Apollo era. It is also possible to re-observe under the same illumination conditions that match those

A.C. Cook, *The Hatfield SCT Lunar Atlas: Photographic Atlas for Meade, Celestron, and Other SCT Telescopes: A Digitally Re-Mastered Edition*, DOI 10.1007/978-1-4614-8639-8_1, © Springer Science+Business Media New York 2014

of past historical sketches so as to make comparisons. Electronic cameras and filter wheels let us detect colors, and by using combinations of filters lunar minerals can be mapped. Very light sensitive CCTV cameras have been used to look for faint impact flashes from meteorites striking the Moon in Earthshine. Observations of occultations of stars disappearing and reappearing on the Moon's limb can be used as an independent check on our existing topographic maps of the limb regions, and sometimes these tell us if the stars are double or multiple because the starlight would fade in short steps.

The darkness of lunar eclipses and the brightness of Earthshine have been shown to provide useful information on the state of our upper atmosphere and also the reflectivity of global cloud cover. This information may be helpful in extrapolating atmospheric data back to earlier eras when lunar eclipses and Earthshine were observed, but when meteorological satellites did not exist.

Finally there are occasional, alleged observations of short-term glows, colors, or obscurations seen against the lunar surface. These Transient Lunar Phenomena (TLP), also known as Lunar Transient Phenomena (LTP), can often be explained away as atmospheric or instrument effects, or even misinterpretations of the lunar surface. Most lunar scientists think that the Moon's surface is geologically inactive apart from small quakes and meteorite impacts. Only a very small number of TLP have ever been independently confirmed or detected with certainty with modern imaging instruments. NASA's Lunar Atmosphere and Dust Environment Explorer (LADEE) may help to eventually confirm or disprove at least some of the explanations for TLP.

So, what improvements have been made in this new edition of the *Hatfield SCT Lunar Atlas?* First, although the atlas photographs taken by Commander Hatfield were state of the art probably until the mid-1980s, the advent of CCD cameras, image stacking software, and sharpening techniques have moved the goal posts considerably in terms of image resolution. Modern amateur astronomer images of the Moon can be nearly diffraction limited, the ultimate resolution only restricted by the aperture size of the telescope. To cater for this, while retaining the familiar appearance of the original book, scanned copies of the Hatfield photographs have undergone a process of digital augmentation, merging a virtual high-resolution view of the lunar surface with the 1960's era photographs. Secondly the charts of lunar feature names have been updated in line with a selection of replacement and new names ratified by the IAU. Computer visualization simulations for sunrise and sunset at 32 different locations have been added. Finally, a lunar observing methods section has been included to assist beginners, and even some more advanced astronomers, with observing ideas.

The Maps: Overlaps, Scales, Grids

This atlas is aimed primarily at northern hemisphere visual observers using a catadioptric telescope equipped with a right-angled star diagonal eyepiece. The image seen with such telescopes will be a mirror image, usually with north at the top, south at the bottom, east on the left, and west on the right; consequently all of the maps and plates contained in this SCT atlas are printed with this same orientation in mind. A sister atlas, *The Hatfield Lunar Atlas: Digitally Re-Mastered Edition*, is available for visual observers using traditional Newtonian, refractor telescopes, or SCT users using cameras without the use of a star diagonal.

The atlas is divided into 16 sections, each of which is made up of a map and several photographic plates. Each map is based primarily on the facing plate. Feature heights are given in meters (m) and distances in kilometers (km). The section Index of Named formations (see Appendix 3) includes many cross references to aid with the location of the most appropriate map.

No attempt has been made to adhere rigidly to the boundaries of the various numbered sections in the Key to Maps and Plates. Indeed, in some cases the boundaries of the plates have purposely been allowed to encroach into neighboring sections. The Key Plate is intended to guide the reader into the right area. The Index of Named Formations lists every map on which a particular feature will be found. The scale of each map, and of the main plates that accompany it, have been adjusted so that the Moon's diameter is nominally 64 cm (any variation in this diameter is stated in the relevant caption).

Despite this, the true scale (the relationship between the diameter of a crater on the Moon and its diameter in this atlas) of each map and plate varies from place to place; furthermore, the north–south scale may well vary in a different way from the east–west scale. This is part of the very nature of the orthographic projection, in which the observer views a globe from a great distance.

In order to give the reader some idea of the true scale from place to place, the diameters of five craters have been noted beneath each map. As the scale varies from plate to plate, so the size and shape of the various photographic images will also vary. The grids on the various maps are intended to be used only for reference purposes. They bear no relationship to lunar latitude and longitude, or to any of the cartographic grids that are in use at the present time. If a formation appears on more than one map, then its grid references on each map will almost certainly vary. An unnamed and unlettered formation on any map may be identified with certainty by quoting the map number and grid reference, e.g., Map 3 Square a4, and then making a small tracing of the square showing the formation.

Nomenclature

Feature names, coordinates, and sizes have been sourced from the official IAU web site on lunar nomenclature, maintained by the U. S. Geological Survey (USGS). There are also some single letters for many craters, and these have yet to be given proper names and have come from the NASA Reference Publication 1097 *NASA Catalogue of Lunar Nomenclature* by Leif E. Andersson and Ewen A. Whitaker (1982). These single-letter names refer to satellite craters, or smaller craters that lie close to larger named craters. A couple of examples of satellite craters are "Plato B" and "Plato C," shown on Map 6 at grid Square c6. There was previously a "Plato A" in the same grid, but the IAU has since changed the name to "Bliss." It may be that eventually all lettered craters will be given proper names, but this will be a long ongoing process.

You will find that although the majority of the official IAU names have been used, it has been necessary to leave some of them out to avoid overcrowding the map, and anyway a few of the named craters would be too small to be seen easily through a telescope. There are, however, also a very few features that are not contained in the NASA catalog, or IAU list, and these have been allowed to retain their earlier names for historical reasons. These names are shown enclosed by brackets on the maps in this book and identified in the Index of Named Formations by an asterisk. An example of an old name is the "Stagg's Horn Mountains," which lies on the southern end of "Rupes Recta."

The Libration Keys

Beneath each main plate there is a small key that shows the numbered area in which the plate lies and the Moon's optical libration when the photograph was taken. The Moon does not present exactly the same face to Earth all the time, but rocks gently back and forth in all directions, so that at one time or another an observer on Earth will see about 60 % of its surface. Referring to these libration keys, the reader should imagine that the Moon has rotated in the direction of the arrow by the amount indicated. Thus in Plate la the movement is 7.3° in a direction a little east of north, and therefore an area just west of the south pole has been exposed, which could not be seen if the Moon were in its mean position. If the blacked in 'square' on these keys lies in the same semicircle as the libration arrow, then the direction of the libration is fairly favorable for the area depicted on the plate. If the arrow passes through the square concerned then the direction is very favorable; an unfavorable libration will exist if the square concerned lies beyond the head of the libration arrow.

The amount of libration can vary between nothing and a maximum of about 7°–8°. The Moon's physical libration and diurnal libration have been ignored, since their combined effect would not generally alter the aspect of the various photographs appreciably.

The Moon's Age and the Sun's Selenographic Colongitude

In previous editions of this atlas, the Moon's age in days was quoted with each plate to signify the lunar phase at which the photograph was taken, e.g., 0.0 days would be New Moon, 7.4 days would be First Quarter, 14.7 days would be Full Moon, and 22.1 days would be Last Quarter. Although this was at one time used to signify the state of illumination of the lunar surface, most amateur astronomy organizations nowadays prefer to quote the Sun's selenographic colongitude.

Table 1.1 This illustrates the relationship between lunar phases and the Sun's selenographic colongitude, or the longitude (old sense) of the sunrise (morning) terminator of the Moon

Phase	Days since new moon	Sub-solar longitude	Sun's selenographic colongitude
New moon	0	180°	270°
First quarter	7.4	90°	0°
Full moon	14.7	0°	90°
Third quarter	22.1	270°	180°

The Sun's selenographic colongitude, more often referred to simply as colongitude, is the longitude of the morning terminator on the lunar surface in the "old sense" of when the western limb (the side that Grimaldi lies on) used to have a longitude of +90° and the eastern limb (Mare Crisium side) used to have a longitude of −90°. Longitude on the Moon's surface now runs from −90° (or 270°) on the western limb to 0° (or 360°), which is at the center of the Moon's disk (assuming no libration), to +90° on the eastern limb. Table 1.1 should help to clarify this.

Any observations made at a telescope should always include the date and UT, but colongitude certainly helps to put the observations in context. Colongitude can be found in the "Handbook of the British Astronomical Association," the circular from the BAA's Lunar Section, or derived from the JPL HORIZONS website, by subtracting the sub-solar longitude (in the modern sense of longitude) from 90°.

The Digital Image Augmentation Process

Although it is often possible to apply filtering to sharpen modern digital images that have been blurred by the atmosphere, this approach will not work very well for original photographic plates by Commander Hatfield because the photographic grain also gets sharpened and enhanced (see Fig. 1.1). Because there was no possibility of stacking images from the original atlas (only one photographic exposure exists for each plate) to improve clarity, a different approach was needed to achieve the higher resolution detail required for this edition of the atlas. This has been accomplished using "digital image augmentation" – a technique of generating a detailed computer visualization of the lunar surface and merging this with the original Hatfield photographs.

Applying digital image augmentation to the atlas images was possible only thanks to results from NASA's Lunar Reconnaissance Orbiter (LRO) mission. The LRO had several excellent instruments on board, one of which was the Lunar Orbiter Laser Altimeter (LOLA), which over the lifetime of the mission has obtained very accurate point height measurements across the entire lunar surface. NASA's Planetary Data System (PDS) has a 1/64th of a degree digital elevation model (DEM) of the entire Moon, extracted from the processed results from this instrument. Using a prototype piece of software called ALVIS (Aberystwyth

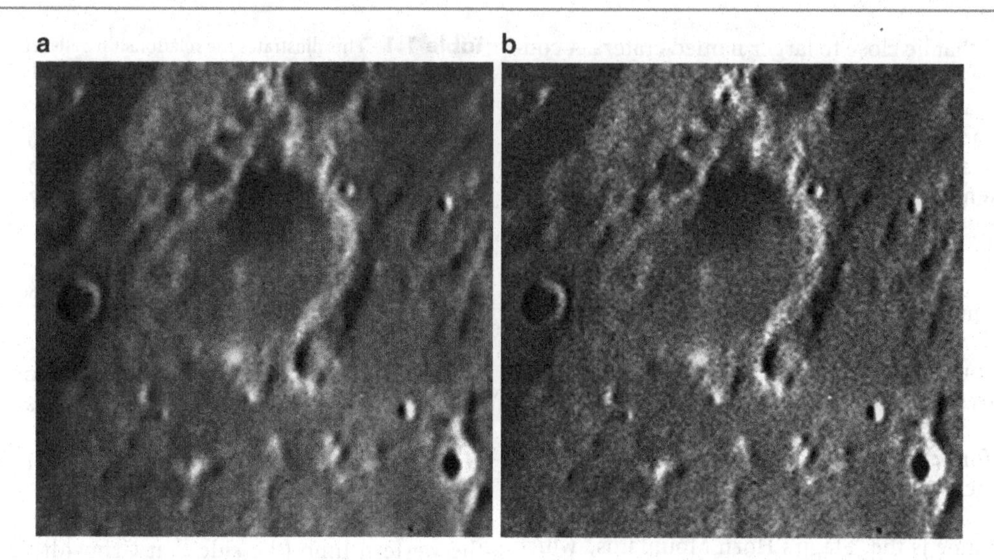

Fig. 1.1 These two pictures illustrate why conventional digital image sharpening does not work well on the photographs used in previous editions of this atlas. (**a**) An enlargement of the Julius Caesar area from Plate 1a (grid d3) from the last edition of the atlas. (**b**) The same area, but after undergoing unsharp masking (radius = 2.5 pixels). Note that although sharper, the photographic grain has also become more noticeable

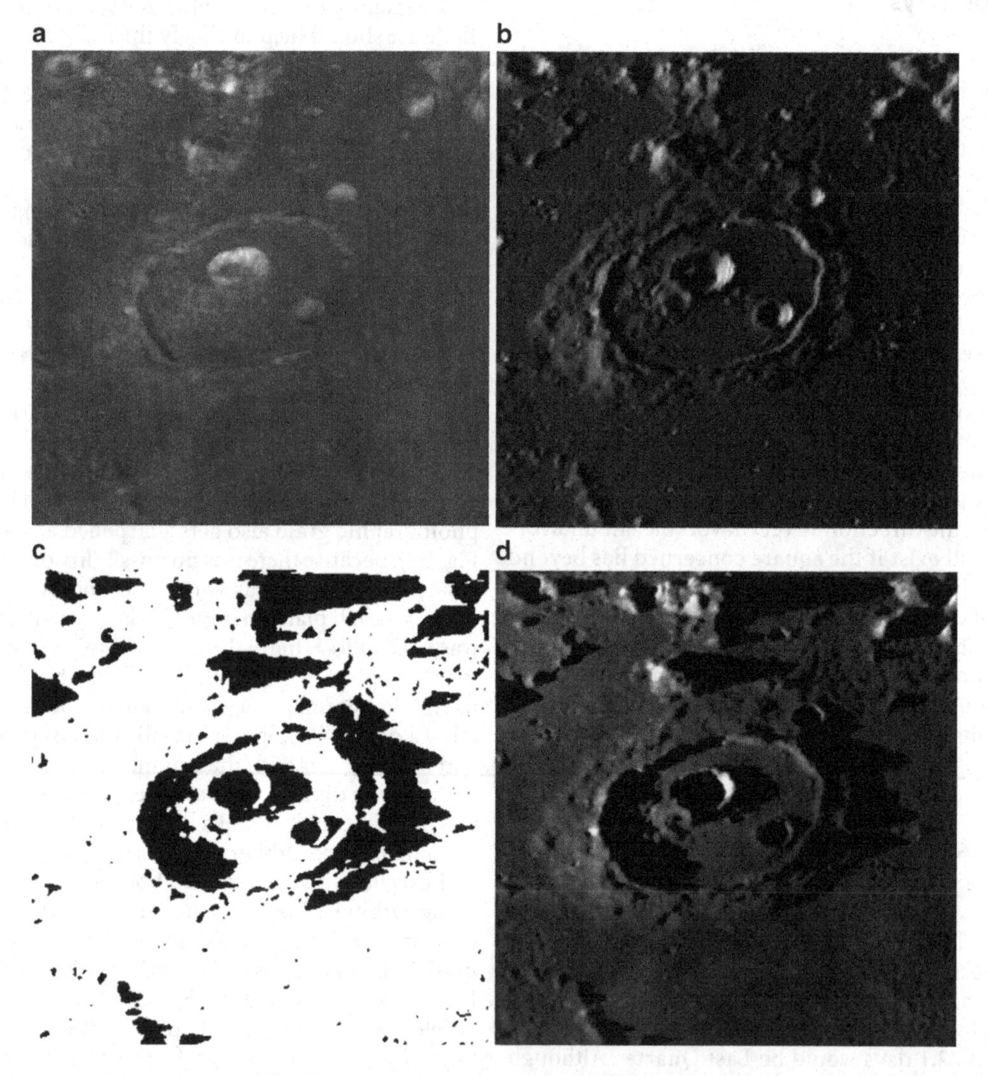

Fig. 1.2 ALVIS generates a virtual image of the Moon using three components: (**a**) an albedo map based upon the Clementine UVVIS mosaic – this represents the reflectivity of the lunar surface. (**b**) A shadings map, to show how the lunar surface reflects light off slopes. (**c**) A shadow map. These images are then multiplied together to produce a virtual image (**d**)

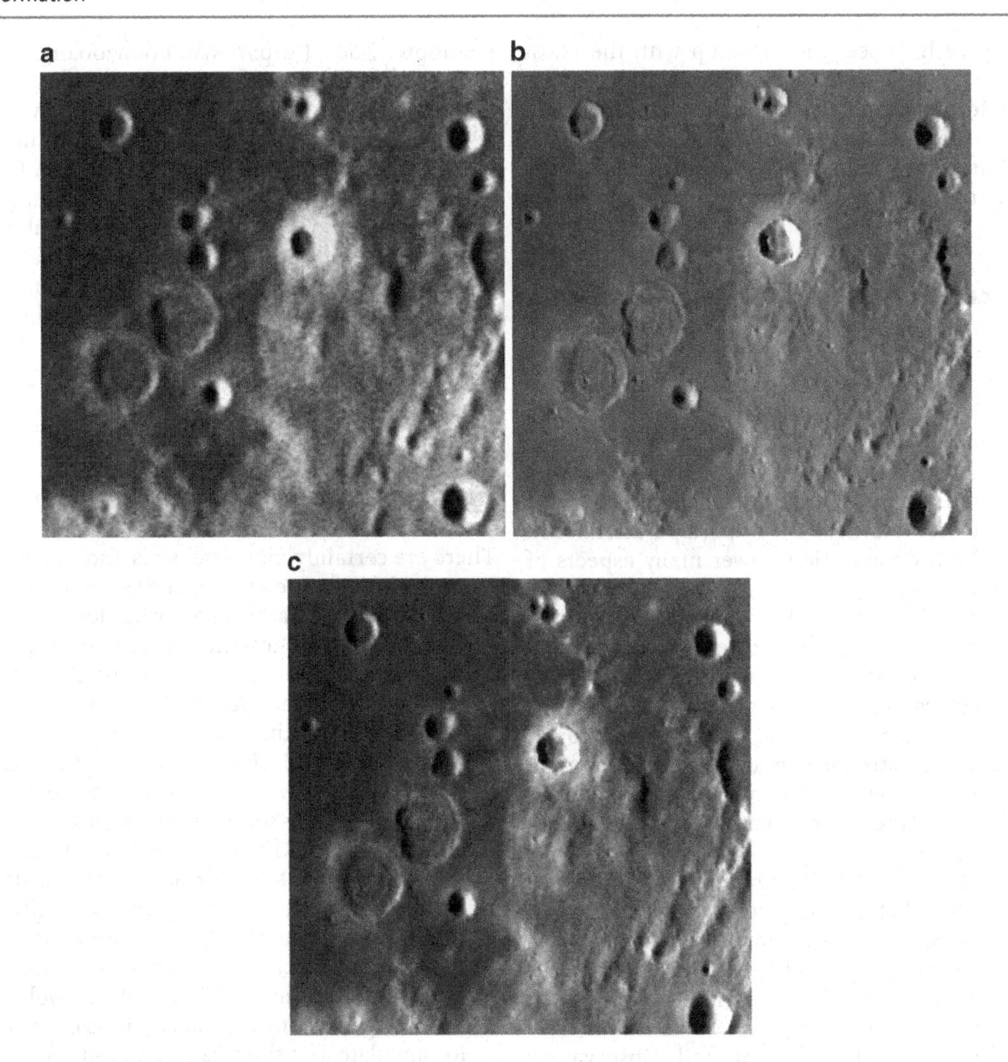

Fig. 1.3 The Hatfield photograph (a) is multiplied by the virtual image (b) to produce a new digitally augmented image (c)

Lunar Visualization and Information System) at Aberystwyth University, it is possible to produce very detailed simulated images of what Commander Hatfield would have photographed in the 1960s if he had had perfect atmospheric conditions and no photographic grain.

There are other software tools available to amateur astronomers to let them generate their own virtual images of the Moon, such as the Lunar Terminator Visualization Tool (LTVT). Unfortunately, despite simulated images of the Moon being very sharp and good at showing where the shadows are, they do not always represent accurately the way sunlight is reflected off the lunar surface by different materials and at different viewing and illumination angles. This means that some parts of the lunar surface, for example ray craters or slopes, may be brighter in real-life images than the computer visualizations show. It is for this reason that the Hatfield photograph images were retained to moderate the ALVIS images for such effects.

To illustrate the digital augmentation procedure, take a look at the steps in Fig. 1.2. ALVIS generates three virtual images of the lunar surface for each photograph used by the atlas, namely a shadow map, an albedo map, and a shading map. These are combined together to form a virtual image of the Moon. The next step is to overlay and align the ALVIS virtual image to the Hatfield atlas photograph, then multiply them together, as is demonstrated in Fig. 1.3. This uses the brightness information held in the original atlas photograph and blends in the detail from the computer model.

Although much of this work can be achieved automatically on a computer, some manual experimentation was still needed to attempt to strike the right balance between how much of the photograph and how much or the virtual image should show through in the final revised atlas image. So on occasions, where ray information has been needed to be preserved, a little less high resolution detail is shown.

For Further Information

With so much information available online, it would be remiss not to list some useful websites that lunar observers may wish to explore, to enhance their knowledge of the Moon. Having studied a particular crater through the telescope you may wish to see it in close up with the latest spacecraft imagery, or learn about the associated geology or age. It should be realized, though, that these are just a snapshot of Internet sites available in 2013, and as time progresses, some of these sites may no longer be in existence, or will have changed their web address.

Astronomical Societies and Organizations

The user of this atlas is strongly encouraged to join his or her local astronomical society, where advice and assistance can be obtained. In addition, consideration should also be given to joining national astronomical societies such as the British Astronomical Association (BAA), http://britastro.org/baa/, or, in the United States, the Association of Lunar and Planetary Observers (ALPO), http://alpo-astronomy.org/. Although both organizations cover many aspects of amateur astronomy, they have thriving lunar sections in which guidance is given by the director and coordinators on many aspects of lunar work for amateurs of all ages and capabilities, such as topographic drawing, imaging, occultations, surface colors, and the study of TLP.

The following are some additional websites for national/international amateur astronomy organizations with strong lunar interest groups. However, this is just a sampling. There are likely to be many other good lunar observing organizations not listed here.

- The Society for Popular Astronomy (http://www.popastro.com/) is a UK-based observing group dedicated to beginners in astronomy, some of whom are quite advanced.
- L'Unione Astrofili Italiani is an Italian astronomy society with an active section of lunar observers (http://www.uai.it/web/guest/home).
- The Wilhelm-Foester Planetarium and Observatory (http://www.planetarium-berlin.de) in Berlin, Germany, although primarily a local astronomical society, has a very influential lunar working group.
- *Selenology Today* is an online periodical produced by the Geologic Lunar Research Group (GLR) and can be found on http://digilander.libero.it/glrgroup/. It carries quite advanced papers on backyard lunar science but will give you an idea of what advanced observing techniques can be achieved using Earth-based amateur-sized telescopes.
- The American Lunar Society (http://www.amlunsoc.org/) is devoted purely to American amateur astronomers dedicated to the Moon.
- The International Occultation Timing Organization (http://www.lunar-occultations.com/iota/iotandx.htm) deals with timing stars disappearing behind the Moon and other objects. Their analyzed results can be used to check heights on the Moon and diameters of asteroids, as well as help with looking for double and multiple stars.
- TLP alerts are available via Twitter (http://twitter.com/lunarnaut), and the initial detection and subsequent observations of these are coordinated by the ALPO and BAA lunar sections.
- Moon Zoo (http://www.moonzoo.org/) and Moon Mappers (http://cosmoquest.org/Moon_Mappers:_The_Mission) are a couple of citizen science websites that let anybody who registers with them online measure craters and identify interesting features in NASA LRO images.
- Finally, International Observe the Moon Night (http://observethemoonnight.org/) is an annual event encouraging anybody who is interested, no matter what their experience level, to observe the Moon, take part in locally organized observing events, and submit images of the Moon.

Lunar Ephemeris Data and Useful Software

There are certainly many websites and much freeware and commercial software suitable for lunar observers. However, here are the ones that this editor has found most useful:

- The Sun's selenographic colongitude can be found for the start of each day in a given month in the lunar section circular of the BAA via the BAA Lunar Section website (http://www.baalunarsection.org.uk/), or you can access NASA's JPL HORIZONS web interface (http://ssd.jpl.nasa.gov/horizons.cgi) and calculate sub-solar longitude, then simply subtract this from 90° to get the colongitude. The HORIZONS website is an excellent way to find the most up to date lunar ephemeris data.
- Lunar eclipse dates, times, and visibilities, including maps, can be found on the Goddard Spaceflight Center website (http://eclipse.gsfc.nasa.gov/eclipse.html).
- The Lunar Terminator Visualization Tool, which can be used for generating simulated views of the lunar surface for any date and time, can be found on http://ltvt.wikispaces.com/LTVT. Alternatively you can use the Virtual Moon Atlas; see http://www.ap-i.net/avl/. These are similar to the prototype ALVIS program used in this atlas.
- The *Registax* image-processing software will transform your web camera, or CCTV, images through the telescope into ultra-sharp views of the lunar surface. This can be downloaded from http://www.astronomie.be/registax/.
- *LunarScan* software can be used to scan automatically through video the night side of the Moon, looking for impact flashes. It can be obtained from the NASA's Marshall Spaceflight Center here: http://www.nasa.gov/centers/marshall/news/lunar/program_overview.html.

Maps, Images and Surface Names

Here are some websites that will let you explore the lunar surface in the form of maps and images.

- The Lunar and Planetary Institute has a very extensive range of maps, telescopic atlas plates, and lunar spacecraft images from the 1960s onwards, and can be found at: http://www.lpi.usra.edu/resources/mapsImagery/index.shtml.

- The USGS Map a Planet (http://www.mapaplanet.org/explorer/moon.html) is a geographical information system for most of the solid planets in the Solar System. It allows you to build up maps with overlays of different data sets. To find out the latest approved names of features on the Moon and where they are located, you can go to the IAU's website at http://planetarynames.wr.usgs.gov/Page/MOON/target. NASA's Planetary Data System (http://pds.nasa.gov/) is a portal to all of the publicly available NASA images and science data sets, though it is rather technical and difficult for non-scientists to understand. The Moon Wiki (http://the-moon.wikispaces.com) is easier to read and provides very comprehensive pages about most named craters on the Moon, some of which feature in the Lunar Picture of the Day (LPOD) (http://lpod.wikispaces.com) website.
- NASA's Lunar Reconnaissance Orbiter website (http://lroc.sese.asu.edu/index.html) lets you zoom into many areas of the Moon down to the meter level of resolution. It also shows you where the spacecraft is currently located and possibly doing its imaging. Lastly, while it is still operational, and there remain areas of the Moon left to image, members of the public can propose target areas of the surface to capture with the high-resolution camera.
- A large part of the Apollo Image Archive can be accessed via http://apollo.sese.asu.edu/, and so, too, the Lunar Orbiter images from the 1960s on (http://www.moon-views.com/). Japan's *Kaguya* lunar mission images can be studied on http://www.kaguya.jaxa.jp/en/index.htm, and ESA's SMART-1 mission on http://www.esa.int/esaMI/SMART-1/index.html.

Data Sets Used in This Atlas

The editor wishes to express his grateful thanks to the Council of the BAA (who own the original copyright) for allowing the use of the Hatfield maps and photographs that form the basis of the present publication. He is also grateful for the public availability, via NASA's Planetary Data System, of the LOLA 1/64th degree lunar digital elevation model (LRO-L-LOLA-4-GDR-V1.0, version 1.06) produced by the Goddard Space Flight Center, and the USGS-produced Clementine UVVIS global image mosaic of the Moon, without which the digital augmentation process would have been impossible.

2 Lunar Observing Techniques

Methods of observing the Moon have changed considerably over the years. Back in Commander Hatfield's time, lunar sketches and photography were the primary techniques employed, the former providing the highest resolution views of the lunar surface through a telescope. Now, however, traditional photography has been replaced largely by digital cameras, and drawings are being equaled, or even surpassed, by digital imaging in terms of resolution.

Although it has never been part of previous editions of this atlas, a few ideas concerning observing methods to help new amateur astronomers get started have been included here, along with some additional directions for the more advanced amateur. Before we get into the project ideas, though, let us consider some important aspects of observing.

Most of us are used to risk and safety assessment in our jobs, and so it is probably a good idea to go through a similar exercise purely in your head before observing, but without the boring paperwork. Unless it is within a few days of Full Moon, moonlight will not supply enough light to see with at night, so a red LED torch, or white torch with a red filter, will provide enough illumination without destroying your dark adaptedness. Remember also that tiredness at night can make you more accident prone, so try to slow down and think before moving around, setting up the telescope, handling eyepieces, etc. Always be aware of your surroundings in the dark, especially any cables from the telescope or tripod legs, which can be trip hazards.

The author has known of several instances where colleagues have dropped expensive eyepieces, or knocked equipment over. If you are using power cables from indoors, it is far safer to use 12-V DC cables than mains supply voltages, although if you have to use the latter then make sure that that there is adequate safety trip protection and keep the cables well away from water, dampness, or anything that could snag or cut the cable. Remember also that as the outside temperature gets near freezing, your brain activity slows down, and therefore you should, too! In freezing temperatures, wear gloves if you are not doing so already; this is because skin can stick to frosty surfaces, like it does inside kitchen freezers.

If your telescope is not in an observatory already, then take it outside and let it cool down for at least half an hour. This can be done just before sunset, if placed under cover and in shade. This will allow the heat from the telescope to dissipate and reach the ambient temperature of the surrounding air before you observe. If you do not do this, then thermal air currents from inside the telescope tube, and around the optics, will distort the lunar image badly, at least until they have all cooled down enough. Finally, when leaving a telescope unattended, make sure you periodically check the weather conditions outside, just in case rain clouds appear unexpectedly.

Projects for Beginners

Here are a few projects that will help beginners learn their way around the Moon and prepare themselves for more advanced work – if they should desire to move on. Whatever project you attempt, it is really important to get out and use your telescope to gain experience, and to share your results with other amateur astronomers.

Now, what you will be able to resolve on the lunar surface all depends upon the aperture, or diameter of the front lens or mirror, of your telescope. Naturally, the larger the aperture, the finer the detail you will be able to see on the lunar surface. But then, of course, you are limited by what telescope you have available to use, or can afford. Improved modern optics means that for beginners, at least, 5-inches (12.5 cm) catadioptric telescopes are probably just about adequate, and the smallest usable aperture refractors would be 3–4 inches (7.5–10 cm).

It is very helpful if your telescope can track or follow the sky as Earth rotates; at least then you will be able to do some of the specialized imaging techniques mentioned in the advanced users section. However there are also non-tracking telescope mounts, such as the Dobsonian, which are both low cost and adequate for sketching, taking single images, and generally learning your way around the Moon. The main disadvantage, though, is that you will have to reposition the 'scope on the Moon at least every 2 min at Full Moon, and more frequently at lesser phases.

Smaller diameter telescopes can be used to help you become familiar with the Moon's geography, but any sketches and images made through these will probably be too low in resolution to show to other astronomers, or publish. Only during lunar eclipses can smaller 'scopes

A.C. Cook, *The Hatfield SCT Lunar Atlas: Photographic Atlas for Meade, Celestron, and Other SCT Telescopes: A Digitally Re-Mastered Edition*, DOI 10.1007/978-1-4614-8639-8_2, © Springer Science+Business Media New York 2014

or binoculars make a useful contribution, because their wide field of view will show the Moon as a whole.

Two other points to consider when choosing a telescope: (1) Ignore the urge to buy a telescope advertised with very large magnification, because it is aperture size that matters. (2) Think about what other uses you may want to put the telescope to; for example, if you intend to look for deep sky objects, such as faint nebulae, then a f/5 focal ratio telescope is better than an f/10 (see below).

The focal ratio number of a telescope corresponds to its focal length divided by the diameter of the 'scope's aperture. A small focal ratio number, e.g., f/5, will give you wider views of the Moon and more sensitivity to faint areas in Earthshine. But a larger focal ratio number, e.g., f/10, will provide high-resolution views of the Moon and planets, albeit less bright and a smaller field of view. Magnification can be calculated from Eq. (2.1), so for a 6 inches (15.2 cm or 152 mm) f/5 reflector with a 6 mm focal length eyepiece this would give us a magnification of ×127.

As a personal rule the highest practical magnification that the author has found helpful is about 30 times that of the telescope aperture in inches, so for a 6-inches reflector this would be ×180 magnification, and for a 12-inches the limit would be ×360. However, in theory, under the very best observing conditions, 50 times the aperture in inches is regarded as the ultimate upper limit.

$$Magnification = \frac{Telescope\ \mathrm{f}\ /\ No. \times Aperture\left(mm\right)}{Eyepiece\ focal\ length\left(mm\right)} \quad (2.1)$$

Sketching the Lunar Surface

The geography of the Moon is especially difficult to get used to because features change quite considerably in appearance over the lunar daytime, or 14.8 days in Earth time. A complete lunar day and night, of course, lasts 29.5 Earth days. Without much shadow it becomes very difficult to find your way around. By making sketches of a lunar feature on several dates, you can build up a portfolio of how features change and will become familiar with the beauty of lunar sunrise and sunset, when the shadows are long and sometimes look like spires. At later stages in the lunar day, outlines of many features in the brighter highlands become difficult to depict, as rays and interior bands dominate many craters. High sun angle (far away from the terminator) phases are a good time to look for small dark halo craters and dark spots from lunar volcanic pyroclastic deposits.

So by sketching several features on the Moon, you will start to gain vital expertise in being able to navigate your way around the lunar surface. Furthermore, the observations that you build up will be welcomed by amateur astronomical societies, who will collect and possibly publish these. Now, apart from studying attractive craters, such as Aristarchus, Copernicus, Gassendi, Proclus, Theophilus and Tycho, try to find and sketch some more obscure features by picking names at random from the index of this atlas. Alternatively, if some astronomical societies have a campaign to observe specific lunar craters, then have a look at these.

Because of the turbulent nature of the atmosphere that we look through, you should spend at least 5 min looking at any given feature on the Moon, waiting to catch brief moments when the atmosphere steadies enough to give crisp views of the surface. Making a sketch will take longer, typically between 10 and 60 min, depending upon the size of the area you are interested in studying.

So how does one go about making a sketch at the telescope? Well, a piece of good paper on a drawing board is a necessity, along with a dark tone pencil, e.g., 2B or 4B, unless of course, you are using an electronic sketchbook on a tablet. Be aware, though, for electronic sketching, that electronics, in particular LCD screens and batteries, may not work so well at the very low temperatures experienced by high latitude astronomers in the winter months.

Start by making a faint outline of the feature and the boundaries of any shadows, and depict these as accurately as possible to define the shape of the feature (see Fig. 2.1a). Now, outline faintly the smaller detail within the sketch (see Fig. 2.1b). You can then pencil in relative intensity numbers (see Fig. 2.1c) into discreet regions. Finally, once indoors, you can erase these numbers and shade in the appropriate densities with a soft pencil, and add dark ink to fill in shadow interiors and white liquid paper/correction fluid to highlight bright spots (see Fig. 2.1d). Instead of this relative brightness scale that is only local to the feature concerned, some amateur lunar observers prefer to adopt the Elger scale to indicate brightness, which is given in Table 2.1. There are alternative styles to shading drawings, too, and Fig. 2.2 shows the same mountain but depicted instead using "stippling," or different sized/density of dots, that reflect gray levels when viewed from a distance.

Before finishing your observation, it is really very important to write the date and Universal Time (UT) on your sketch, along with information about telescope aperture, magnification, your name and location, and atmospheric seeing and transparency. Make sure that you do not use local dates or times, or indeed daylight saving times, as it is important for astronomers to be able to compare your observations with others from around the world. You can obtain the correct date and UT from the U. S. Naval Observatory (USNO) clock website [1]. It is also vital to include a direction indicator to show which way is north, south, east, and west, other wise it may become difficult to compare with sketches made by other astronomers. Indeed if you are planning on sending your observations in to an astronomical society, or a web page, try to adopt the points of the compass orientations that they use.

Concerning atmospheric seeing and transparency conditions on the North American continent and some other countries, the atmospheric seeing, or steadiness of the image through the telescope, is described on a scale of 1–10 (Table 2.2). Similarly the transparency of the atmosphere can be described on a scale from 1 to 10, but in this case the number refers to the magnitude of the faintest star that you can see. So in both cases the higher the number the better the observing conditions.

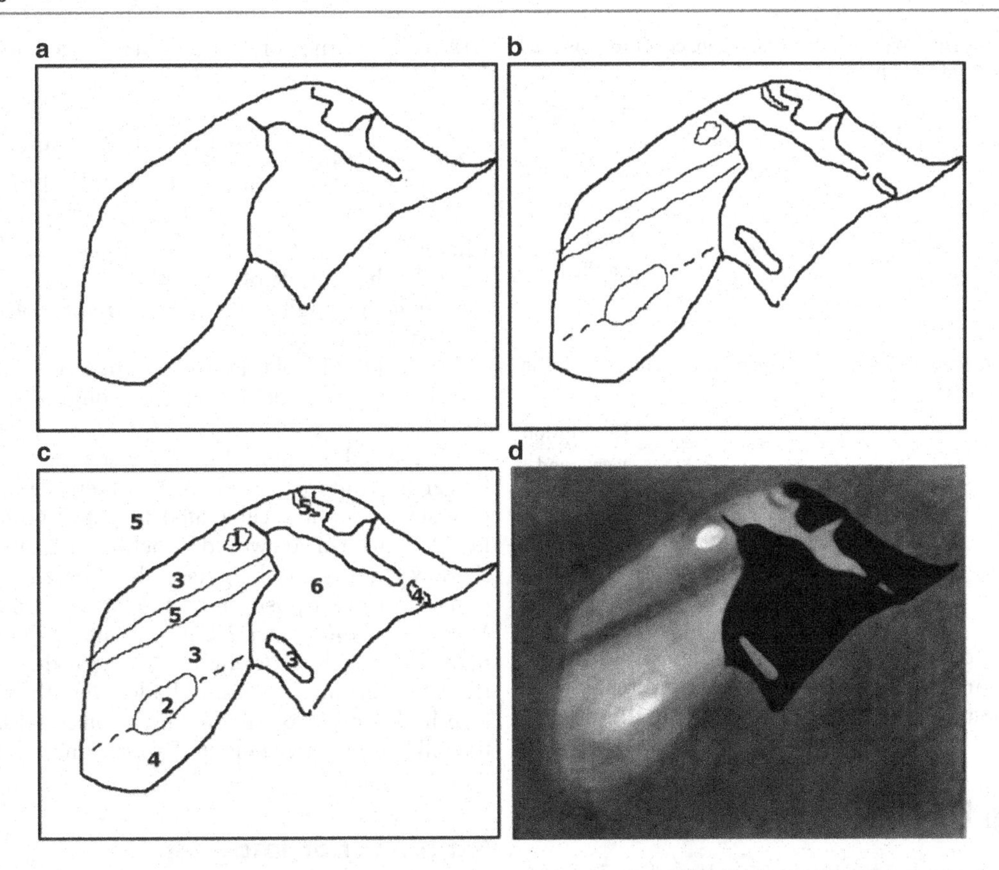

Fig. 2.1 This figure illustrates the development of a drawing of Mons Pico that the editor made on May 17, 1978, at 21:00 UT using a 12 inch reflector at ×260 under Antoniadi III seeing conditions. Moon's age = 10.7 days, Sun's selenographic colongitude = 36.5°, south is toward the bottom. (**a**) Outline sketch to constrain the shape of Mons Pico. (**b**) Interior detail outlines added. (**c**) Relative intensities of regions penciled: 1 = very bright to 6 = black shadow. Note that this is not same as the Elger scale. (**d**) Pencil shaded to reflect relative intensities

Table 2.1 The Elger scale of visual comparative intensity on the lunar surface

Intensity	Description
0	Pure black shadow
1	Very dark gray, e.g., on dark mare just after sunrise
2	Dark gray, e.g., southern floor of Grimaldi
3	Medium gray, e.g., northern floor of Grimaldi
4	Medium-light gray, e.g., west of Proclus
5	Light gray, e.g., Archimedes floor
6	Brighter white gray, e.g., Copernicus ray system
7	The very light grayness of a crater ray, e.g., Kepler's rays
8	Very bright white, e.g., Tycho's rim
9	Pure white, e.g., southern floor of Copernicus
10	Brilliant white, like one gets on the central peak of Aristarchus

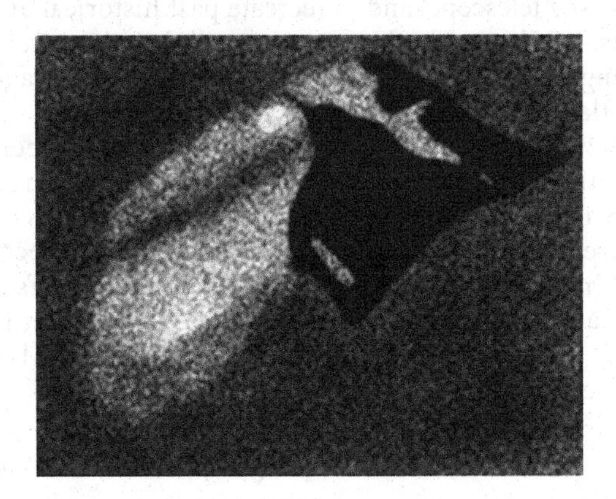

Fig. 2.2 This illustrates the alternative "stipple" approach, a popular way to shade drawings. It is best viewed from a distance

In European countries and many other places, the Antoniadi scale is used for atmospheric seeing (Table 2.3). Atmospheric transparency for European observers is commonly just descriptive words, e.g., very good, good, average, poor, very poor, or even other descriptions such as haze, fog, cirrus, etc. So what happens if conditions change during an observation? In this case just include a dash to show that there has been a change, e.g., seeing = 5–7 or III–II.

Lastly, it is very useful to write the Moon's Age, or more helpfully the Sun's selenographic colongitude, onto any sketch. At least then if you obtain a sequence of drawings

Table 2.2 A description of astronomical seeing conditions as used in North America and elsewhere

Seeing	Description
1–2	Very poor
3–4	Poor
5	Average
6–7	Good
8–10	Excellent

Table 2.3 The Antoniadi scale of astronomical seeing, as used in Europe and elsewhere

Seeing	Description
I	Very good
II	Good
III	Moderate
IV	Poor
V	Very poor

from different dates, you can arrange them by order of increasing lunar phase, or you can compare directly with images of similar selenographic colongitude in this atlas.

Measuring Heights on the Moon

Establishing how high a mountain is, or how deep is a crater, is now no longer a scientifically useful pursuit for amateur astronomers. This is because the entire lunar topography has been measured by laser altimeters flown aboard several spacecraft. Nevertheless it is satisfying for any amateur astronomer starting out to try this as an exercise at the telescope, and to recreate past historical experiments. If it is cloudy then you can still do this exercise by making use of the ample supply of shadows in images in this atlas.

Now the relative height H of a mountain between the tip of the shadow on the ground and the mountain top is given in Eq. (2.2). However this applies to non-curved surfaces, and so can only be used on the Moon effectively over short distances, and well away from the Moon's limb. More advanced equations are available to cope with surface curvature, but are outside the scope of this beginners' section.

$$H = L \times \mathrm{Tan}(\mathrm{Alt}) \tag{2.2}$$

where:

L is the shadow length in kilometers.
Alt is the altitude of the Sun at the feature concerned.

You can obtain the length of the shadow in kilometers by comparing it to the east–west diameters of craters, finding the most similar one, then looking up the diameter from the feature index in this atlas. So, for example, if a shadow was 1/10 the diameter of Plato crater, and Plato has a diameter of 101 km, then the shadow length is approximately 10 km. The altitude of the Sun at a lunar feature comes from Eq. (2.3):

$$\mathrm{Alt} = -\mathrm{ArcSin}\begin{pmatrix} \mathrm{Sin}(\mathrm{Flat}) \times \mathrm{Sin}(\mathrm{Slat}) + \mathrm{Cos}(\mathrm{Flat}) \\ \times \mathrm{Sin}(\mathrm{Slat}) \times \mathrm{Sin} - (\mathrm{Col} + \mathrm{Flon}) \end{pmatrix} \tag{2.3}$$

where:

Flat is the latitude of the feature.
Flon is the longitude of the feature as measured positive towards the east.[1]
Slat is the sub-solar latitude of the Sun.
Col is the Sun's selenographic colongitude.

Be warned, though, that if the shadow length is much longer than approximately 30 km then the Moon's curvature starts to affect calculated heights significantly using Eq. (2.2). Instead you would be better off measuring relative shadow lengths and heights with respect to other mountains at a similar position on the terminator. For example, Mons Pico is just over 2 km high above the surrounding mare, so if a hill somewhere else, but at the same distance from the terminator, has a shadow length of a tenth the length of Mons Pico's shadow, then you could safely assume the hill has a relative height of approximately 200 m.

Simple Lunar Imaging

Having learned your way around the Moon, it is personally very satisfying to be able to take images and show your friends just what you have been up to on dark nights. Fortunately this does not involve any expensive equipment. Just make use of any digital camera, web camera, or cell phone that you have. You will need to point the telescope at the Moon and hold the camera up to the eyepiece so that it is looking optically straight into the eyepiece and not tilted. You should see an image on the camera screen, but it will probably be out of focus, and so you will need to experiment with adjusting the eyepiece and the distance of the camera away from the eyepiece, until you get sharp images that show craters. Sometimes the field of view may be small, in which case try changing the eyepiece.

If you can, take your pictures on a time delay, so as to avoid camera shake that would otherwise occur when you press the button. A camera tripod can also help to keep the camera steady. Finally, there are also camera adapters sold by astronomical equipment suppliers that will help you attach digital cameras with removable lenses, or web cameras, directly to the telescope. By fixing the camera to the telescope, this allows you to use a shorter focal length eyepiece, or a Barlow lens, to yield higher resolution images – providing that the atmospheric seeing allows.

Because many digital cameras are sensitive to near-infrared light, the use of a near-IR blocking filter to remove

[1]Proclus at 46.8° E would have a longitude of +46.8°, whereas Aristarchus at 47.4° W would have a longitude of –47.4°.

Table 2.4 Lunar eclipse dates and times [2] from November 2012 until November 2020. The second column describes the state of the eclipse: penumbral only, partial umbral, or total umbral. For penumbral only eclipses, the first and last contacts refer to when the penumbral shadow first touches the lunar limb, and when it finally leaves the lunar limb. For partial and total umbral eclipses, first contact is defined similarly, except that it refers to the dark umbral, and not the lighter penumbral shadow. For total umbral eclipses, 2nd and 3rd contact refer to the start and end of totality. Please note that you will need to check that the Moon has risen from your geographical location between the times given for each eclipse date in order to ensure that it is visible

Date	Eclipse type	First contact UT	2nd contact UT	Mid UT	3rd contact UT	Last contact UT	Amount
2013 Apr 25	Partial	19:54		20:08		20:21	1 %
2013 May 25	Penumbral	03:53		04:10		04:27	2 %
2013 Oct 18/19	Penumbral	21:51		23:50		01:50	77 %
2014 Apr 15	Total	05:58	07:07	07:46	08:25	09:33	100 %
2014 Oct 08	Total	09:15	10:25	10:55	11:24	12:34	100 %
2015 Apr 04	Total	10:16	11:58	12:00	12:03	13:45	100 %
2015 Sep 28	Total	01:07	02:11	02:48	03:23	04:27	100 %
2016 Mar 23	Penumbral	09:39		11:47		13:55	77 %
2016 Sep 16	Penumbral	16:55		18:55		20:54	91 %
2017 Feb 11/12	Penumbral	22:34		00:44		02:53	99 %
2017 Aug 07	Partial	17:23		18:22		19:18	25 %
2018 Jan 31	Total	11:48	12:52	13:31	14:08	15:11	100 %
2018 Jul 27	Total	18:24	19:30	20:23	21:13	22:19	100 %
2019 Jan 21	Total	03:34	04:31	05:12	05:43	06:51	100 %
2019 Jul 16	Partial	20:02		21:31		23:00	65 %
2020 Jan 10	Penumbral	17:08		19:10		21:12	90 %
2020 Jun 05	Penumbral	17:46		19:25		21:04	57 %
2020 Jul 05	Penumbral	03:07		04:30		05:52	35 %
2020 Nov 30	Penumbral	07:32		09:43		11:53	83 %

this light is highly recommended, especially for refractors. Without this, the images may be slightly blurry and low contrast, because near IR has a different focal point to visible light once the light has passed through most optical glass, and the effect is especially acute in refractors.

Post-observation processing with image software such as Adobe *Photoshop*™ can help improve image contrast, enhance color, and sharpen. You can then compare with images in this atlas. One final tip. Be sure to set the date and time to UT in your camera set up; at least then you can recover the correct date and UT from file properties at a later date. Alternatively keep a record of the dates and times that you took the exposures and rename the image files to contain these.

Lunar Eclipses

Lunar eclipses happen fairly infrequently [2], being visible between 0 and 3 times per year from a particular geographical location (Table 2.4). They can be total, making the whole Moon almost vanish to the eye; partial; or penumbral (where the eclipse shadow is barely perceptible).

Over many decades astronomers have estimated the darkness of a total eclipse by its Danjon or L number, which lies between 0 and 4 (see Table 2.5). Because the state of Earth's atmosphere can be affected by regional weather, volcanic eruptions, forests fires, and possibly even the solar cycle, the analysis of the reported darkness of lunar eclipses over the centuries can give us important information on the state of the upper atmosphere long before the weather satellite era.

Cameras and webcams, with telephoto lenses, are excellent ways to image lunar eclipses, although a tripod is essential if you want to have the eclipsed disc filling much of the screen and wish to avoid blurred images resulting from camera shake. You can try a range of exposures by switching the camera off its automatic exposure meter; this will help you to find the optimum setting and consider whether you want to show detail in the dark umbra or in the lighter penumbra. Adjust the exposure accordingly. If you take some nice-looking images, send them into amateur astronomy societies or astronomical magazines, as they are always pleased to receive colorful illustrations. To identify features, refer to near Full Moon images in this atlas taken between 13.2 and 16.7 days, or between 70° and 110° in selenographic colongitude.

Earthshine

Small telescopes and binoculars can be used to monitor the brightness and color of Earthshine and could give a useful yardstick for comparing present-day estimates of Earth cloud cover reflectivity with those from past centuries. Earthshine is best seen between New Moon and First Quarter, or between Last Quarter and New Moon. All you need to do is to record the date and UT, your location, the diameter of telescope that you are using, magnification, atmospheric transparency, and then what you can identify in Earthshine in order of brightness, e.g., Aristarchus, Copernicus, Tycho, Kepler, and some of the dark features that you can see as well, such as Plato and Grimaldi. Also

Table 2.5 The Dajon value of how dark a lunar eclipse is. This is commonly referred to as the L value

L value	Description
0	A dark eclipse, with a dark gray or brown shadow and most features invisible
1	A dark eclipse, with a dark gray or brown shadow and most features invisible
2	A deep red eclipse, where the main shadow is a deep red or brown colour
3	A brick red central eclipse with a bright or yellow ring on the umbral shadow edge
4	Bright eclipse with an orange umbral shadow often with a very bright bluish edge

take a note of the color, though this may be affected by the altitude of the Moon.

Occasionally you may also see bright stars and planets occulted by the Moon. For beginners it is just worth the effort to watch these vanish abruptly behind the Moon in the case of stars, and fade in brightness in the case of planets. On the rare occasions of alignment, where you see stars being occulted near the poles of the Moon, you can see these flashing off and on as they graze the limb of the Moon, behind mountains, reappearing in gaps between valleys. Occultations can make nice photogenic digital images, and video of grazing occultations can be even more spectacular.

Disproving Past TLP Observations

TLPs, or Transient Lunar Phenomena, are enigmatic glows, colors, and other phenomena that some astronomers claim to have seen on the Moon; these will be covered in more detail in the advanced section. Looking for TLP with small telescopes is frowned upon because of resolution issues, and they are exceedingly rare anyway. However, one useful task that observers are encouraged to do is to re-observe a TLP site under very similar illumination conditions to what existed at the time of the original TLP. In doing so we occasionally find that the original TLP descriptions, for example "a floor of a crater looking like an obscuration of detail," is actually what the crater should really look like at that stage in illumination. In such cases the original observer made a mistake, but it is only by re-examining the crater that we learn of the error of calling this a TLP. Be sure to use this atlas to make correct identification of the lunar features concerned. The BAA and Aberystwyth University organizes a website [3] that carries predictions for when to examine certain lunar features to check out past TLP sites at similar illuminations.

Lunar Citizen Science

Even when it is cloudy, it is still possible to continue to contribute some lunar science. MoonZoo [4], which is part of Zooniverse, was formed to encourage members of the public to help scientists measure crater diameters, search the Moon for boulder fields, and report interesting and unusual lunar features. A similar site is part of Cosmoquest and is called Moon Mappers [5], but this site lets users check up on computer vision estimates of crater locations and diameters in images. At the time of this writing, in 2012, another website [6] allowed members of the public and scientists alike to specify targets on the Moon for NASA's Lunar Reconnaissance Orbiter (LRO) to point its camera at. Careful planning is needed to select a site that has some real scientific interest and has not been imaged already. Many months may follow, waiting to see if a requested area has been targeted with the camera, but this editor and other amateur astronomers have had parts of the Moon imaged in this way and acknowledged in e-mails from the NASA LRO team.

Advanced Techniques

In order to achieve lunar science these days, amateur astronomers need to make maximal use of a wide range of computer tools, online resources, and electronics. We will take a brief look at high-resolution imaging issues before going onto some potential advanced projects. Again, this atlas will be of great use for comparing your observations with, and also for lunar feature identification purposes.

High-Resolution Imaging

Although spacecraft imagery surpasses the resolution that we can obtain with Earth-based imagery, telescopic observing still has some advantages. For example you can view the Moon under illumination conditions that were not recorded in the limited set of spacecraft images of a given area. Secondly, although we have simulation software, such as LTVT [7] the very best Earth-based imagery is still slightly better in resolution, especially near the terminator. Also, LTVT-like software cannot simulate rays and other high albedo areas with the same accuracy as shadows. Therefore, high resolution telescopic imagery can still be of great importance in some of the advanced projects described below.

Amateur astronomy societies are very happy to receive spectacular-looking lunarscapes for publication in their magazines and journals. Some societies have regular observing campaigns on specific features that you can follow. Alternatively pick features at random out of the index and build up a portfolio for different stages of illumination.

Clearly large aperture telescopes help here in terms of angular resolution. The best diffraction-limited resolution that can be achieved with a telescope is given by Eq. (2.4).

$$R = 0.025 \times \lambda / D \tag{2.4}$$

where:

R is the resolution in seconds of arc.
λ is the wavelength of light being used in nm.
D is the telescope aperture in cm.

So, for a 30-cm refractor (12 inches) at a wavelength of 600 nm, it should be possible to resolve an angular separation of 0.5 seconds of arc, or 1 km on the lunar surface. Unfortunately, Earth's atmosphere does not let one see such crisp, sharp views of the surface very often, and even when it does, it may be only for brief fraction-of-a-second views. So if one were to take, say, 1,000 images of the same part of the surface, then maybe less than 10 % (less than 100) would be sharp enough to see the most detail. This threshold will vary depending upon the astronomical seeing and gets smaller the worse the seeing becomes.

Software such as *Registax* [8], which takes a large sequence of images and selects the sharpest ones to use, are based on this principle. The few remaining images can then be automatically aligned, combined (stacked) to reduce image noise, and sharpened (using wavelets) to produce high-quality clear views of the lunar surface.

There is, however, one further factor that defines what a camera can see, namely the image scale. If you want to achieve 0.5 arc second resolution, then the image scale needs to be 0.25 arc second per pixel or better. This is due to the Nyquist Sampling Theorem, which in effect means that you cannot resolve anything smaller than two pixels across.

Actually, though, the situation is not quite as simple as this because the theorem applies to obvious star-like points, and most of the Moon is covered in shades of gray. If you had a crater, then could you be sure to recognize it as a crater if it were only five pixels across? Probably not; in reality something like 10 pixels across is the minimum size regime you'd need to start to recognize discrete objects on the surface of the Moon. Exceptions to this rule are linear features such as rilles and fractures that can have widths of less than a pixel but can be still seen and recognized for what they are.

Image scale can affect the area of the Moon that you are trying to capture with a camera, as well as exposure time. Say that you have a 1,024 × 1,024 CCD chip, and this has an image scale of 1 arc second per pixel. Then the total area of the Moon covered would be 1,024 × 1,024 in seconds of arc on each side, or approximately 2,000 km × 2,000 km at the Moon's distance. Now if you tried an image scale of 0.1 seconds of arc per pixel, then the area of the Moon goes down to approximately 200 km × 200 km, or roughly two Plato lengths on each side. This may sound wonderful, but as the pixels are looking at 1/10th × 1/10th = 1/100th of the Moon's surface, as it was at the previous scale, there will now be 1/100th less light falling on each pixel, and so the exposure will need to be correspondingly a hundred times longer. In the turbulent Earth's atmosphere, longer exposures are not always possible, and anything longer than say 0.2 seconds is almost certainly going to be blurred, except under very good atmospheric conditions. Consequently, even with *Registax* [8] exposure will impose a limit on the best image scale that you can achieve. Because all cameras are different, it is up to the observer to find an optimum exposure range and image scale through experimentation at different lunar phases.

Another resolution-limiting factor is color or wavelength, as was shown in the telescope diffraction limit equation. In theory, blue light will give you higher resolution of the lunar surface than red, or the near IR. There may also be focal point issues because any transmission glass between the camera and the Moon will lead to chromatic aberration, or the focal point, and indeed image scale being slightly different according to spectral sensitivity. This effect is more acute in refractors (even through achromatic optics in the near IR), and less so in reflectors, but as Barlows or eyepieces are likely to be used, and there is a glass window over the camera chip, then there will always be some chromatic aberration effect present. The use of a near IR blocking filter or narrow band filters will reduce this effect considerably.

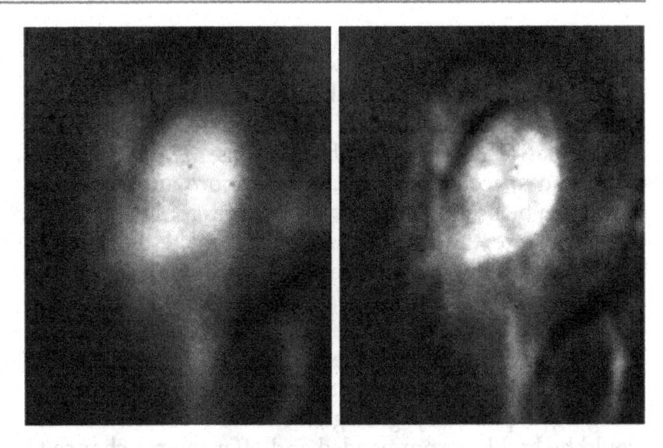

Fig. 2.3 Aristarchus crater captured by the editor on November 11, 1989. *Left*: A raw 1/25th second frame. *Right*: A result after aligning and stacking 110 frames in *Registax*, and sharpening

Finally, to illustrate the effects of using *Registax*, but without frightening off newcomers with staggeringly detailed images that will take a lot of practice to achieve, this editor has applied it to some old VHS video of Aristarchus taken back in 1989, using just over a hundred frames. Figure 2.3 (left) above shows one of the raw images looking rather blurred by atmospheric seeing conditions. Figure 2.3 (right) shows the result of applying *Registax* and sharpening. If you capture imagery of Aristarchus, or any other crater, after reading the online instructions and comments on astronomy forums about the use of this software, then you should be able to achieve far superior results than Fig. 2.3 under good conditions. Also many people advise that you check that your telescope optics are well aligned; otherwise image quality will drop off considerably. An excellent guide to high resolution astronomical imaging can be found in Martin Mobberley's book *Lunar and Planetary Webcam Users Guide* [9].

Lunar Features and Cataloging

One task that professional planetary geologists do not have much spare time to do is to go around looking at every part of the Moon's surface and viewing every available image in order to build up catalogs of features. Automated image searching software has been tried but is not as reliable as human vision, so far.

Fortunately, amateur astronomers have more time on their hands, and if organized properly and working in groups, can build up scientifically useful catalogs, as has been demonstrated already with domes. Here we include some suggestions of objects that amateur astronomers can catalog, measure, and quantify. This can be achieved using a combination of telescopic observations, *LTVT* [7] simulations, and spacecraft images, two or more approaches of which can be used to confirm any features found.

Please note, though, that in order to make the catalogs useful, it is really very important to publish these, or at least make them available to other astronomers or planetary scientists, perhaps via an on-line spreadsheet file or on a DVD upon request. Secondly, some verification process must be gone through to independently check each feature found. Bringing a planetary geologist onboard to offer advice would be highly recommended and offer them a chance to become involved in public outreach activities. If you do discover new features, send them to lunar sections of astronomical societies, or forums on MoonZoo, for second opinions from others and possible publication.

Visual observing and imaging are serendipitous methods of discovering previously unknown features on the lunar surface, when the illumination is just right, and also for confirmation of features found with the other two methods, so long as the feature size is not beyond the resolving capability of the telescope. Furthermore, observations through the telescope allow you to look at any part of the illuminated surface without waiting for visualization software to finish, or having to search through millions of spacecraft images.

Images are especially useful, as they have a known pixel scale, thus allowing direct measurements; the images can also be enhanced to make faint features easier to see. Visualization software, such as *LTVT* [7], although taking time to set up and run, do allow illumination to be controlled, and tried from solar altitudes and directions that never happen in nature. The latter, for example, is useful in making east–west aligned ridges on the equator, which are normally shadowless and difficult to detect, stand out with prominent north or south shadows. Visualization software, and especially telescopic observations, are limited in resolution; therefore spacecraft images are vital if we wish to look into the realm of sub-kilometer-to-meter scale lunar feature types. However, because of issues of interpreting the lunar surface under a single illumination, the availability of multiple illumination angle imagery, or stereo imagery, would help enormously in their interpretation.

Below are some detailed ideas for cataloging craters, domes, and rilles. Then examples of other features that you could catalog will be given. But the suggested descriptive categories, or measurements, will be left up to you. Hopefully the crater and dome sections should give you some ideas.

Craters and Impact Basins

Craters are everywhere on the Moon and by far the most prominent geological feature. They were formed when high-velocity objects struck the lunar surface at anywhere between a few kilometers per second up to many tens of kilometers per second. They range in size from microscopic up to hundreds of kilometers across. There are many more smaller craters than larger craters, and these are generally younger because any old small craters get eroded by subsequent impacts and are more difficult to see. Larger craters are deeper and less easily eroded, and therefore are older.

Geologists find that craters come in three size classes. "Simple craters" are smaller than 15–20 km in diameter and are bowl shaped. "Complex craters" are larger than simple craters and often have terracing in the walls, as well as central peaks. Around 300 km in diameter and above, craters transition into "basins," have outer rings, and sometimes have inner rings, too. Some of these more ancient basins are so highly degraded that they may only show up as partial circular arcs of rim topography. But this is something that amateur astronomers can look for through a telescope, or seek evidence for in the plates in this atlas or through *LTVT*-type [7] simulations at non-standard illumination directions.

Both visual observing near the terminator and *LTVT* [7] are also useful for finding evidence for ghost craters, which are craters that have been almost rendered invisible because they were highly degraded in the first place and were subsequently covered over by mare material. Their presence is seen only at low illumination angles near the terminator, where shadow and shading extenuates their underlying topographic relief on the mare covering. Many ghost craters remain to be discovered.

Apart from these simple size classifications of craters, there are three other classifications schemes, and these are based upon the morphology, or visual form, of the craters. In1963 Ralph Baldwin [10] devised a five-class scheme for crater classification according to roughness (Table 2.6), and he applies these to 199 craters in his *Measure of the Moon* book. Chuck Wood has another five-class classification scheme [11] that uses five named craters as examples, but has not published his catalog (Table 2.7). Florensky et al. [12] define crater morphologies based upon the following six craters: Schmidt, Dawes, Rümker, Tycho, Copernicus, and Plato (Table 2.8).

So, as you can see, there are many classification schemes. However, the classifications may be influenced by the illumination angle at which the craters have been studied. Higher latitude craters are seen more frequently in shadow, and therefore will tend to look rougher. Software such as *LTVT* [7] allows anyone who uses it the ability to illuminate a crater from any direction and at a fixed solar altitude, say 10°. This removes any effects of latitude that makes craters look rougher than they are.

Crater rays are another important aspect of craters and are best seen towards Full Moon, but can start to appear earlier. They are the result of ejecta thrown out during the formation of craters. The ray material must be a thin coating, because it does not usually show up in topography and is only visible because of its compositional and optical properties being different from pre-existing underlying lunar soil. Rays fade over millions of years due to space weathering and the constant bombardment of micrometeorites plus solar radiation. Hence it is only geologically young craters that we can now see with rays. The subject of rays have not been studied much by amateur astronomers,

Table 2.6 The Baldwin crater classification scheme

Class	Description	Example crater	Map location
1	Young craters with sharp rims lying on top of mare	Tycho	10 d5
2	Intermediate craters that may have been formed before mare flooding and have slightly degraded walls	Clavius	10 d3
3	Older craters with rims that have been degraded by subsequent impacts	Cyrillus	13 a5
4	Elderly craters that are heavily degraded	Walther	10 b7
5	Craters obscured partly by mare	Grimaldi	7 g1

Table 2.7 The wood crater classification scheme

Class	Description	Size range	Example crater	Map location
ALC	Bowl shaped	<=20 km	Albategnius C	13 f5
BIO	Bowl shaped but with a flat floor over part of bowl	<=20 km	Biot	15 d3
SOS	Shallow flat-floored craters	5-35 km	Sosigenes	1 d3
TRI	Collapses on interior rim that partly obscure an otherwise flat floor	15–50 km	Triesnecker	1 g2
TYC	Craters with different terrace levels on the inside rim and a partly flat floor	30–175 km	Tycho	10 d5

Table 2.8 The Florensky et al. crater classification scheme

Class	Description	Map location
Schmidt	Simple bowl-shaped craters	1 d1
Dawes	Craters with a knobbed and ridged bottom	1 b4
Römer	A pronounced central peak with the lower parts having a crater wall, terraces being near to the base. The floors are not typically flat.	1 a6
Tycho	A central peak that dominates over the impact melt/lava flooded floor	10 d5
Copernicus	Similar to the Tycho class, but has more flatness to the floor	5 e4
Plato	No central peaks or flat floors	2 h5

perhaps because they are difficult to map, there are so many of them, and because their edges are not always defined very well.

Below is a list of the many things that you can record about a crater. Do not be put off by the number of items. You could pick all, or just some, of these to start to build up a catalog. Of course in any spreadsheet catalog that you are building up, it would by necessity include the crater name (if it has one), the latitude, the longitude, and maximum diameter. ACT-REACT Quickmap [13] can be used to estimate crater diameters and rim widths, with the ACT-REACT Quick Map path tool. Other items you might like to record are:

1. What is the crater's morphological class? (Use some well defined example of morphological class craters for comparison, as mentioned above, or define your own.)
2. Does it have a flat floor, and if so of what area?
3. Does it have a dark lava-filled floor?
4. How many sinuous rilles does the crater interior have?
5. How many domes does the crater interior have?
6. How many graben does the crater interior have?
7. Does the crater have floor fractures?
8. Does it have any central peaks(s), and if so how many?
9. If the crater has rays:
   - What is the earliest range of selenographic colongitudes in which the rays first become visible, and the latest in which they disappear?
   - List the color seen in the rays in color images.
   - What is the length of the longest ray?
   - What is the length of the shortest ray?
   - What percentage of the rays would you say are dark rays?
   - Is there an obvious direction to the rays, and if so, on what azimuth of the lunar surface?
   - At what distance from the crater rim do the rays start?
10. How many light interior bands does the crater have?
11. Does the crater have a exterior dark halo? (This is best detected when the Sun is high above the lunar surface and there is little shadow.)
12. Does it have wall terracing?
13. Does it have a spiral pattern to its interior and walls?
14. What is the longest and shortest diameter of the crater?
15. If the crater is basin class in size (greater than or equal to 300 km diameter):
   - How many interior rings does it have, and what diameters?
   - How many exterior rings does it have, and what diameters?
16. Any other measurement or description you can think of?

Domes

Lunar domes are often smooth mounds on the lunar surface that may be volcanic in origin (see Mons Rümker in Plate 8e). It is not always feasible to get a definitive identification of whether they are volcanic or simply hills draped by basalt flows or impact melt; nevertheless it is possible to gather supporting evidence, e.g., if the topographically measured slope lies within the accepted range of volcanic domes, whether the domes have a color different to their surroundings, whether there is a craterlet on top of the dome and whether there are other domes in the area.

Extensive dome catalogs have been produced already, first by ALPO and subsequently by GLR [14], but it is always worthwhile to continue looking for previously undiscovered domes using a telescope, or spacecraft imagery. Visualization software can be used to illuminate the lunar surface from any direction; thus users can experiment to bring out even the slightest hint of a dome. Heights and topographic profiles can also be measured using QuickMap [13]. Descriptors that you can record about domes should definitely include its name (if it has one), its latitude and longitude, and its maximum diameter. Other items might include:

- What is the morphological class of the dome?
- Any summit craterlet?
- What is the height above the surrounding terrain?
- What is the average slope angle?
- Are there any other domes nearby forming a group, and if so what is the minimum and maximum distance to its closest and furthest neighbor?
- Is the dome in mare, highlands, or inside a crater?
- Are there any rilles nearby?
- Any other measurement or description you can think of?

Sinuous Rilles

Rilles are exciting features to find on the lunar surface, being more spectacular near the terminator. They appear to be winding former riverbeds but were carved out by lava flowing from high to low areas, the most famous example being Vallis Schoteri (see Map 8 e2).

You should not confuse sinuous rilles with graben. The latter are straight, mostly uniform in width, and have little respect for the topographic highs and lows that they pass through. Sinuous rilles can be singular, or multiple, and sometimes appear to start from an obvious outflow vent in the lunar surface.

The best way to catalog rilles is to describe them in the following ways:

- Rille name, or name of the nearest feature.
- Longitude and latitude limits of a rectangular box that would entirely enclose the rille.
- Minimum and maximum widths of rille.
- Minimum and maximum depths of the rille above the Moon's surface.
- Length of rille.
- Maximum number of levee levels inside the rille.
- Number of connecting rilles.
- Any other measurement or description you can think of?

Other Features

Rather than going into very specific descriptive and measurement details for every class of lunar feature, what follows below is a general summary of the common known classes.

GRABEN look like straight-sided rilles, but instead of flowing clearly downhill and winding around topography, they carry straight on through mare, crater floors, mountains, etc., before flattening out. A graben usually has a uniform width but can sometimes slowly broaden. Graben can be parallel to each other and on other occasion cross over each other. They happen when the crust of the Moon is pulled apart slightly under stress. The surface fractures in two parallel cracks, and the central block sinks slightly. Graben can be categorized in a similar way to rilles.

SCARPS are cliff-like structures such as Rupes Recta (see Map 9 b2-b3). Lobate scarps, a sub-class, occur where the crust is being forced together and one part moves over the other part. A lobate scarp has a topographic profile with a shallow slope on one side and a steeper slope on the other side that looks naturally lobate in appearance. One geologist described lobates as gradually advancing avalanches. Scarps in general can be categorized in a similar way to rilles, except that they are one-sided relief features and do not have levees. You can also include information on whether the scarps curve, and whether they have a lobate appearance.

WRINKLE RIDGES are best seen in mare close to the terminator (see Plate 6a, to the south and west of Mons Pico), as they are fairly low-lying features. They can occur where the mare lava has flowed over underlying crater rims, or where the edge of cooling lava has overflowed previous deposits. There may be some descriptive parameters, as used with rilles and scarps, that are applicable to these, although wrinkle ridges have less well defined edges.

SWIRLS are made of a light material with a swirl-like appearance. A good example is the Reiner Gamma formation (see Map 7 f4). At the time of this writing, the origin of these features is unknown, although it is speculated that they lie on the opposite side of the Moon to large impact basins and have magnetic fields strong enough to deflect low energy-charged solar wind particles. They have no measurable topography, and in a few cases cross highland and mare, though so far most have been found in mare areas. A catalog of names (or nearest features) and longitude and latitude limits would be useful to describe the locations of these features, as well as what selenographic colongitude limits they are visible over.

HOLLOWS are a relatively recent category of feature, discovered during the Space Age, because the high resolution from spacecraft imagery was needed to appreciate their nature. They are basically irregularly shaped holes in the surface that are of the order of a kilometer in size or smaller, do not have raised rims, and have floors that differ in appearance from the surrounding terrain. Often they appear as though the mare lava had avoided the holes, or some geological process had excavated away the material from within. Two well-known examples are Ina and a collection of hollows on the floor of Hyginus crater.

Geological Mapping and Stratigraphy

Unless you are already a geologist, or have studied geology at a university, this task is best achieved by working in a group to check each other's work and agree by consensus.

The idea is to draw boundaries around regions of uniform textured areas to represent different geological units. The number density of craters on each unit must be visually similar. Older terrain has more craters per square kilometer than younger terrain.

On an image mosaic, overlay your boundaries and color in regions of similar age and texture. This in a very crude sense is how planetary geological maps are made. For each colored geological unit, provide a key or legend that describes the morphology and texture. For a better understanding of what you would be expected to do, take a look at some example USGS geological maps of the Moon [15], and more importantly read a good lunar geology book [16]. In this modern era we now have mineral composition maps produced by spacecraft using different filter wavebands. These can aid with the interpretation of units that make up geological maps.

Lunar geological maps were produced mostly in the 1960s and 1970s, and relatively few have been published officially since then. Therefore any maps made of new areas, and at higher resolution, may be of benefit to the planetary science community, so long as these are checked by geologists and made available to a wider audience online.

Another obvious task that could be accomplished by amateurs, and which goes hand in hand with the crater catalog and geological mapping, is to work out which feature overlies another feature, in a sense stratigraphy. This can give a relative "for and after" chronology, for example, if a small crater lies on top of a big crater, then clearly the small crater must be younger.

Occultations

Observations of these were originally intended to get the shape of the Moon around its limb, and to map out the polar regions around the edges of permanently shadowed crater rims. However, now that we have LOLA (Lunar Orbiter Laser Altimeter), this mission has largely been fully accomplished. Nevertheless, it is still worth observing to act as an independent check on the LOLA results.

Occultations also provide the opportunity to look for evidence of double or multiple stars. You can find out about when occultations are about to happen at your observing site by downloading the *Occult* software from the IOTA website [17]. Occultations of stars as faint as magnitude 10 are possible to see against the dark limb of the Moon. Reappearances on the bright limb are especially tricky to observe, though, because of stray light and not knowing exactly where to look along the limb. Although we have not covered lunar eclipses much in this advanced section, these do at least offer an opportunity to search for fainter occultations if the eclipse is especially dark.

In an ideal situation one would use a photometer attached to the telescope focal plane and center this on the star, and sample the light at, say, a thousand times per second, storing the results digitally on a computer. At this speed you should be able to detect Fresnel diffraction fringes off the edge of the lunar limb a fraction of a second before occultation.

A fiber optic at the focal plane, coupled to a high-speed light-sensitive photodiode, can be used to measure the light output. Older photomultiplier tubes, or indeed photon counting devices, could be utilized but are expensive and can be easily swamped by the glare of Moonlight. Recording of good low-noise occultation light curves can be used to look for very close double stars, or measure stellar angular diameters down to 1/1,000 of a second of arc. Stars with measurable stellar diameters tend to be of spectral type K or M. Use of filters helps to isolate the Fresnel diffraction fringes.

Slower sampling speeds at TV frame rates of 30 frames per second in the United States and 25 frames per second in Europe are found on low light CCTV cameras such as those described in the impact flash section. By de-interlacing odd and even lines you can achieve sampling rates of 60 fields per second in the United States or 50 fields per second in Europe, but this is still not fast enough to resolve most stellar diameters and many close double stars. An exception occurs, though, when the occultation is close to the polar areas of the Moon. Then the relative approach rate to the limb is slowed down from the usual 0.5 seconds of arc per second (approximately 1 km/s at the Moon's distance), and this will make diffraction fringes, close doubles, and diameter effects last longer, although irregularities in limb topography will start to contribute more noise.

For the brightest occultation (magnitude 6 and brighter) it should be possible to undertake imaging spectroscopy by placing a low-resolution diffraction grating, e.g., 50–200 lines per mm, in front of the CCD camera window. An example spectra, albeit not from a star, is shown in Fig. 2.4. The spectra produced will have a profile that varies according to the temperature of the star. A cooler star will be brighter in the red and near infrared part of the spectrum. As the occultation takes place, bands may appear in the spectra, associated with the Fresnel diffraction fringes.

It is possible that different parts of the stellar atmosphere might also be revealed as the last parts of the star are eclipsed by the lunar limb, and some strong spectral lines may show up, too. However your best chance of seeing these would be if the star was undergoing a near grazing occultation close to the lunar poles, so that events are slowed down. Needless to say, some standard spectra should be taken of some non-occulted stars of known spectra and temperature in order to calibrate the spectral dispersion.

Earthshine

An interesting project would be to conduct time lapse imaging of Earthshine. Observational studies by professional astronomers prove that there can be significant changes in brightness of Earthshine occurring in the order of tens of minutes, as cloud banks rotate around the limb of Earth. There are also landmass and ocean effects according to which part of Earth's hemisphere is facing the Moon, and which fraction of this is illuminated. Seasonal effects have also been detected.

A longer term study by amateur astronomers could replicate some of this work if the calibration and stray light

Fig. 2.4 A 1/60th second de-interlaced video frame showing spectra with absorption bands from a bright strobe light from an aircraft that happened to be passing through the line of sight between the telescope and the Moon. Captured by the editor, on February 25, 2001, using a holographic diffraction grating

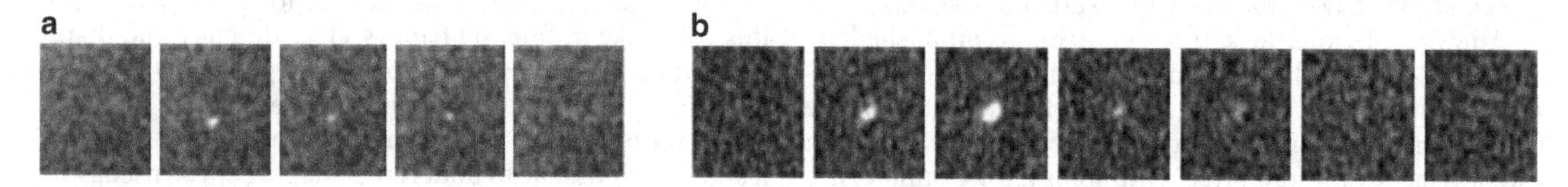

Fig. 2.5 A 1/60th second de-interlaced video sequences of lunar impact flashes from the 2001 Leonid meteor shower as videoed by the editor. (**a**) November 18, 2001, UT 23:19:15, magnitude 6. (**b**) November 19 UT 00:18:58., magnitude 5

issues can be resolved. Advanced users could attempt to make quantitative measurements of this by taking two exposures, one for Earthshine and one for the sunlit side. They should then be able to compare the brightness of features in Earthshine with those on the daylight side of the Moon for specific phases. Filters would be used to ensure that the effects of atmospheric absorption can be compensated for, as is done in astronomical photometry. What would be more of challenge would be to calibrate out scattered light effects from the bright side, perhaps by using a mask in the telescope focal plane?

Impact Flashes

Surprisingly, impact flashes [18] from meteorites striking the surface of the Moon can be detected from Earth (see Fig. 2.5). They occur at a rate of one every few hours on average, more frequently during meteor showers and less frequently at other times. These are best searched for with a low-light CCTV camera, for example the Watec 902H. Impact flashes as faint as magnitude 10 can be detected with amateur equipment, although it may take as long as 10 h before you detect one, and they are rarer still for brighter flashes. For maximum sensitivity, work in white light, without any filters, and use either a small f/No telescope, or a focal reducer with your present telescope to bring the f/No down. This will both increase image contrast on the night side and expose the camera to a larger part of the Moon's surface. Multi-station observing is very important, because cosmic ray events can look very similar to impact flashes; however, if video recordings from two or more observing stations have been captured, then cosmic rays will be visible in one camera only, whereas impact flashes will of course be visible in both.

Experience has shown that impact flashes are more readily detectable between New Moon and First Quarter than between Last Quarter and New Moon. At phases in between the brightness from the illuminated side makes detection of faint impact flashes on the night side of the Moon more difficult. Also, due to atmospheric absorption and scattering at low altitudes above the horizon, impact flash detection is optimal at altitudes of greater than 15° above the horizon, although if an observer is desperate for clear sky sessions then lower altitudes are practical so long as this is for small crescent phases, and when plenty of detail remains visible in Earthshine other than just the limb.

If possible, observers at different sites should video through different wavebands. Analysis of such results would enable the temperature of the flash to be determined using the black body radiation law. Red and near-IR filters are better than blue or UV because more of the radiated light is at the long wavelength end of the spectrum. For astronomers with a lot of time on their hands, and spare telescopes, using a low resolution and low dispersion diffraction grating in front of the camera might make an interesting experiment to see if you can capture spectra (See Fig. 2.4) of an impact flash. You would do well, though, to focus and calibrate on stars beforehand to make sure that you know the faintest magnitude that you are likely to record a spectra of during the night concerned, and to calibrate the spectral dispersion. This editor has recorded video spectra off stars as faint as magnitude 6, and so it is not inconceivable that if you observe Earthshine for many hundreds of hours that you could capture a spectrum of an impact flash. This would be a quite a scientific achievement because it has not been accomplished at the time of this writing.

In a similar context, impact flash observing has so far been attempted on the night side of the Moon. Well, if you consider the brightness of flashes against the daylit side of the Moon, near the terminator, then the lunar surface is still significantly darker compared to the brightness of the Full Moon, and some of the brightest flashes may be seen against the lunar surface.

To improve chances of detection on the day side, observations could be tried in the near IR, where the impact flash is brightest, or through a narrow band Sodium D line filter (589 nm). The latter has been chosen because it is quite common for meteors in our own atmosphere to exhibit this spectral line, and so one might expect a concentrated sodium flash on the lunar surface during an impact. Needless to say the narrower the spectral width of the sodium filter, the better the chances of detecting a sodium flash over general sunlight.

If using software such as *LunarScan* [19] to detect flashes on the dayside of the Moon, near the terminator, it would be important to mask out bright and dark crater edges to avoid atmospheric seeing effects and triggering false detections of impacts. Another good opportunity to observe impact flashes is during a lunar eclipse, assuming that the umbra is sufficiently dark. However, experience has shown that the eclipse is bright in the near IR, and so observers might consider using a near IR blocking filter to darken the background Moon, or try out a green or red filter.

Video is best recorded in raw format, e.g., monochrome TV frame size running at 30 or 25 frames per second with 8 bits per pixel. These will acquire data at a rate of approximately 26 GB per hour. However, most amateur astronomers use digital video format, where each frame is compressed into a JPEG image and stored in an AVI file. This will accumulate data at a rate of approximately 13 GB per hour. MPEG recordings should be avoided because although these acquire data at only 4.7 GB per hour, they do suffer considerable compression artifacts due to interpolation, making analysis of any impact flashes unreliable.

Impact flash recording should be time stamped, either using the file creation time on the computer (if this can be proven to be reliable), or by using audio time signals from a radio clock such as the station WWV in the United States on 2.5, 5, 10, 15 and 20 MHz (120, 60, 30, 20 and 15 m) short wave, station CHU in Canada on 3.33, 7.335, and 14.67 MHz (90.1, 40.9, and 20.45 m), and from MSF in the UK on 60 KHz.

After an impact flash observing session, software such as *LunarScan* [19] can be used to search for flashes rather than manually watching frame after frame in slow motion. If you discover any flashes, send the results to any astronomical society you belong to, to encourage others to observe, but more importantly send them to NASA's Marshall Space Flight Center, where they catalog all such events. You should find this atlas helpful in identifying the nearest lunar crater to the impact flash site if you compare Full Moon-type images with the Earthshine video. What you see in Earthshine is essentially what you would see on the Full Moon, only considerably less bright.

Transient Lunar Phenomena

This is a potential minefield to study, and at the time of this writing many astronomers and planetary scientists treat this topic with disdain. Their suspicions about TLP form around three main arguments. First, most planetary scientists will tell you that the Moon is a geologically dead world with the last major lava flows occurring thousands of millions of years ago. Second, even if there were remnant outgassing, then this must not contribute much to the already tenuous entire lunar atmosphere, which is so thin that it is equivalent to approximately the amount of gas present in a Zeppelin airship here on Earth. The six Apollo lunar landers actually contributed significant amounts of gas to the previously thin atmosphere, as confirmed by surface experiments left behind. Because no further detectable large boosts to the lunar atmosphere were found over the years that the experiments operated, the amount of outgassing elsewhere must be insignificant. Finally, critics point to the unreliability of at least some of the observed TLP reports.

Given these criticisms it may seem odd that the subject is even mentioned in this atlas and is suggested as being worthy of study. However the topic of TLPs is given some space in this chapter because it is important to train amateur astronomers in recognizing effects that are not TLP, so that they do not repeat the mistakes made by some earlier observers. Another reason for covering the subject in detail is that there have been several publications in highly regarded scientific journals of theories to explain TLPs [20] as well a few published accounts of TLP observations that passed the rigorous refereeing process that such journals adhere to.

Most of the proposed theories involve dust particles that are kicked up above the surface from underground gas reservoirs by relatively small releases of radiogenic and other gases from quakes, landslides, impacts, or tectonics. The theories suggest that dust can either raise the surface brightness temporarily, absorb, or scatter light, producing colors against the surface. Other theories argue that because dust particles can become charged in darkness, they repel and can levitate in clouds until their charge is dissipated in sunlight. These explanations do not account for all TLP sightings, but the investigation of the suspension of dust above the lunar surface is one of the key exploration aims of NASA's LADEE mission. Studies on the statistics of TLP all suggest that they are exceedingly rare, and the chances of seeing a good event would be one per several thousands of hours of observing for a typical amateur astronomer.

In view of the likely rarity of TLP, a practical search strategy might be not to bother to look for them at all, but to be prepared and know exactly what to do in the event that you see something suspicious. Below are types of possible TLP events that people claim to have seen. To assist you in determining whether an appearance of a lunar feature is abnormal, and merits even being considered as a TLP, some flow chart-style checklists have been included in Appendix 2 to help minimize your chances of making the same mistakes that some observers have made in the past.

- Obscurations are where there is an apparent lack of detail on part of the Moon, and you would expect perhaps to see more detail. Be warned, though, that Earth's atmosphere is almost certainly the culprit, or it may have something to do with that specific area of the Moon being smooth anyway (see Appendix 2, Fig. A.1).
- Colored TLPs are sometimes seen on the Moon but can be caused easily by atmospheric effects, optics, or even natural color on the Moon (see Appendix 2, Fig. A.2). Aristarchus has a natural faint blue tint, and an even fainter yellow-orange compositional color can sometimes be seen on the floor of Bullialdus when the atmospheric conditions and illumination are right.
- Brightenings are when an area of the Moon looks brighter than it should be, or where it appears to have changed in brightness during the time that the observer was studying the Moon. Brightenings can happen on the day side of the Moon (see Appendix 2, Fig. A.3) and on Earth's lit side. As you can see from Appendix 2, Fig. A.4, there are many reasons to be extra careful with TLP sightings on the night side of the Moon.
- Shadow-related TLP This can take the form of detail seen inside shadow when the Sun is too low to light up even raised topography, or transient gray shadows (See Appendix 2, Fig. A.5). Be warned, though, that for some of the brighter ray craters, scattered light off the inside of a sunlit rim can illuminate the floor weakly, so only repeat observations at similar lunar phases can prove or disprove whether the effects seen are abnormal.

Although not depicted in the tables, a third false TLP effect has sometimes been seen, namely faint spur or comet-like tails coming off peaks, or other topography protruding from shadows. These are atmospheric seeing flare effects, and if you see one, there are likely to be others if you look elsewhere on the Moon.

- Flashes of light have sometimes been seen on both the day and night sides of our Moon (see Appendix 2, Fig. A.6 in this book), but again there are many straightforward reasons or explanations to account for these, unless they are impact related.
- Moving objects are occasionally seen passing across the Moon, or noticed at its limb. These are almost certainly chance alignment effects, along the line of site, of bugs, birds, aircraft (see Fig. 2.4), and meteors in our atmosphere, or satellites. Also, if you work out the distances involved, then the speed of passage across the lunar disk (approximately 0.5° in diameter), it is always way too fast for large meteorites or impact ejecta at the Moon's distance, and so therefore must be closer to us. You can pretty much classify any object that you see moving across the lunar surface through a telescope as a non-TLP.

In all of the flowcharts (Appendix 2 in this book), the rules given are only guidelines, and it is really common sense to check other features to see if these show similar effects. Once all the checks have been gone through to establish some credibility of any suspected TLP, then further courses of action are needed.

First you should alert other observers to check out the feature concerned for independent confirmation. However, you must never tell them what they should expect to see, lest this leads to biased reporting. So either phone or text to your astronomical colleagues, or contact the coordinator of the ALPO and BAA Lunar Section TLP team. Second, get back to observing as soon as possible and be quantitative in recording what you see. If you are observing visually then try to make some sketches, estimate the size and position of the TLP, and note down every conceivable change that you detect, including color if relevant, or brightness compared to other parts of the Moon. Checks should be made regularly on other features for similar effects.

Always note down the time that the TLP is no longer visible. If you are taking images, capture these at regular intervals so as to monitor the development of the TLP, and again if it is a color TLP, try different colored filters.

So you should now be primed and ready to react in the unlikely event that you do witness a TLP. What else can be done? As was discussed in the beginners' section, it is a very good idea to re-examine former TLP sites under the same illumination conditions as past TLPs – and take high resolution images that you can submit to astronomical societies. Doing so may help discover that what the observers saw was just the normal appearance of the feature. In any case it will give you a good impression of what may have caught observers out in the past, and tune your own interpretation of the lunar surface.

For die hard enthusiasts who want to search methodically for TLP [21], then they could try monitoring color on the Moon in real time using a color webcam, or a color CCTV camera. These devices are more sensitive to color than the human eye, especially if you increase the contrast and color saturation on your video capture window, or TV. Alternatively amateur astronomers could just try time lapse imaging of the Moon over many hours, on either the day or Earthshine sides, and then play back the recordings at high speed, looking for brightness changes either manually or using computer software. Even if they do not detect any TLPs, the passage of the terminator and changing shadows should look impressive.

References

1. U. S. Naval Observatory (USNO) clock (2012) (http://tycho.usno.navy.mil/cgi-bin/anim).
2. Espenak, F. (2012). NASA Lunar Eclipse website (http://eclipse.gsfc.nasa.gov/lunar.html).

3. BAA Lunar Section/Aberystwyth University website for TLP repeat illumination predictions (2012) (http://users.aber.ac.uk/atc/tlp/tlp.htm).
4. Moon Zoo website (2012) (http://www.moonzoo.org/)
5. Moon Mappers website (2012) (http://cosmoquest.org/mappers/moon/).
6. NASA's Lunar Reconnaissance Orbiter LROC Target Acquisition website (2012) (http://target.lroc.asu.edu/output/lroc/lroc_page.html).
7. Lunar Terminator Visualization Tool website (2012) (http://ltvt.wikispaces.com/LTVT).
8. Registax website (2012) (http://www.astronomie.be/registax/).
9. Mobberley, M. (2006). *Lunar and Planetary Webcam Users Guide*, Springer-Verlag, London, pp. 79–174.
10. Baldwin, R. B. (1963). *The Measure of the Moon*, University of Chicago Press.
11. Wood, C. A., and L. Anderson (1978). New morphometric data for fresh lunar craters. Proceedings of the 9th Lunar and Planetary Science Conference. pp. 3669–3689.
12. Florensky, C. P., A. T. Basilevsky, and N. N. Grebennik (1976). The Relationship between Lunar Crater Morphology and Crater Size, *The Moon*, pp. 59–70.1
13. ACT REACT Quick Map website (2012) (http://target.lroc.asu.edu/da/qmap.html)
14. Lena, R., C. Wöhler, J. Jim Phillips, and M. T. Chiocchetta (2012) Lunar Domes: Properties and Formation Processes, Springer-Verlag, Italy.
15. USGS Atlas of the Moon website (2012). (http://www.lpi.usra.edu/resources/mapcatalog/usgs/)
16. Wilhelms, D. E. (1987). The Geologic History of the Moon. USGS Professional Paper 1348.
17. International Occultation Timing Association website (2012) (http://www.lunar-occultations.com/iota/iotandx.htm)
18. Cudnik, B. (2009) *Lunar Meteorite Impacts and How to Observe Them*, Springer, New York.
19. *LunarScan* software website (http://www.gvarros.com/lunarscan15.zip)
20. Crotts, A. P. S. (2009) "Transient Lunar Phenomena: Regularity and Reality," *Astrophysical Journal*, 697(1), pp 1–15.
21. British Astronomical Association Lunar Section Transient Lunar Phenomena (2012) (http://www.baalunarsection.org.uk/tlp.htm).

Maps and Plates

Keys to Maps and Plates

0240 UT Aug 05 1963. Moon's age 15.2 days. Sun's selenographic colongitude = 91.5°.

Map 1

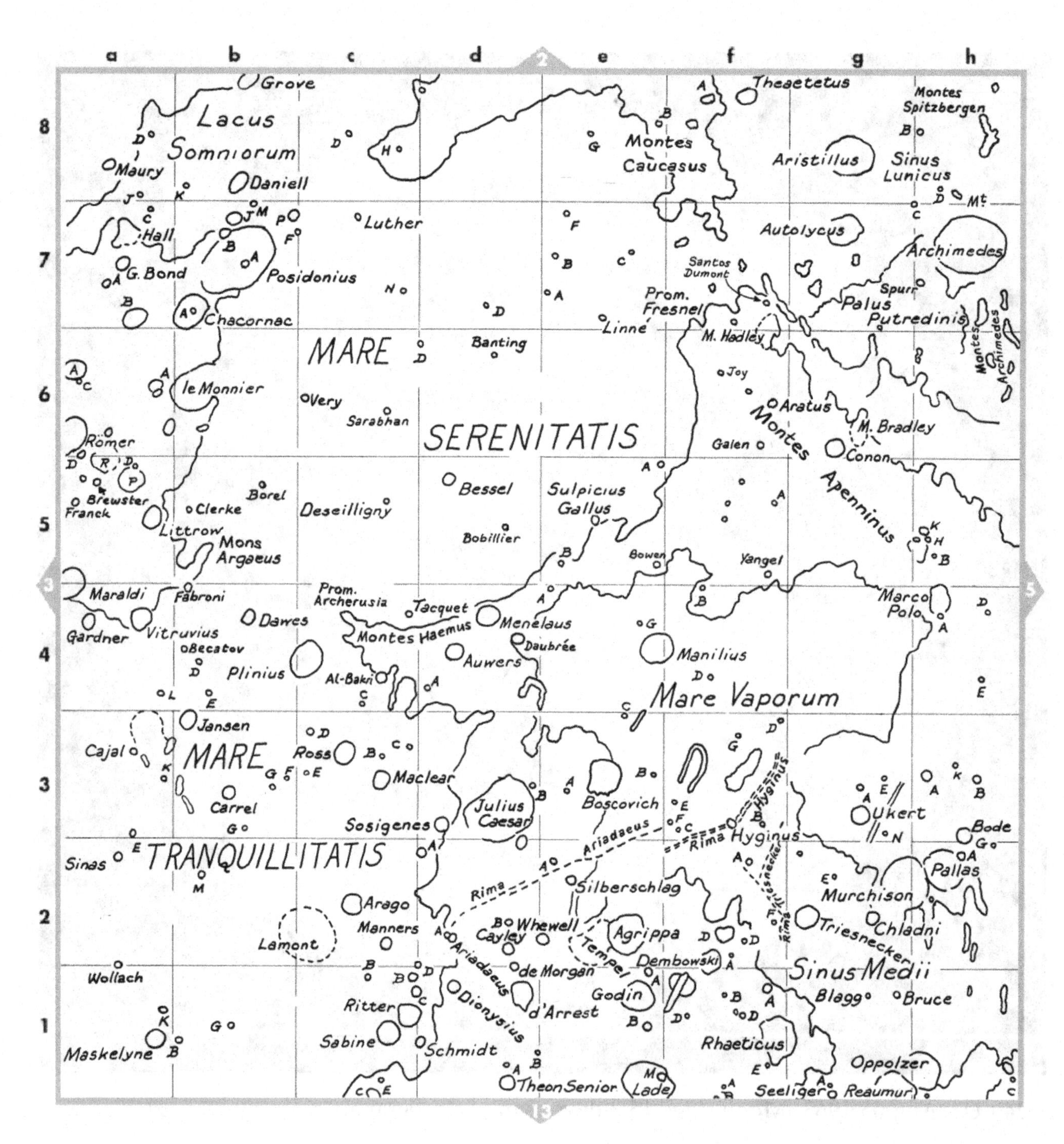

Crater Diameters

Archimedes	81 km (h7)
Posidonius	101 km (b7)
Manilius	38 km (e4)
Maskelyne	22 km (a1)
Bruce	6 km (g1)

This map has been prepared from Plate 1a. Plates 1b, 1c, and 1d show the same area under different lighting. Plate 1e shows larger scale photographs of the Triesnecker (g2) and Linné (e7) areas.

Plate 1a

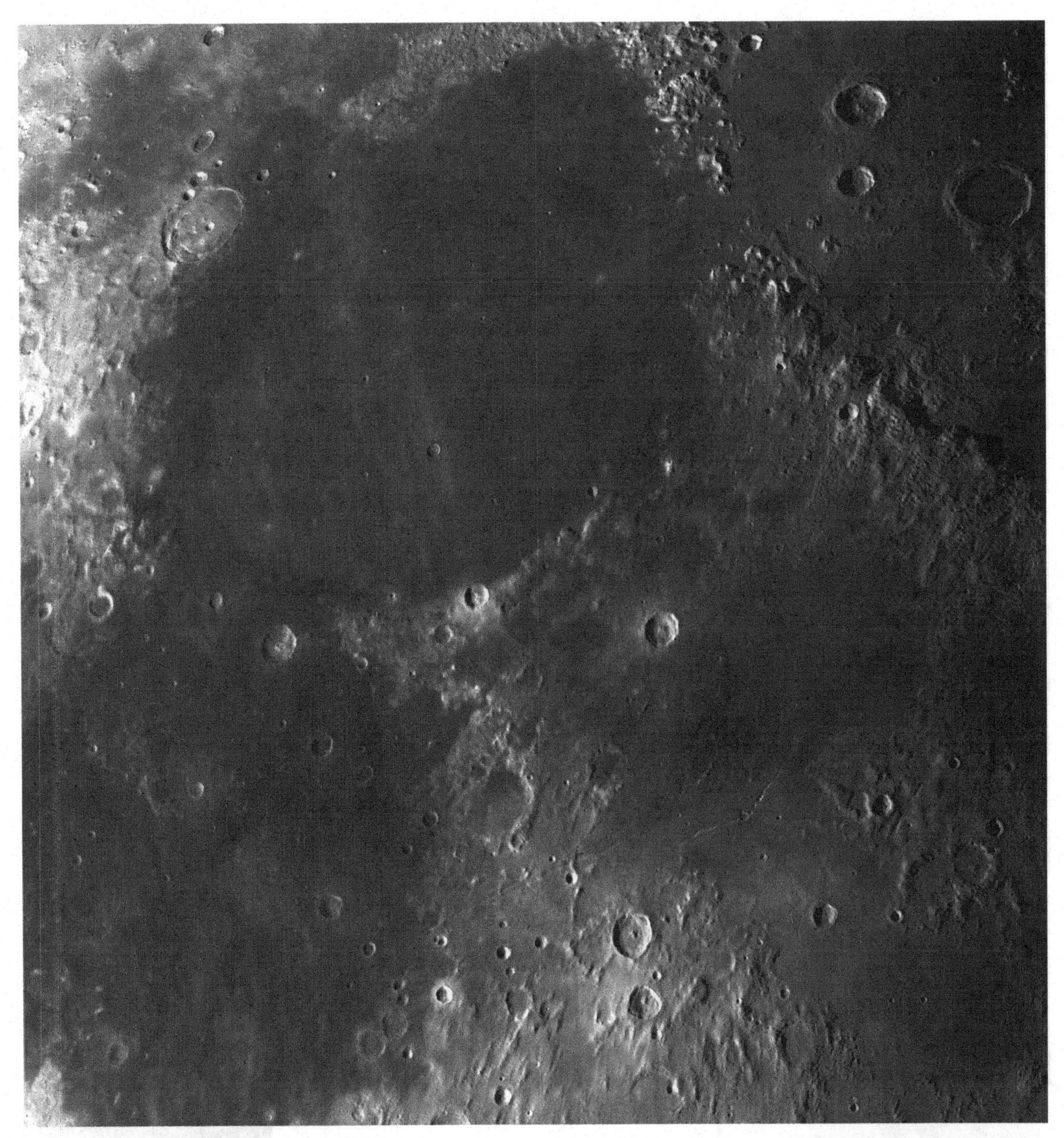

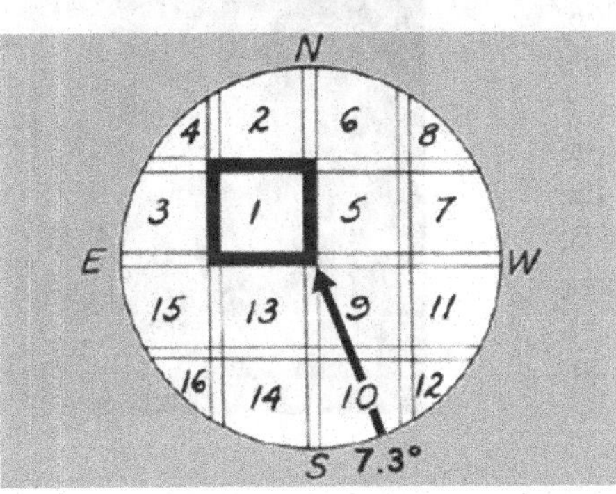

2046 UT Apr 28 1966. Moon's age 7.9 days. Sun's selenographic colongitude = 12.3°. Diameter 64 cm. The Montes Appeninus (g5) rise to 4,900 m in places.

Plate 1b

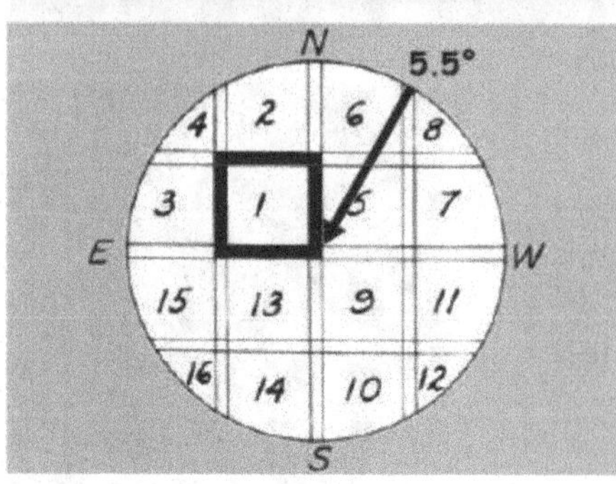

0215 UT Aug 06 1966. Moon's age 18.9 days. Sun's selenographic colongitude = 144.5°. Diameter 64 cm. Note the ridges on the mare to the southwest of Posidonius (b7), and the domes near Arago (c2). These are shown on a larger scale on Plate 3f.

Plate 1c

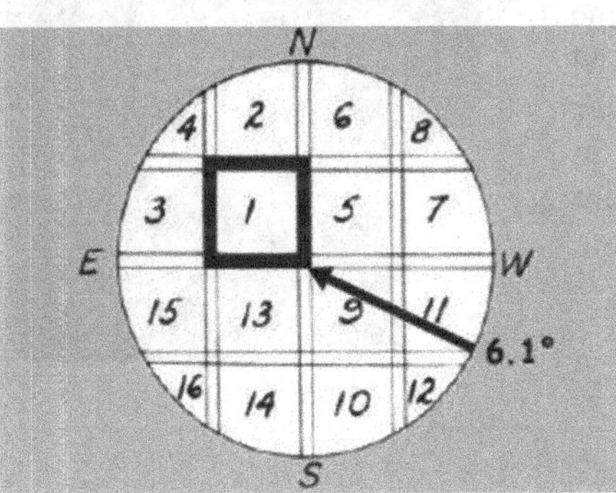

2137 UT Dec 25 1966. Moon's age 13.8 days (2 days before Full Moon). Sun's selenographic colongitude = 72.7°. Diameter 64 cm. Compare this with Plate 1a, which shows almost the same area.

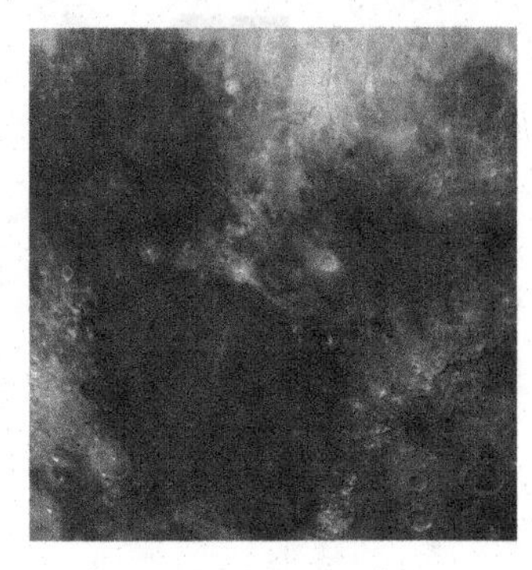

Plate 1d

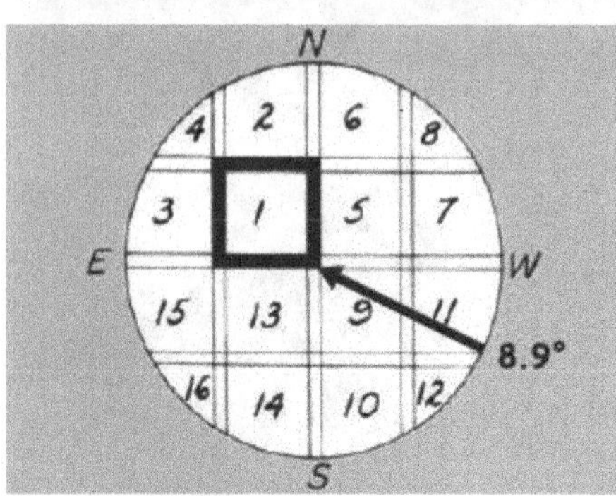

1835 UT Mar 18 1967. Moon's age 7.6 days. Sun's selenographic colongitude = 0.9°. Diameter 64 cm. Although the Moon's age here is almost the same as in Plate 1a, the aspect is different because of the different libration.

Plate 1e

Triesnecker, Hyginus and their systems of rilles. 2016 UT May 16 1967. Moon's age 7.2 days. Sun's selenographic colongitude = 1.5°. Diameter 94 cm approx. Triesnecker lies in Map 1 g2.

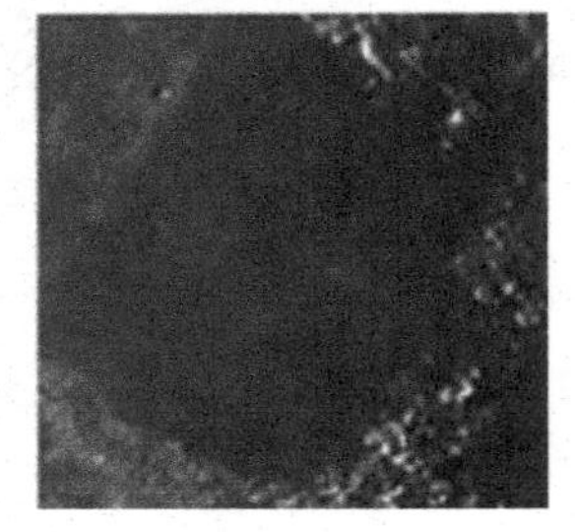

Linné, Bessel and Sulpicius Gallus in the Mare Serenitatis. 2017 UT May 16 1967. Moon's age 7.2 days. Sun's selenographic colongitude = 1.5°. Diameter 94 cm approx. Linné lies in Map 1 e7.

Map 2

Crater Diameters

Autolycus.................................... 39 km (f1)
Daniell .... 28 km (a2)
Egede .... 34 km (e4)
Strabo .... 55 km (b7)
Gioja .... 42 km (f8)

This map has been prepared from Plate 2a. Plates 2b, 2c, and 2d show the same area under different lighting. Plate 2e shows larger scale photographs of the Vallis Alpes (f4) area.

Plate 2a

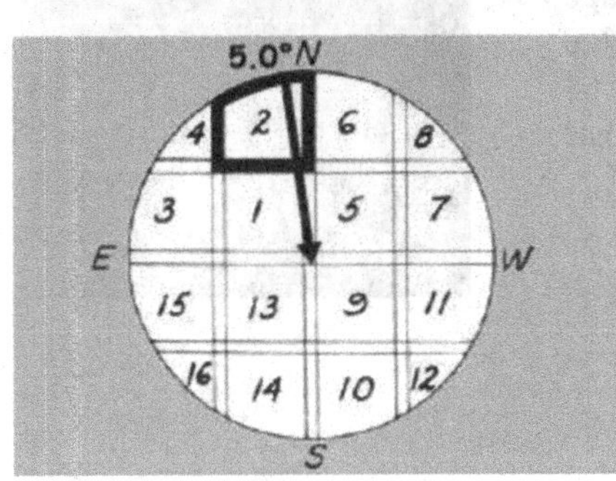

1814 UT Nov 22 1966. Moon's age 10.2 days. Sun's selenographic colongitude = 29.7°. Diameter 64 cm. This is a very favorable northerly libration, which is not likely to be seen very often. Note that Plato (h5) is very nearly circular. The Montes Alpes are about 3700 m high in places.

Plate 2b

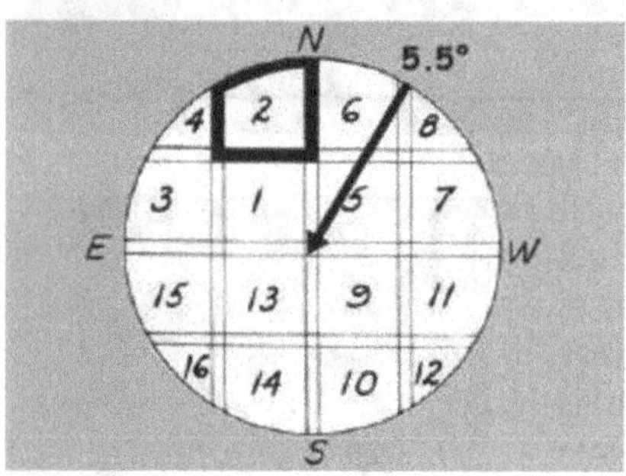

0211 UT Aug 06 1966. Moon's age 18.9 days. Sun's selenographic colongitude = 144.4°. Diameter 64 cm. Compare this libration with that of Plate 2c.

Plate 2c

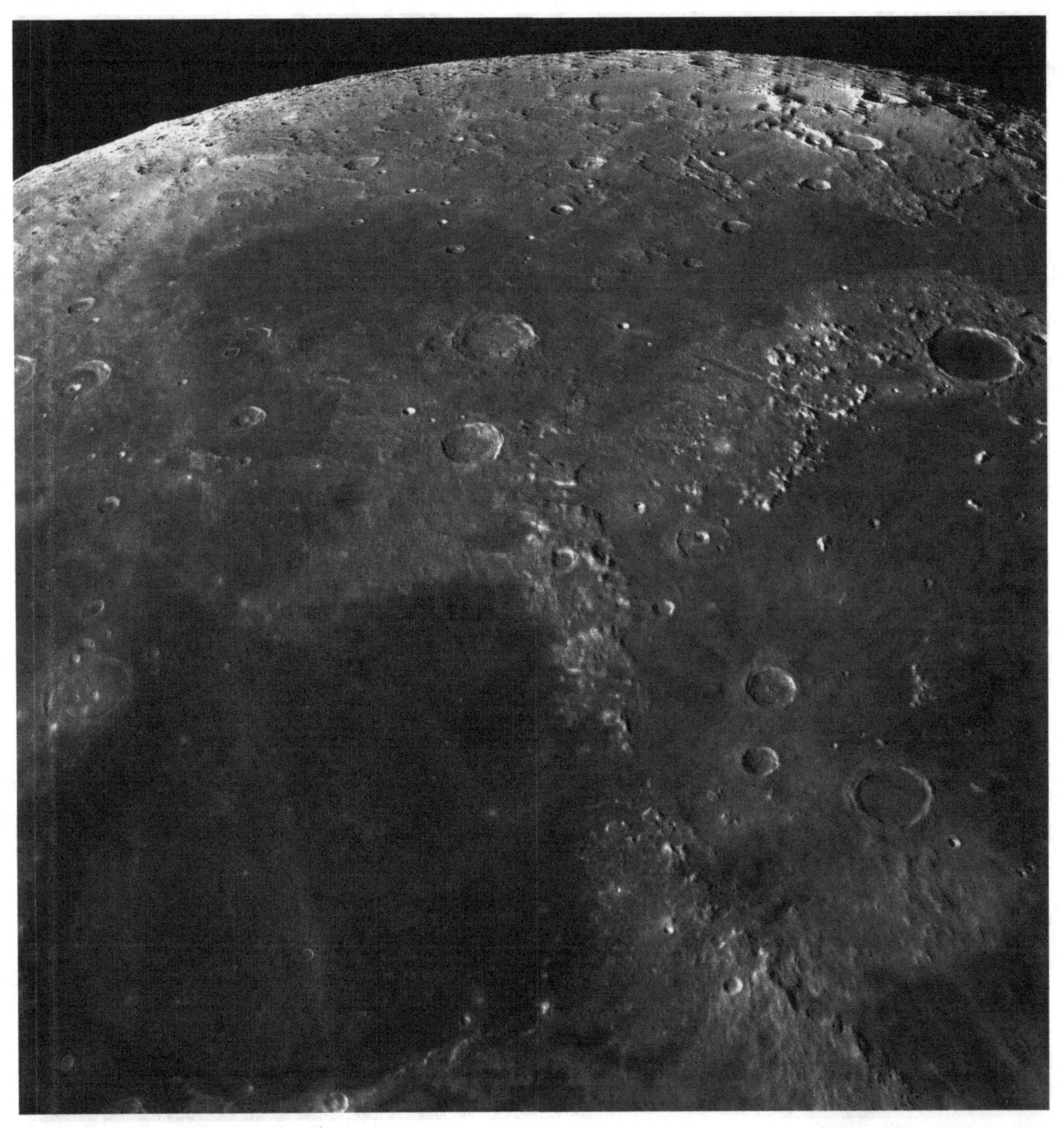

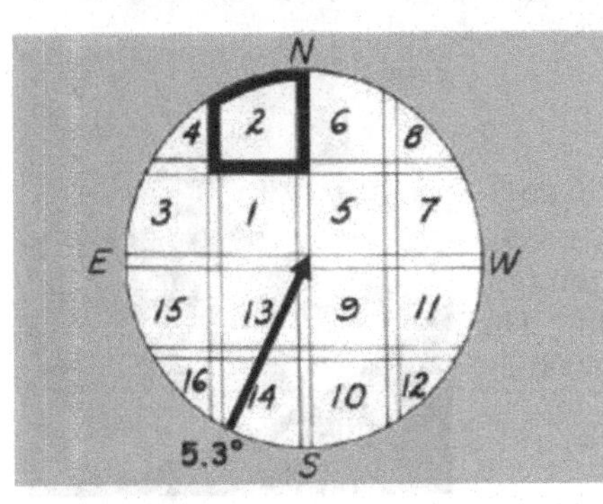

2103 UT May 29 1966. Moon's age 9.4 days. Sun's selenographic colongitude = 31.0°. Diameter 64 cm. This is a relatively bad libration for this area. Note how much closer Plato (h5) is to the limb here than in Plates 2a and 2b.

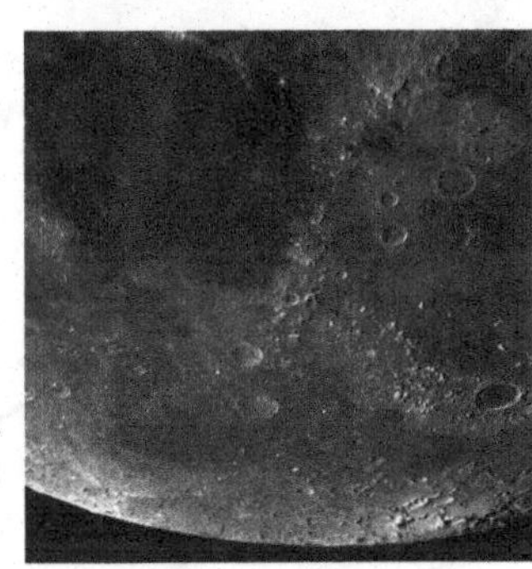

Plate 2d

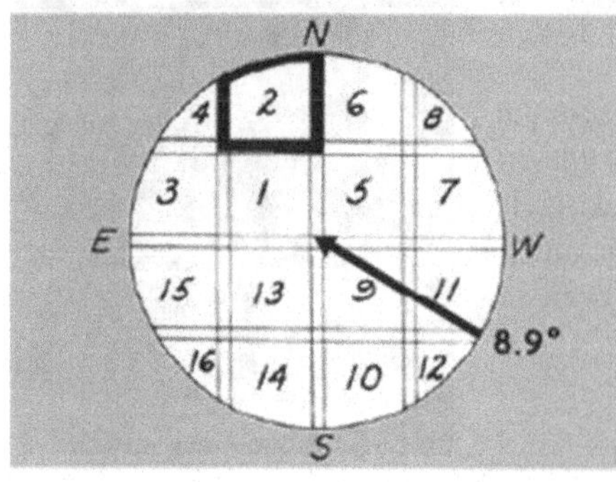

1835 UT Mar 18 1967. Moon's age 7.6 days. Sun's Selenographic Colongitude = 0.9°. Diameter 64 cm. Note how Mons Piton (g3) catches the early morning sunlight. The Montes Caucasus (e2) rise to about 6000 m.

Plate 2e

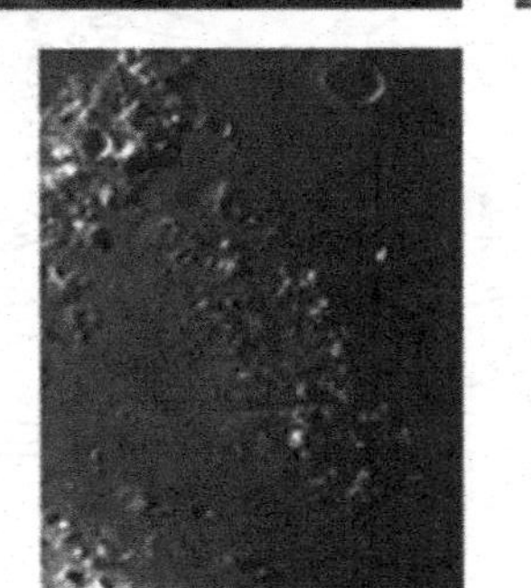

Aristillus, Cassini and the Vallis Alpes. 2012 UT May 16. 1967. Moon's age 7.2 days. Sun's selenographic colongitude = 1.4°. Diameter 94 cm approx. The Vallis Alpes lies in Map 2 f4.

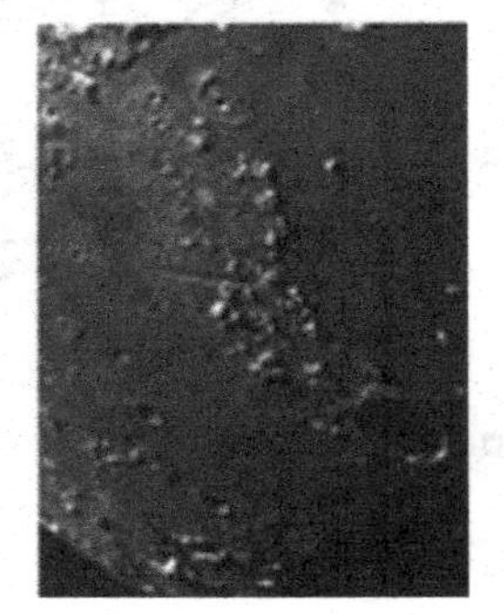

Cassini, the Vallis Alpes and Plato. 1925 UT Mar 19. 1967. Moon's age 8.6 days. Sun's selenographic colongitude = 13.5°. Diameter 91 cm approx. Note the "ghost" crater ring just South of Plato. This has been called "Ancient Newton".

Map 3

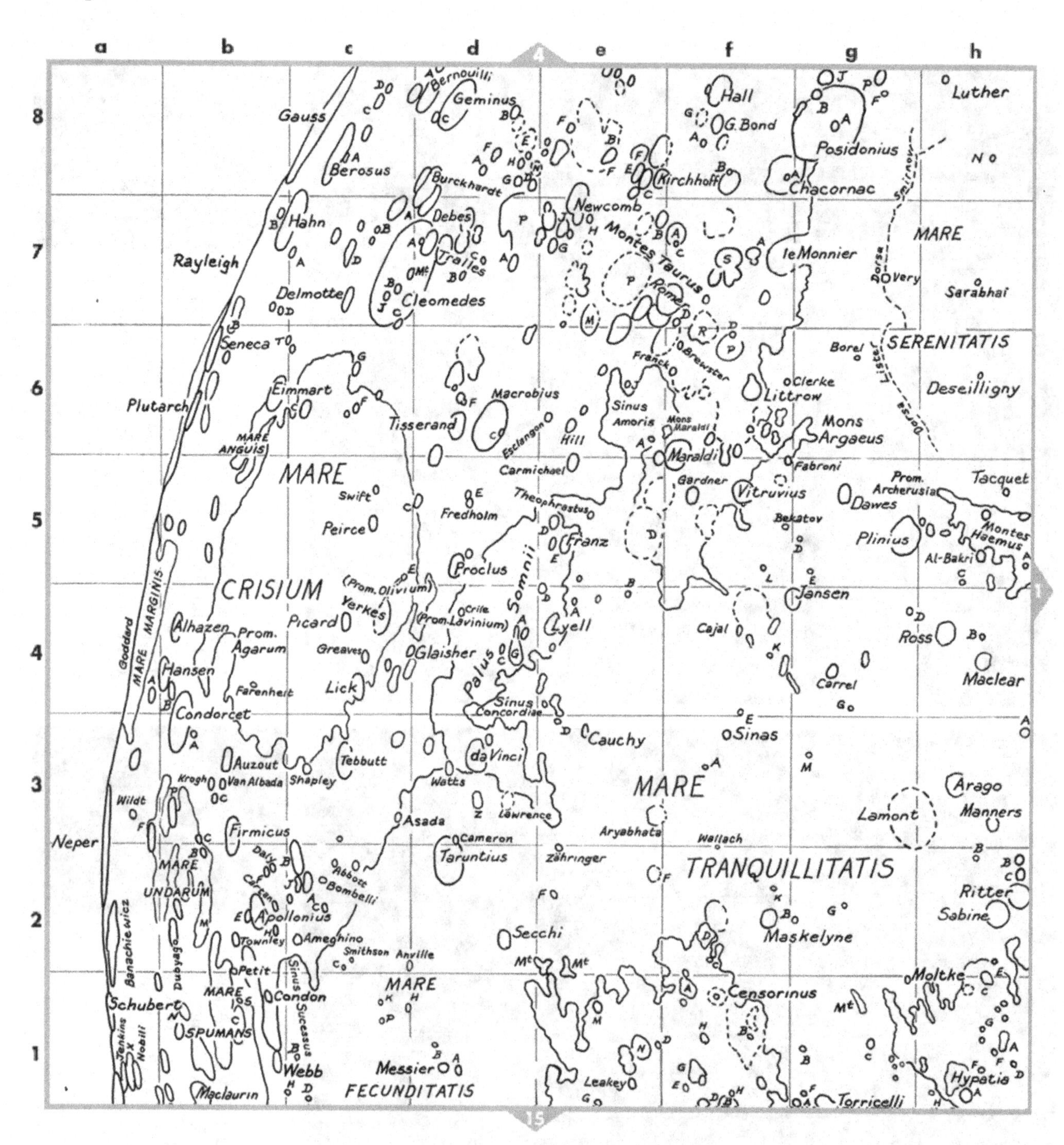

Crater Diameters

Luther	9 km (h8)
Berosus	75 km (c8)
Lyell	31 km (e4)
Sabine	30 km (h2)
Maclaurin	54 km (b1)

This map has been prepared from Plates 3a and 3d. Plates 3b, 3c, and 3e show the same area under different lighting. Plate 3f shows larger scale photographs of the area near Arago (h3), and the wrinkle ridges east of le Monnier (g7). Plate 3g shows Messier and Messier A (d1). *Note:* In Plate 3c d4 are shown the two Promentaria Olivium and Lavinium, which were once thought to be spanned by a natural arch. This is now known to be a shadow effect, and the names are not in current use.

Plate 3a

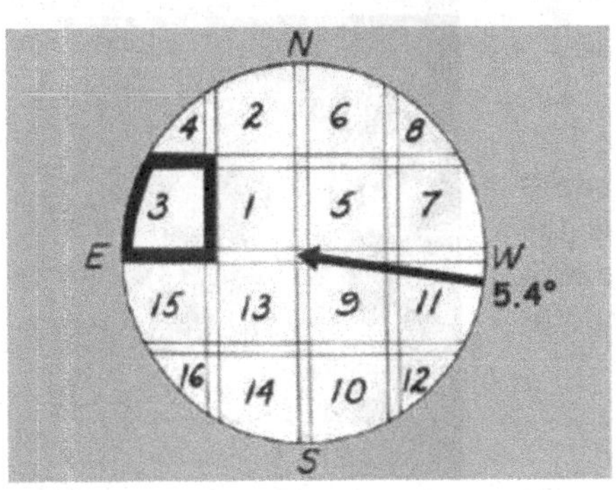

1757 UT Feb 16 1967. Moon's age 7.3 days. Sun's selenographic colongitude = 355.4°. Diameter 64 cm. Note the rays around Proclus (d5). There are several "domes" north and west of Arago (h3).

Plate 3b

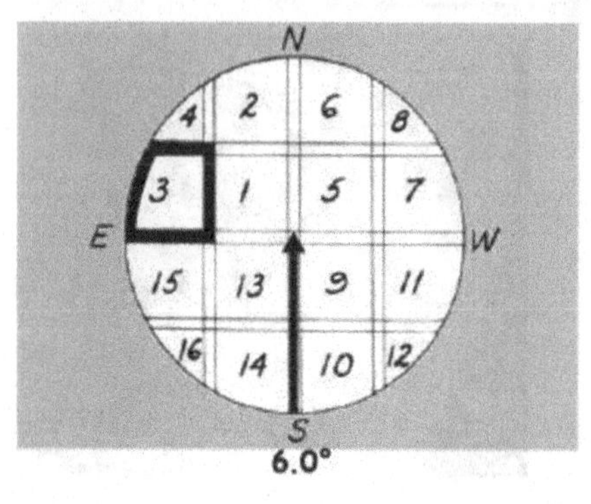

2228 UT Jan 08 1966. Moon's age 17.0 days. Sun's selenographic colongitude = 113.6°. Diameter 64 cm. Compare the shape of the Mare Crisium (c5) here with that in Plate 3a and the left hand photograph in Plate 3e.

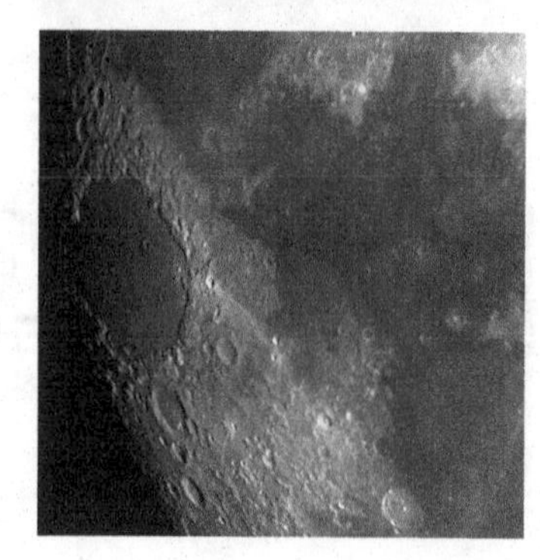

Plate 3c

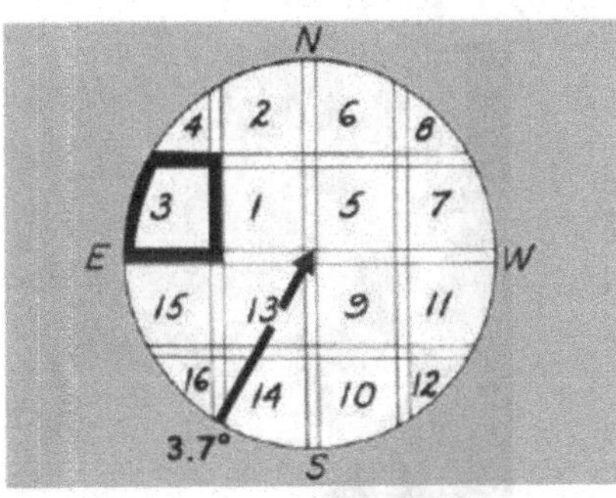

2336 UT Feb 26 1967. Moon's age 17.6 days. Sun's selenographic colongitude = 119.7°. Diameter 64 cm. Compare the Mare Tranquillitatis (f3) here with the same area in Plate 3a.

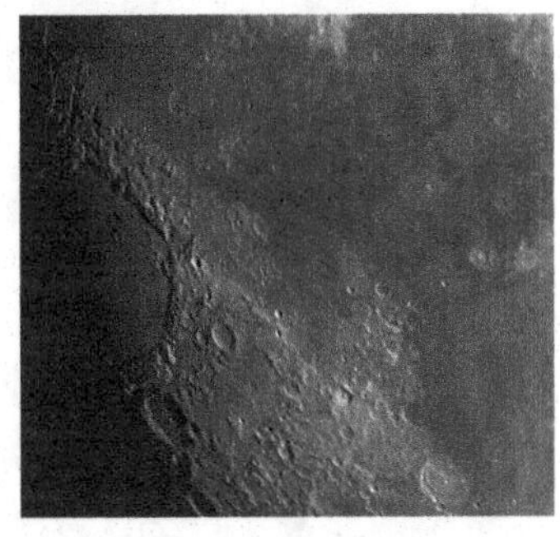

Plate 3d

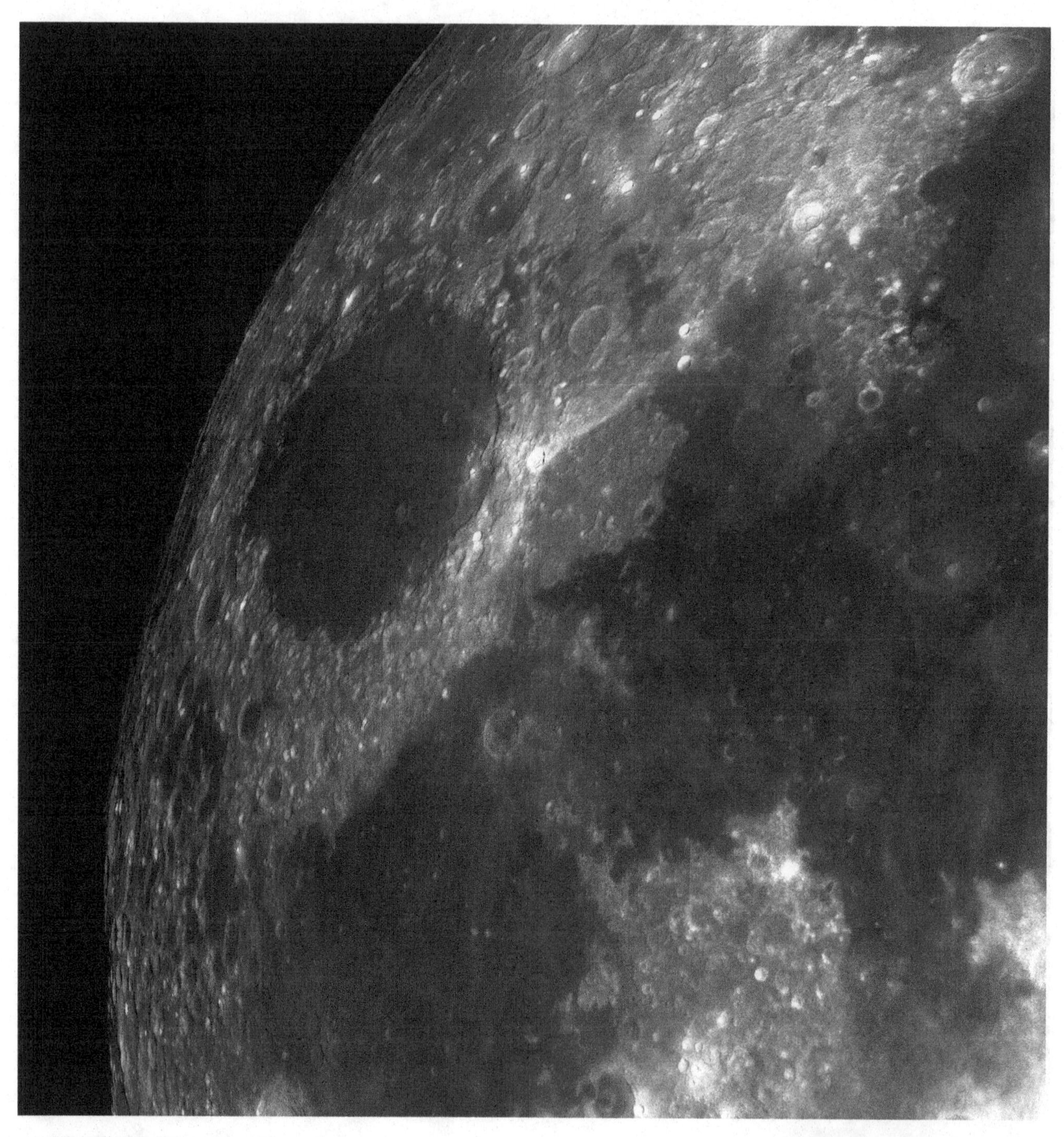

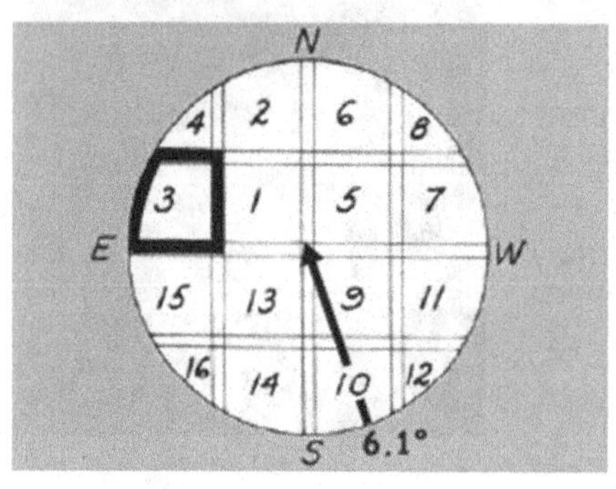

2353 UT Feb 24 1967. Moon's age 15.6 days. Sun's selenographic colongitude = 95.6°. Diameter 64 cm. This was taken 6 hours after Full Moon.

Plate 3e

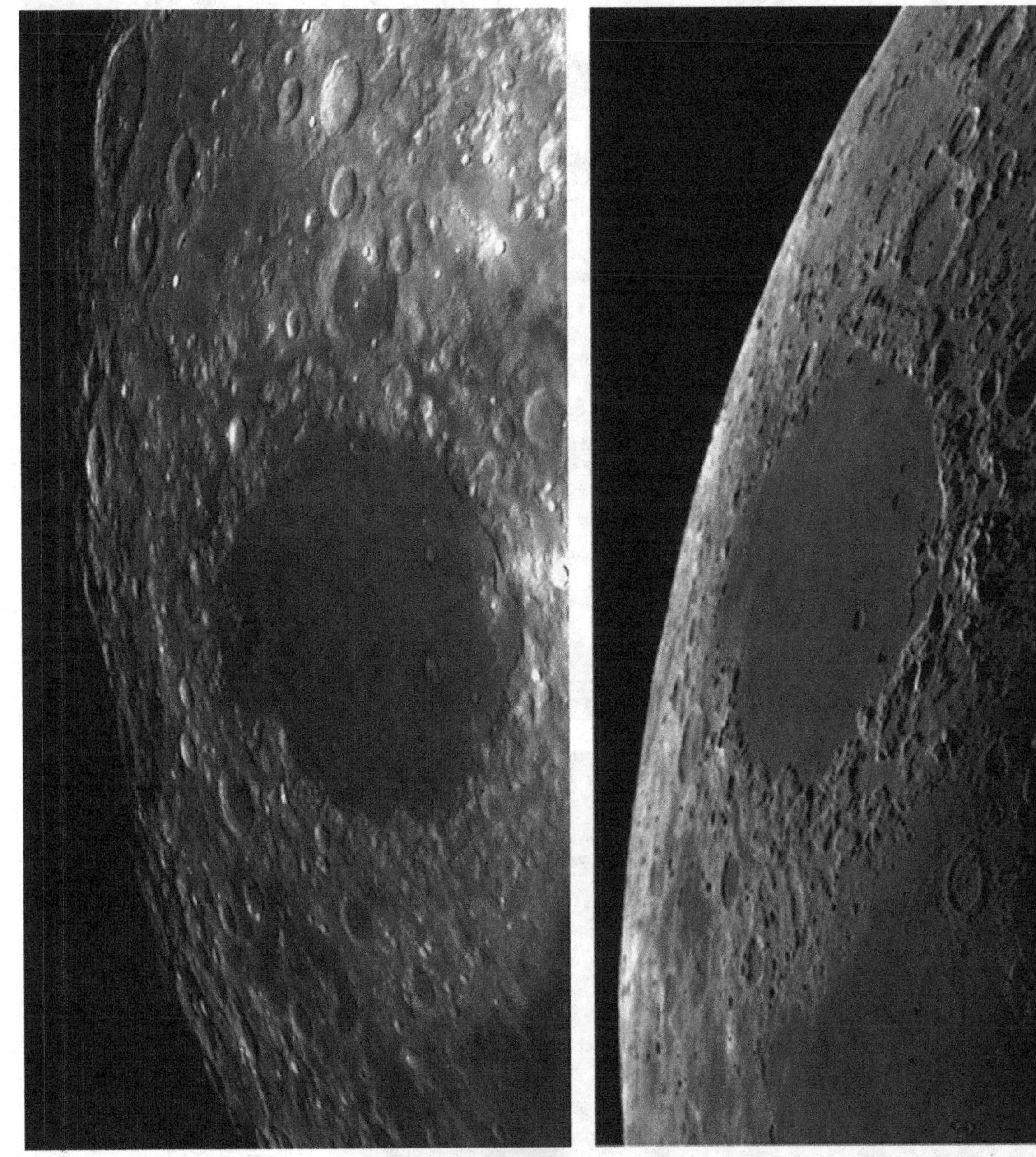

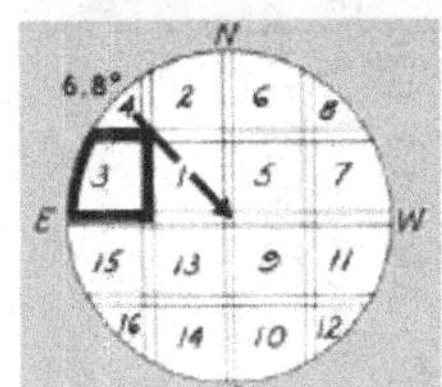

2304 UT Aug 20 1967. Moon's age 14.9 days. Sun's selenographic colongitude = 95.8°. Diameter 64 cm. Gauss (c8) is near the limb and Neper (a2) is on the limb.

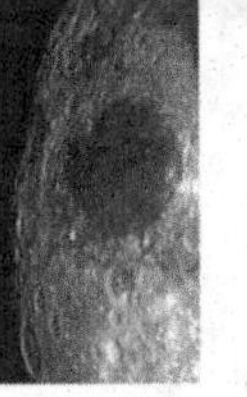

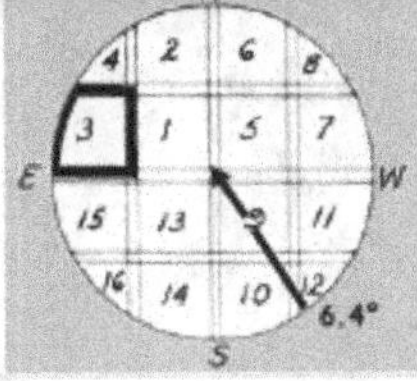

2034 UT May 23 1966. Moon's age 3.4 days. Sun's selenographic colongitude = 317.5°. Diameter 64 cm. Compare this with Plate 3b and with its neighbor here.

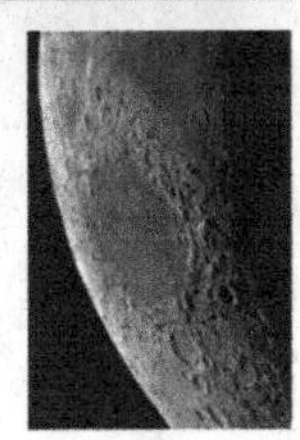

Plate 3f

Arago, its "Domes" and Rima Ariadaeus. 2000 UT May 15. 1967. Moon's age 6.2 days. Sun's selenographic colongitude = 349.1°. Diameter 94 cm approx. Arago lies in Map 3 h3 and Map 1 c2. The domes lie about its own diameter North and West of it. Rima Ariadaeus lies in Map 1 d2.

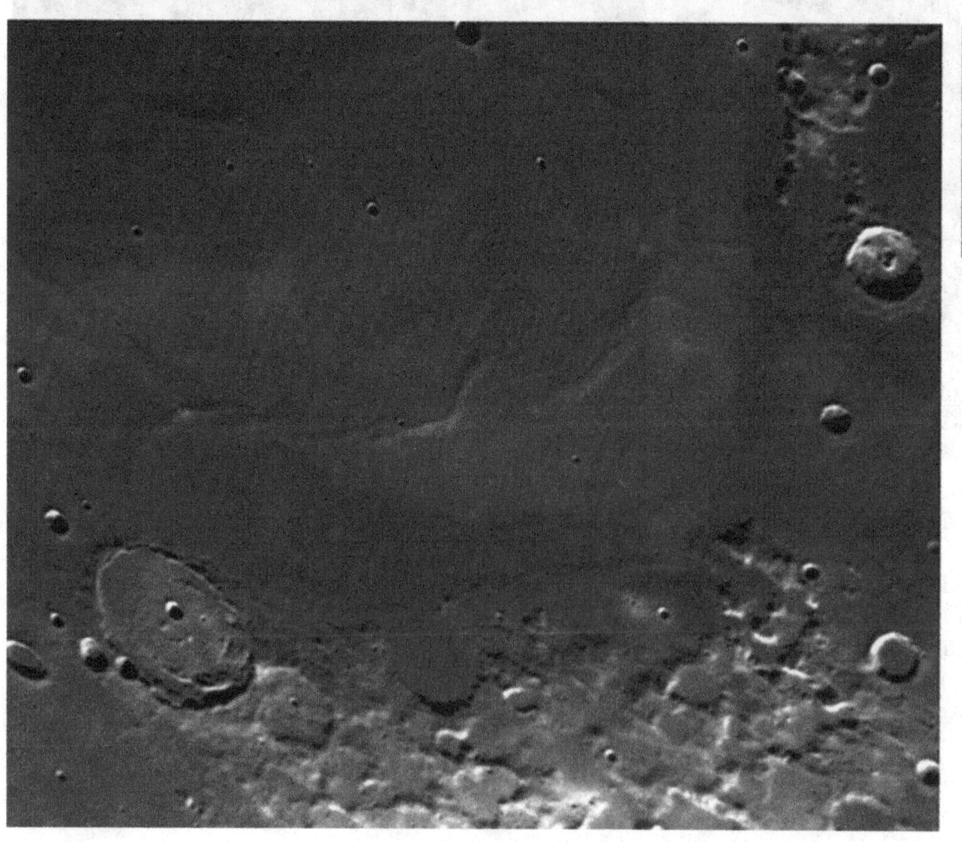

Plinius, Posidonius, Bessel and the "Wrinkle Ridges" in the Mare Serenitatis. 2015 UT May 15. 1967. Moon's age 6.2 days. Sun's selenographic colongitude = 349.3°. Diameter 94 cm approx. Plinius lies in Map 3 g5.

Plate 3g

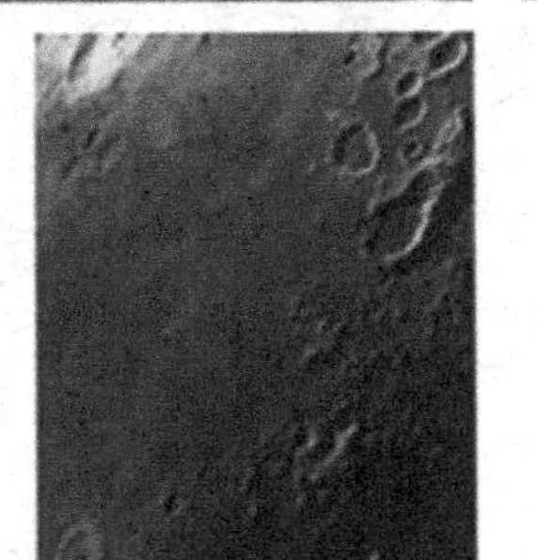

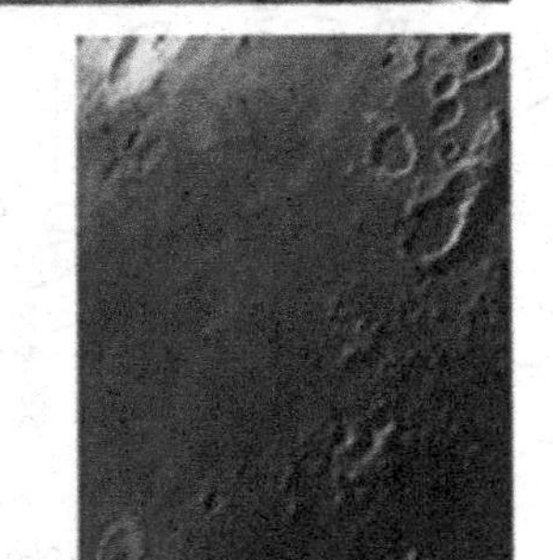

Messier and Messier A. 2002 UT May 13 1967. Moon's age 4.2 days. Sun's selenographic colongitude = 324.7°. Diameter 94 cm approx.

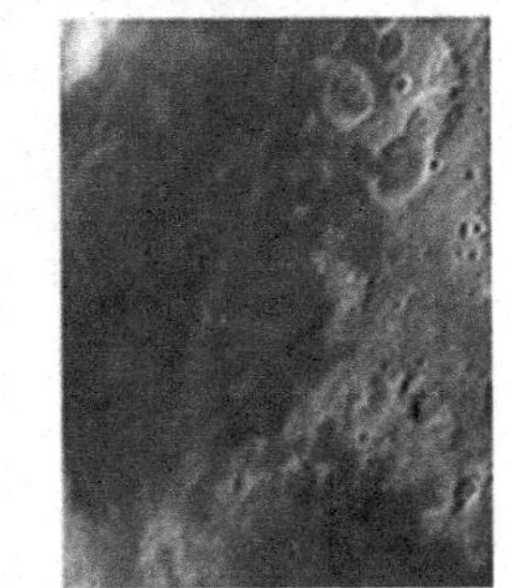

Messier and Messier A. 2008 UT May 15. May 1967. Moon's age 6.2 days. Sun's selenographic colongitude = 349.2°. Diameter 94 cm approx.

Note how these two craters change their appearance in two days. They lie in Map 3 d1.

Map 4

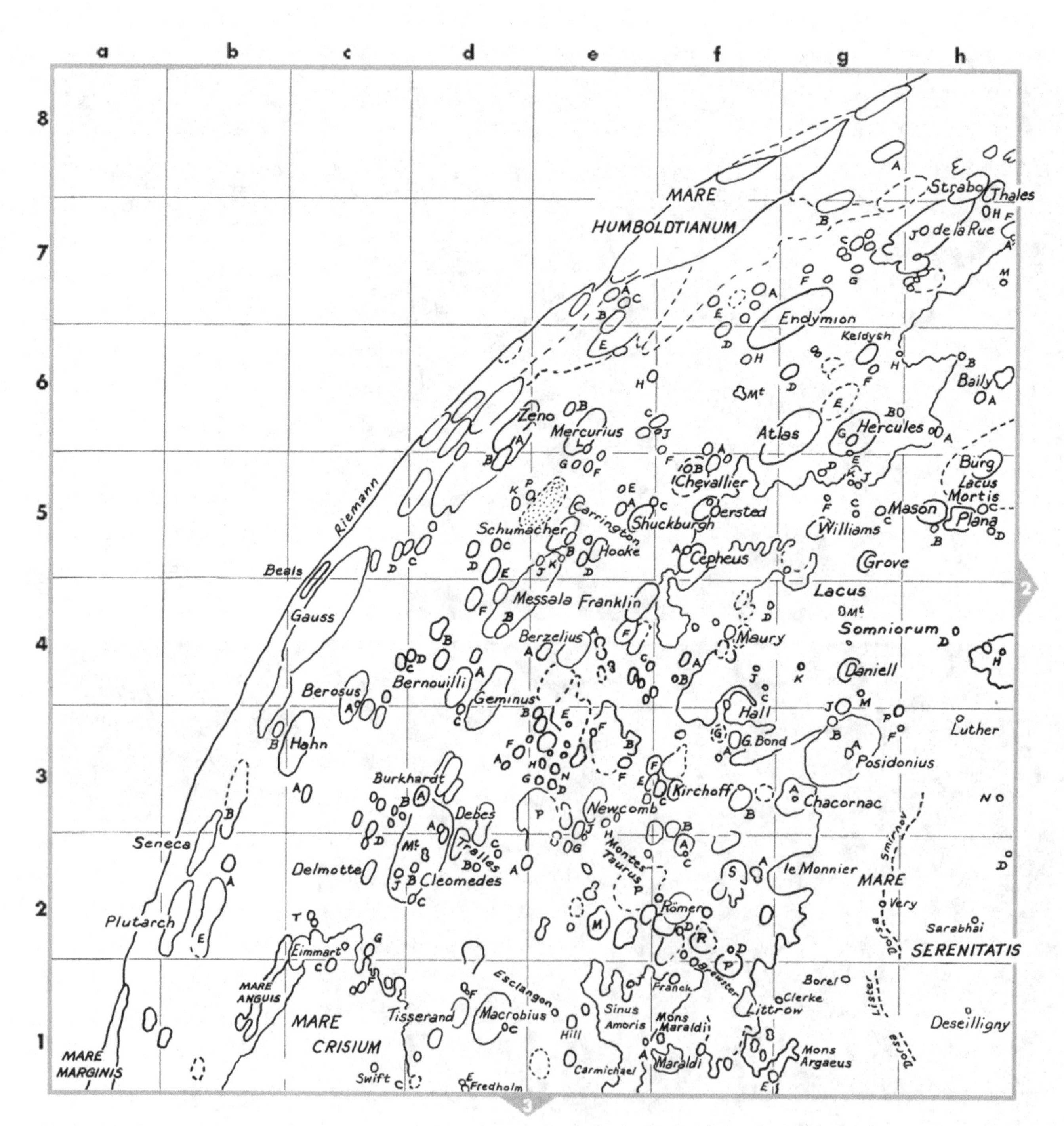

Crater Diameters

Endymion	122 km (g7)
Mason	33 km (h5)
Gauss	171 km (c4)
Römer	41 km (f2)
Tisserand	35 km (d1)

This map has been prepared from Plates 4a and 4e. Plates 4b, 4c, and 4d show the same area under different lighting. *Note:* Plate 4e shows quite a good libration for this area. The Mare Humboldtianum (f7) and Gauss (c4) do not often appear like this.

Plate 4a

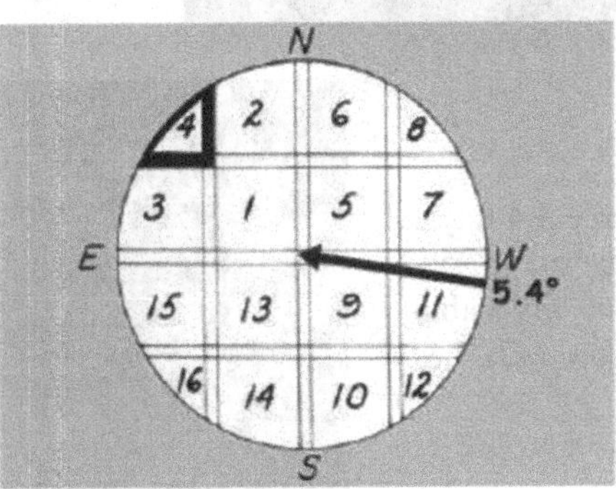

1757 UT Feb 16 1967. Moon's age 7.3 days. Sun's selenographic colongitude = 355.4°. Diameter 64 cm.

Plate 4b

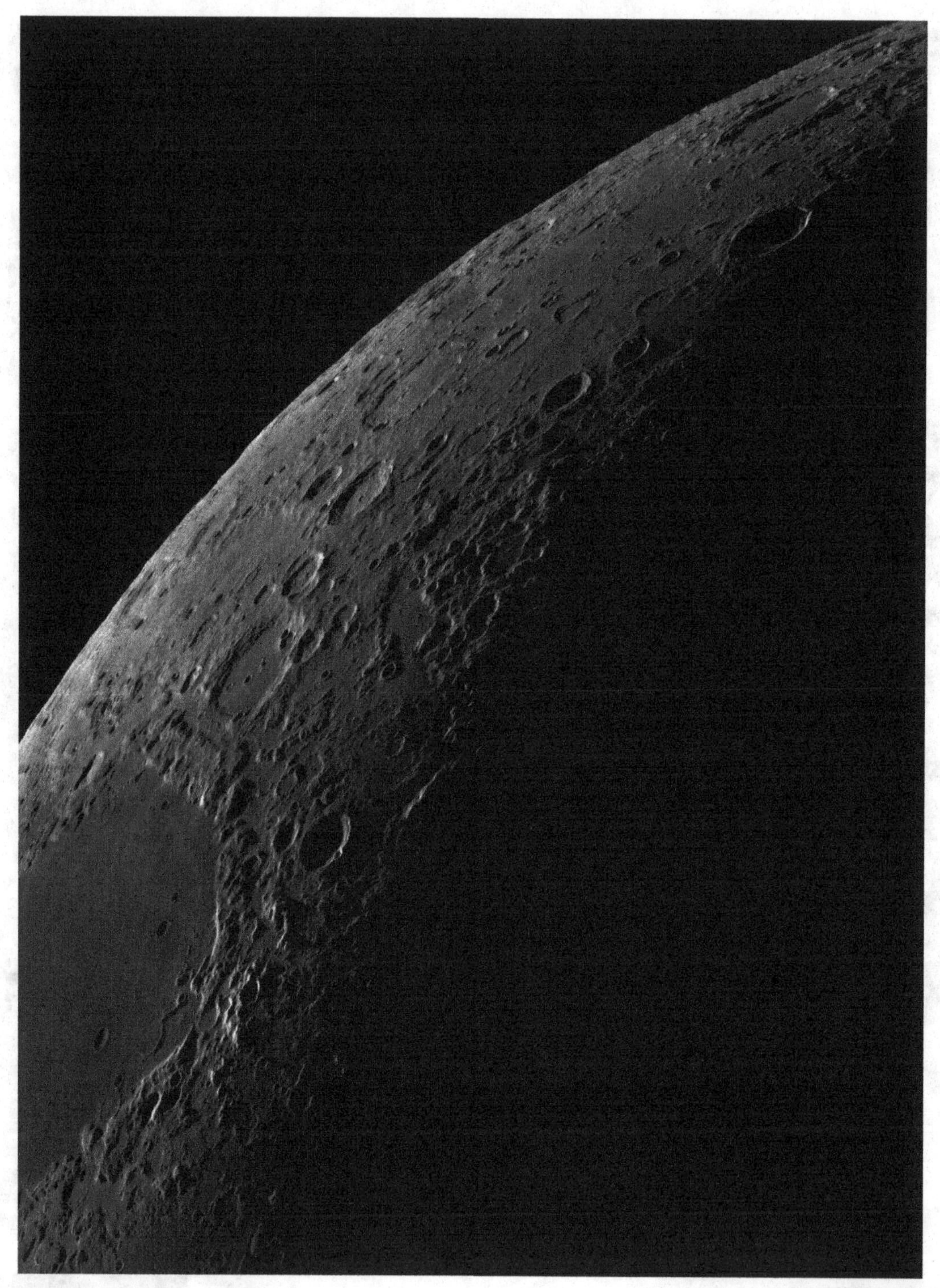

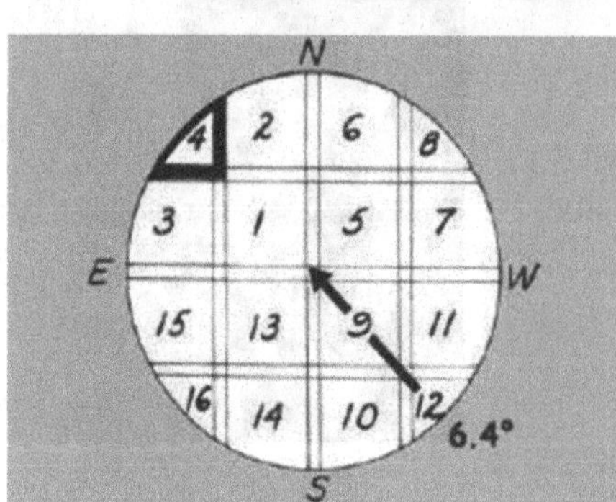

2034 UT May 23 1966. Moon's age 3.4 days. Sun's selenographic colongitude = 317.5°. Diameter 64 cm. This is not a favorable libration for this area. Endymion (g7) lies almost on the limb.

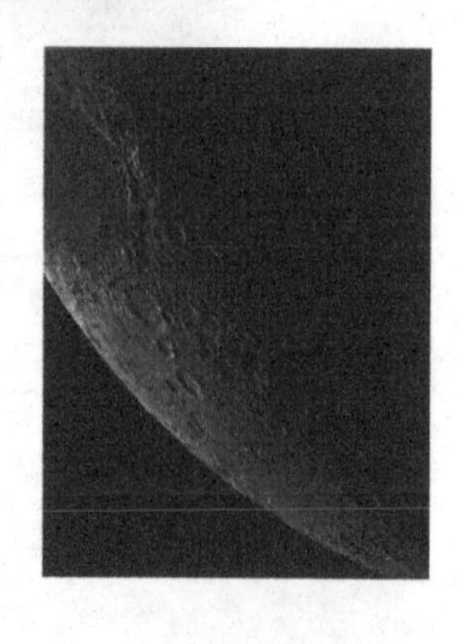

Plate 4c

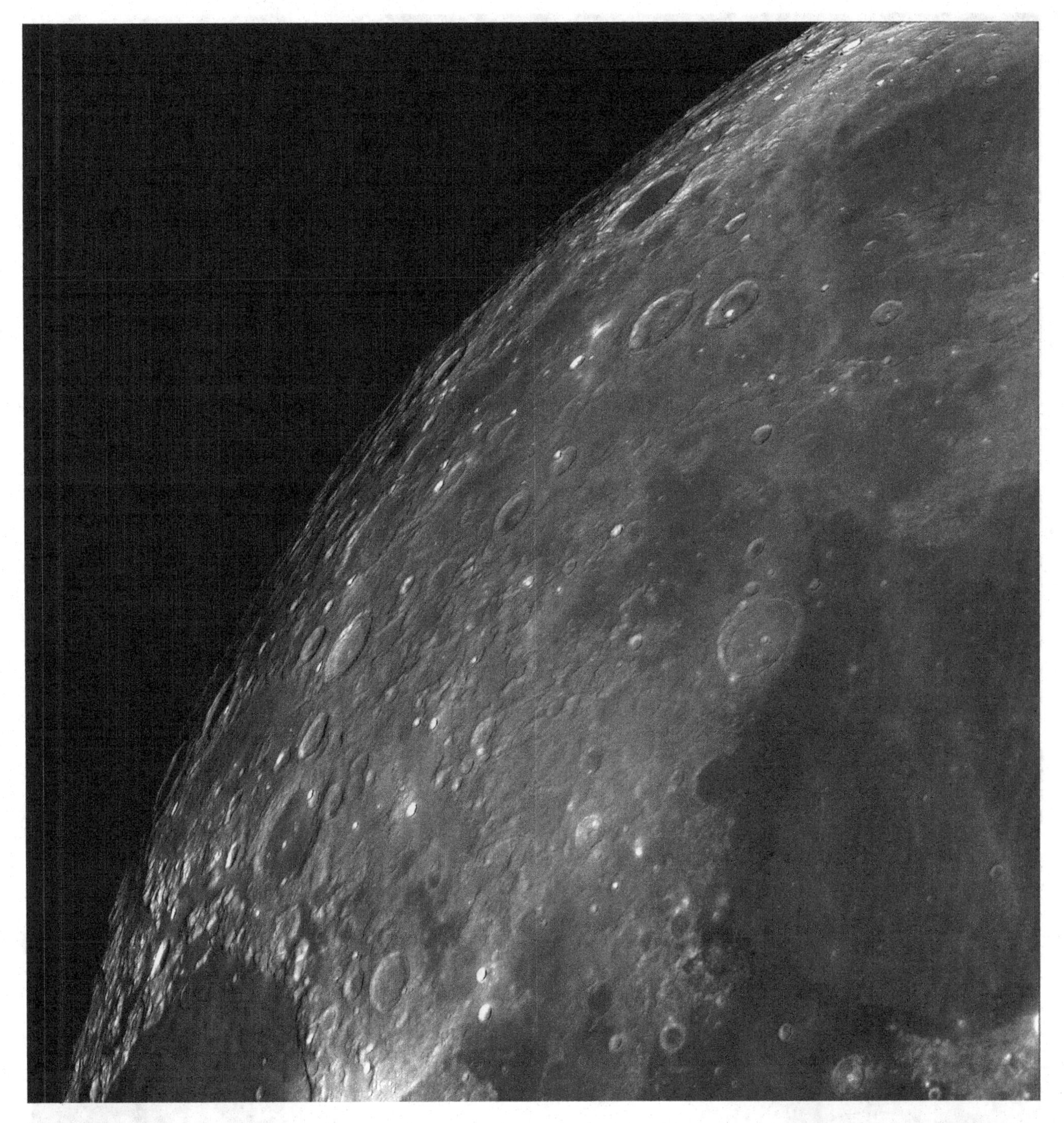

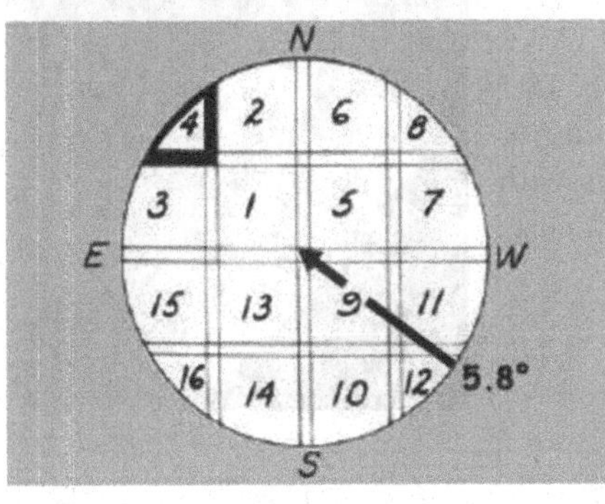

2127 UT Nov 28 1966. Moon's age 16.3 days. Sun's selenographic colongitude = 104.2°. Diameter 64 cm. Libration is similar to that in Plate 4b.

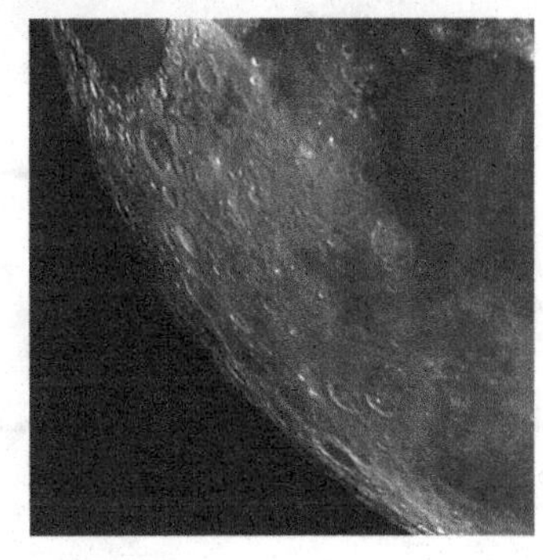

Plate 4d

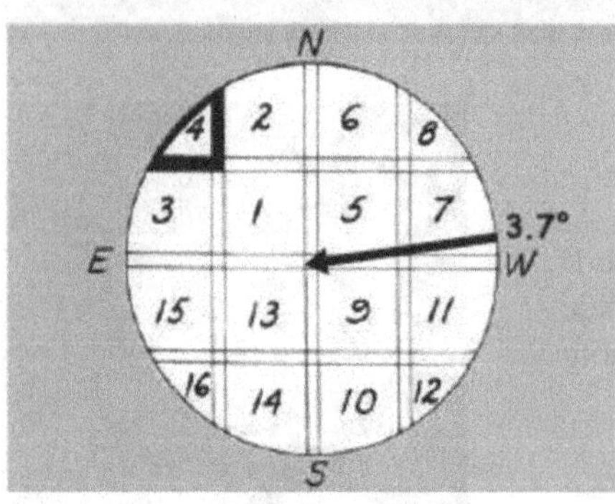

2248 UT Oct 29.1966. Moon's age 15.7 days. Sun's selenographic colongitude = 99.8°. Diameter 64 cm. Compare this with Plate 4e. The area is the same in each case, but the libration is very different.

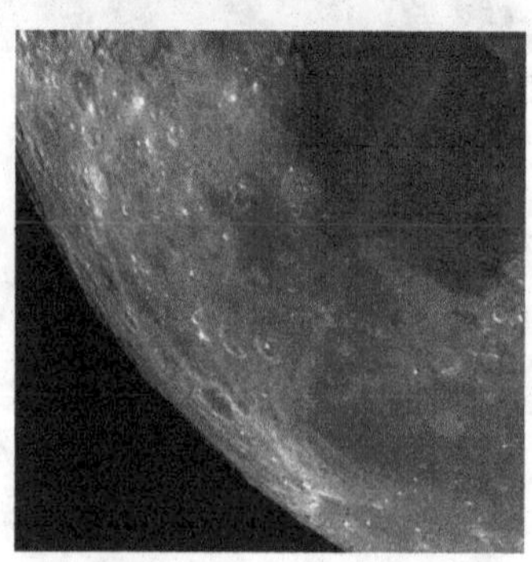

Plate 4e

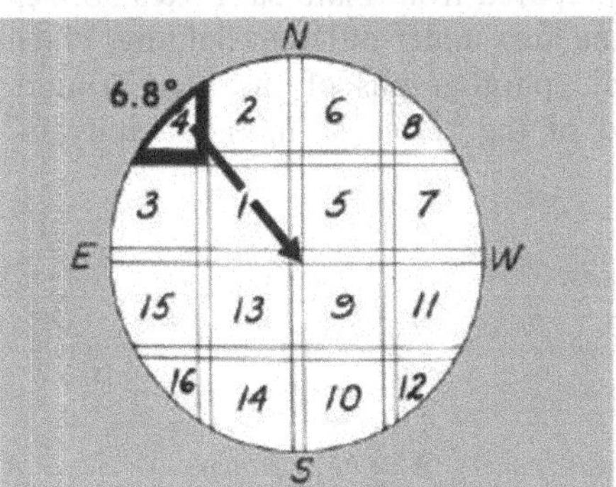

2304 UT Aug 20 1967. Moon's age 14.9 days. Sun's selenographic colongitude = 95.8°. Diameter 64 cm. This is quite a good libration for this area. The Mare Humboldtianum (f7) and Gauss (c4) do not often appear like this.

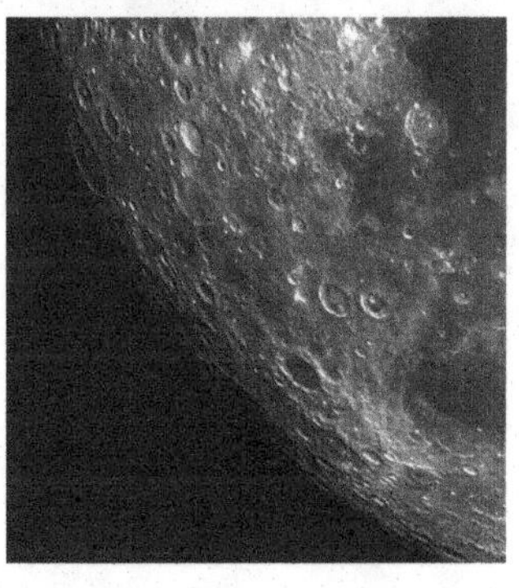

Map 5

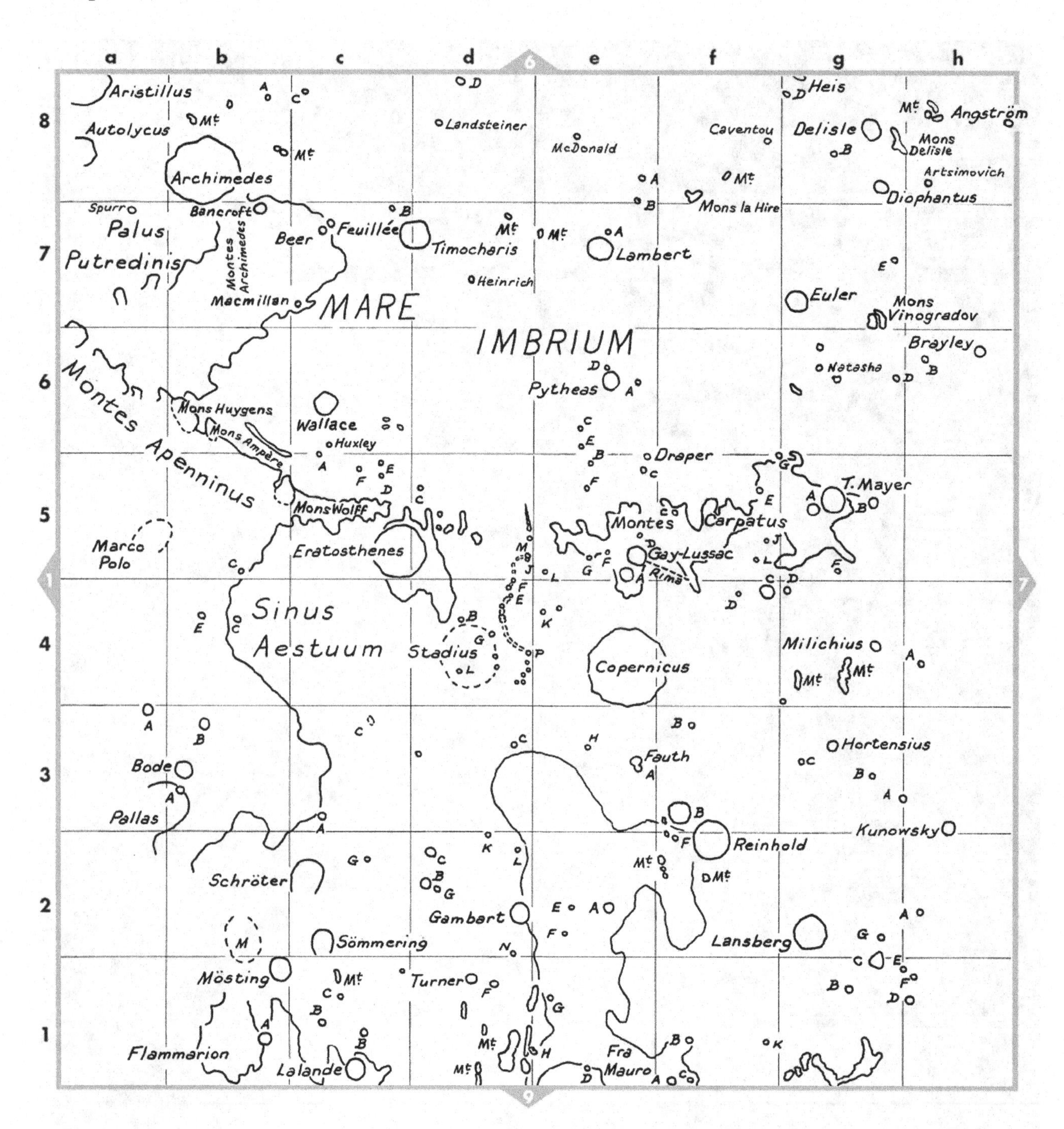

Crater Diameters

Delisle	25 km (g8)
Beer	9 km (c7)
Copernicus	96 km (e4)
Lansberg	40 km (g2)
Mösting	25 km (b1)

This map has been prepared from Plate 5a. Plates 5b, 5c, and 5d show the same area under different lighting. Plate 5e shows larger scale photographs of the area around Copernicus (e4) and of the "Domes" in the vicinity of Milichius (g4).

Plate 5a

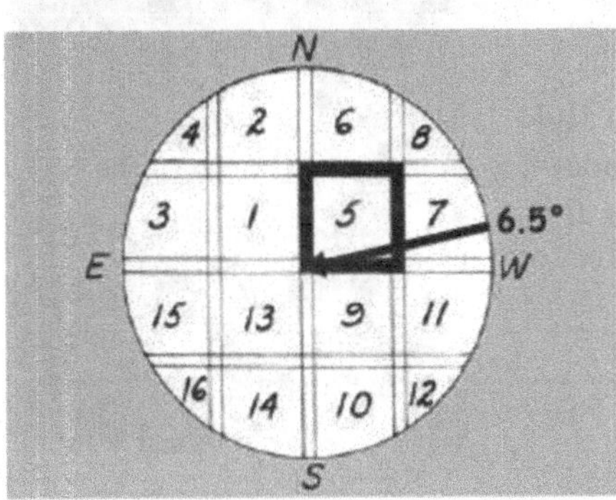

0315 UT Aug 09.1966. Moon's age 21.9 days. Sun's selenographic colongitude = 181.6°. Diameter 64 cm. Note the ridges in the Mare Imbrium and the Sinus Aestuum (c4). The Montes Carpatus rise to about 2,100 m in places.

Plate 5b

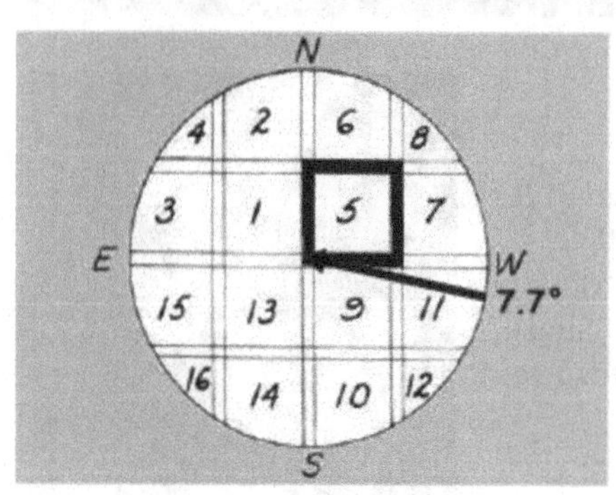

1947 UT Jan 31 1966. Moon's age 10.1 days. Sun's selenographic colongitude = 32.1°. Diameter 64 cm. Compare this with Plate 5c, which was taken nearer to Full Moon.

Plate 5c

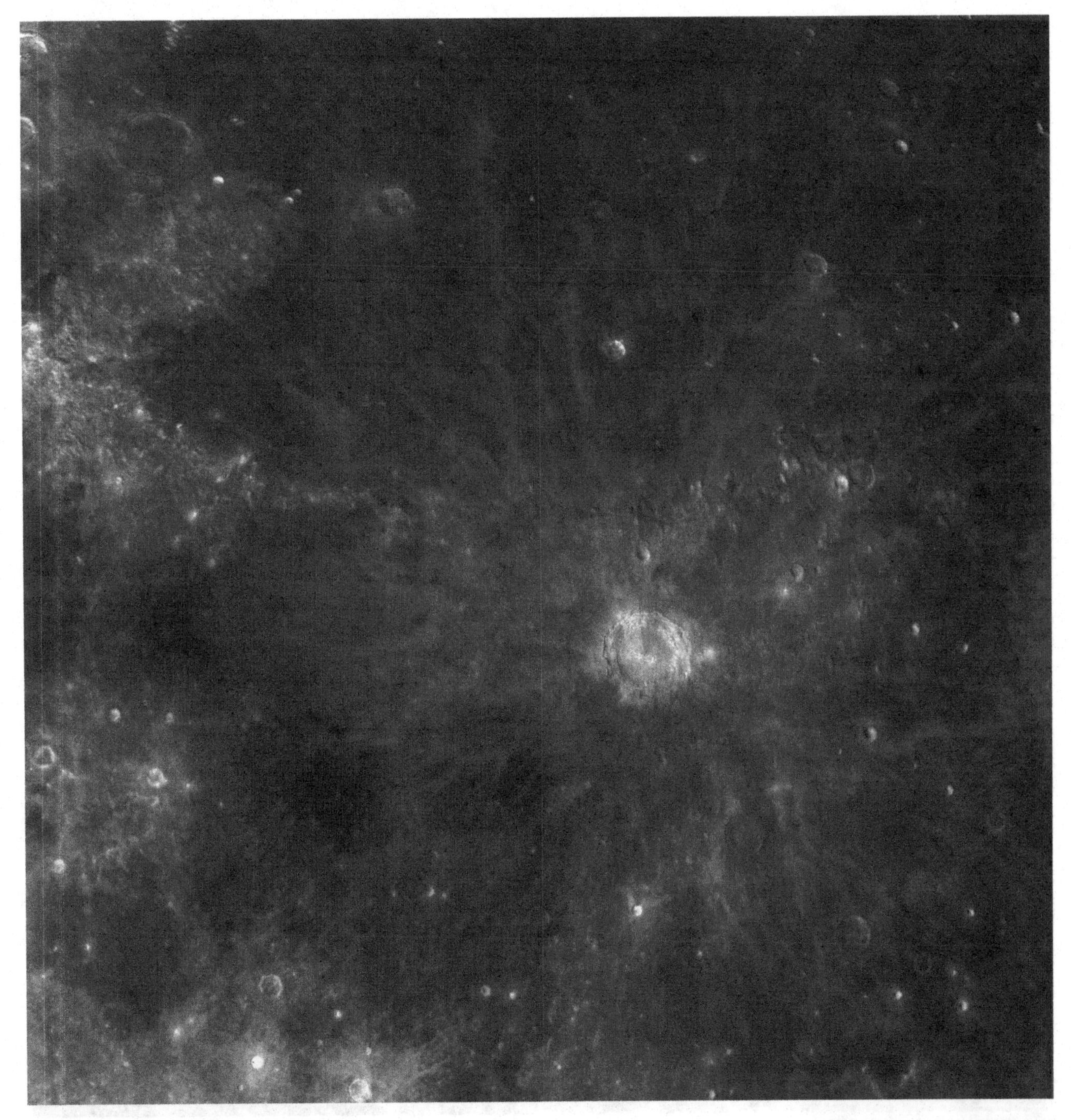

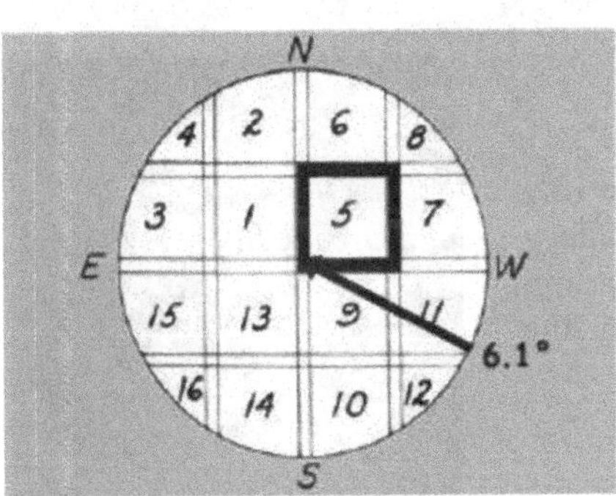

2135 UT Dec 25 1966. Moon's age 13.8 days. Sun's selenographic colongitude = 72.7°. Diameter 64 cm. This photograph shows almost exactly the same area as Plate 5b.

Plate 5d

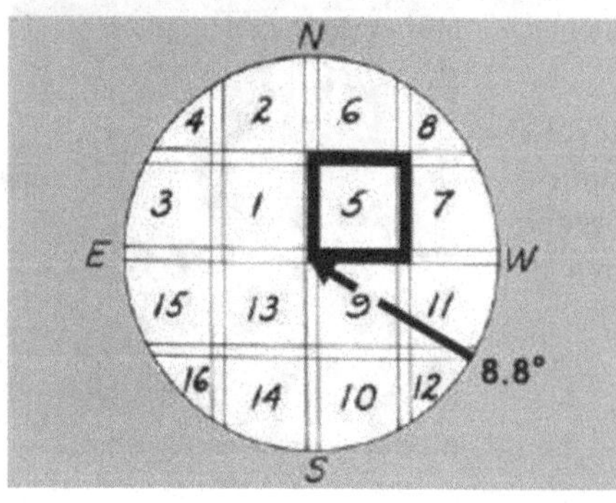

1801 UT Feb 19 1967. Moon's age 10.3 days. Sun's selenographic colongitude = 31.9°. Diameter 64 cm. This photograph extends further to the east than the others in this group.

Plate 5e

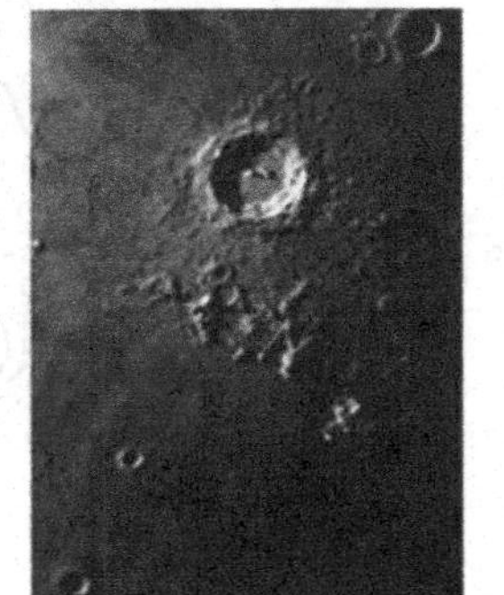

The Copernicus area soon after sunrise. 2041 UT Mar 20 1967. Moon's age 9.7 days. Sun's selenographic colongitude = 26.4°. Diameter 91 cm approx. Draper (e5, e6) is 8 km in diameter.

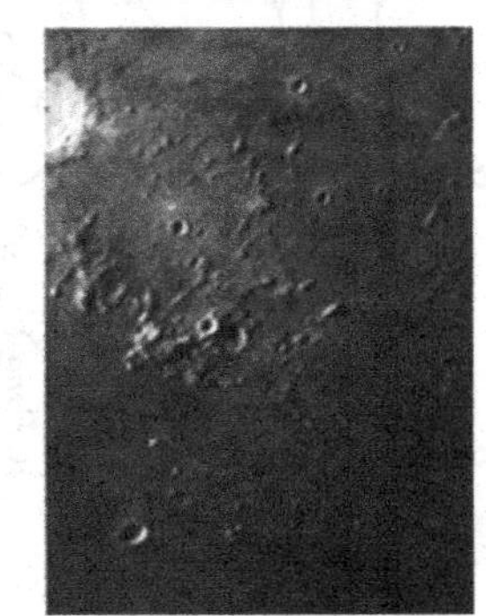

Euler to Hortensius, with the "Domes" near Milichius. 1908 UT Mar 21 1967. Moon's age 10.6 days. Sun's selenographic colongitude = 37.7°. Diameter 91 cm approx. Milichius lies in Map 5 g4; there are several domes near it here, mostly to the northwest.

Map 6

Crater Diameters

Cleostratus	64 km (h7)
Epigenes	55 km (b7)
Le Verrier	21 km (d4)
Angström	10 km (h2)
Timocharis	34 km (c1)

This map has been prepared from Plate 6a. Plates 6b, 6c, and 6d show the same area under different lighting. Plate 6e shows larger scale photographs of the Montes Recti (d5) and surrounding country, and of the Sinus Iridum (e4) soon after sunrise.

Plate 6a

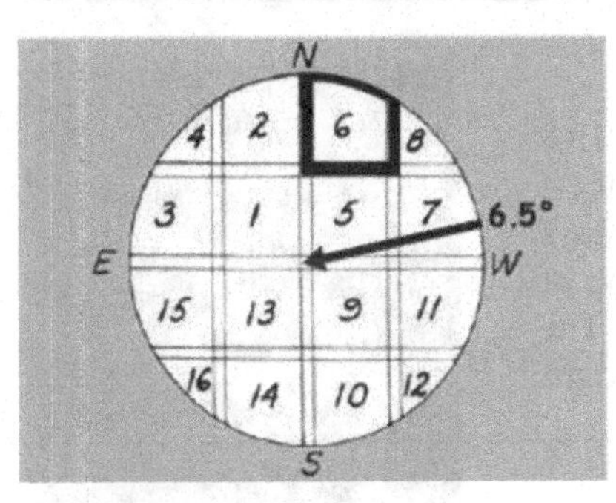

0315 UT Aug 09 1966. Moon's age 21.9 days. Sun's selenographic colongitude = 181.6°. Diameter 64 cm. Note the "ghost" craters and low ridges near the eastern border of the Mare Imbrium. Mons Pico (b4) is about 2,400 m high and is not nearly so steep as it looks.

Plate 6b

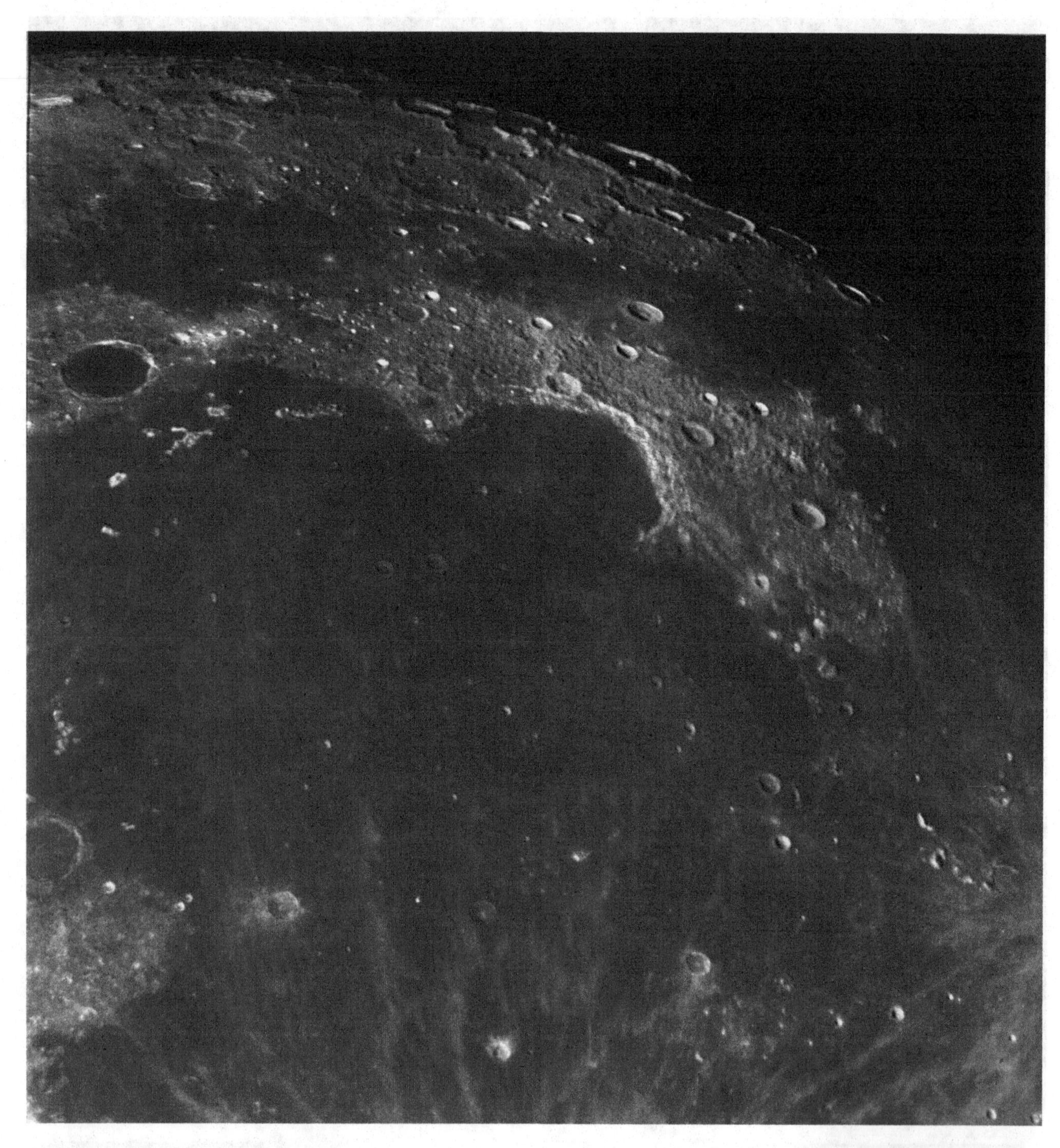

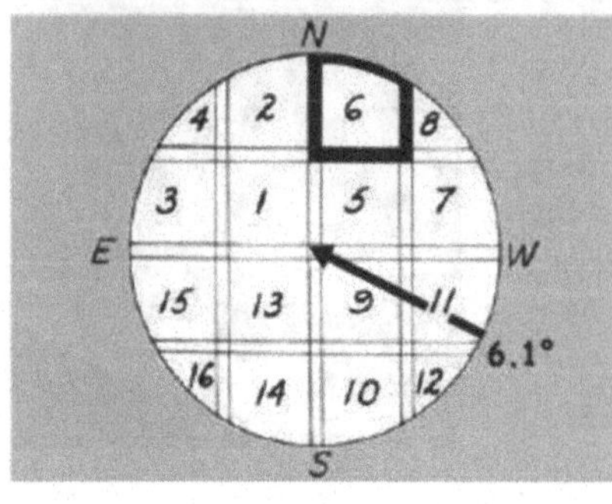

2040 UT Dec 25 1966. Moon's age 13.8 days. Sun's selenographic colongitude = 72.2°. Diameter 64 cm. This was two days before Full Moon. Compare the shape of Plato (b5) with that shown in Plate 6a. The large crater with a central mountain right on the terminator is Pythagoras (f7).

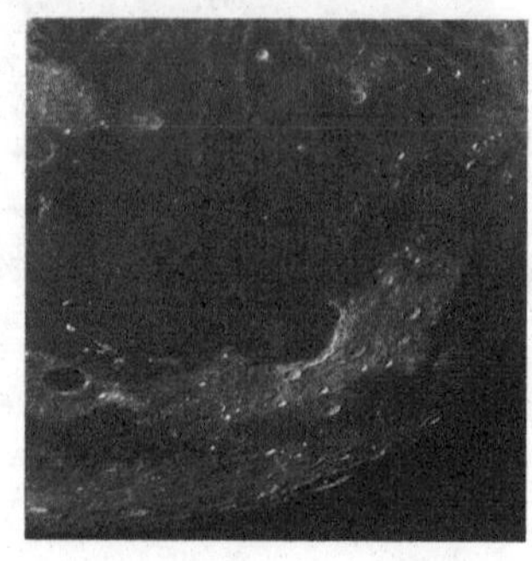

Plate 6c

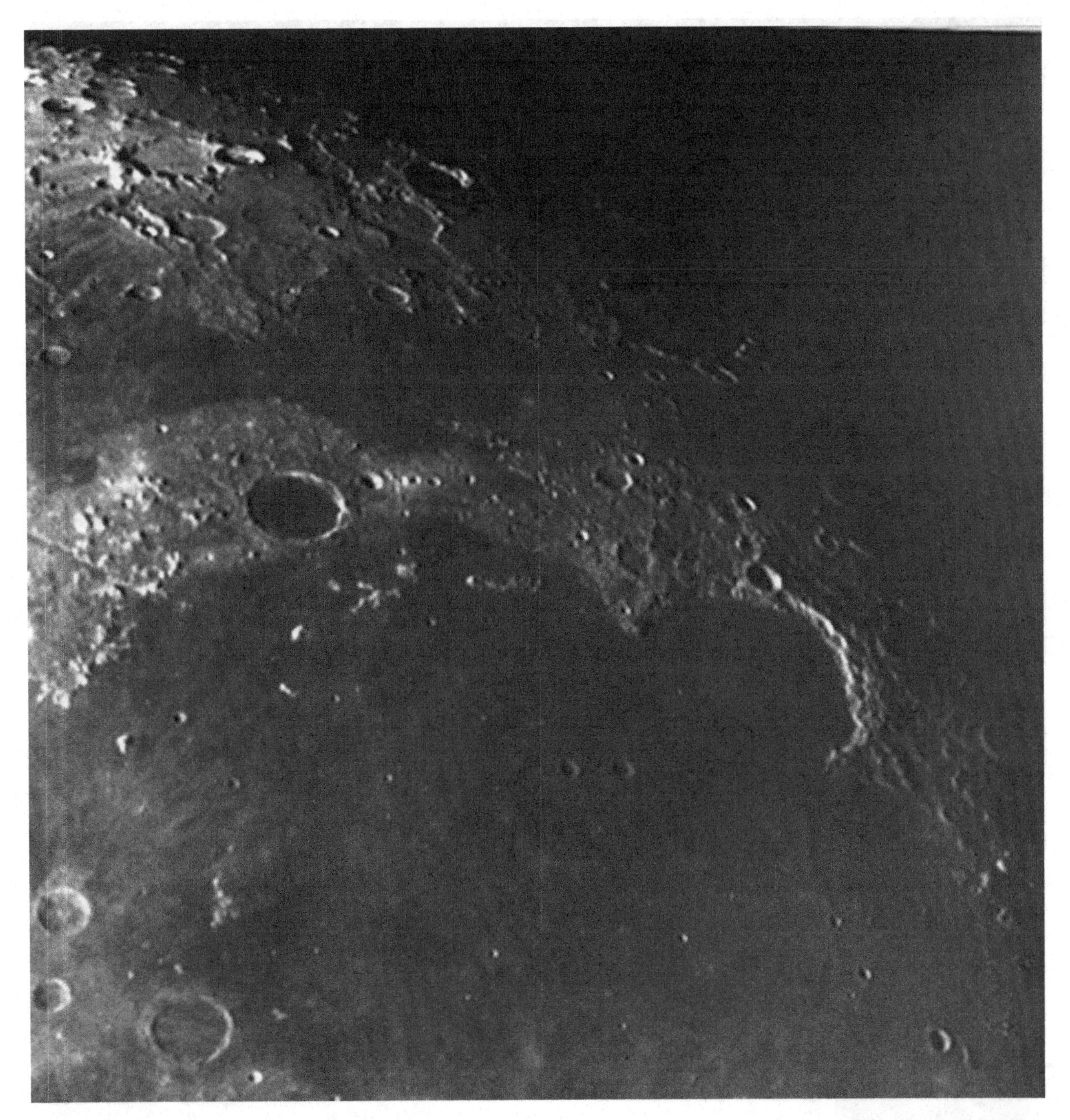

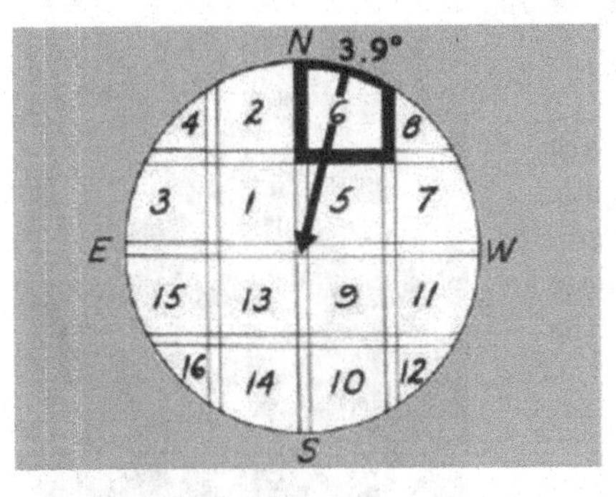

2142 UT Nov 23 1966. Moon's age 11.3 days. Sun's selenographic colongitude = 43.6°. Diameter 64 cm. This is quite a good libration for the north polar regions. The top left-hand (NE) part of this photograph extends beyond Map 6; it is shown on Map 2.

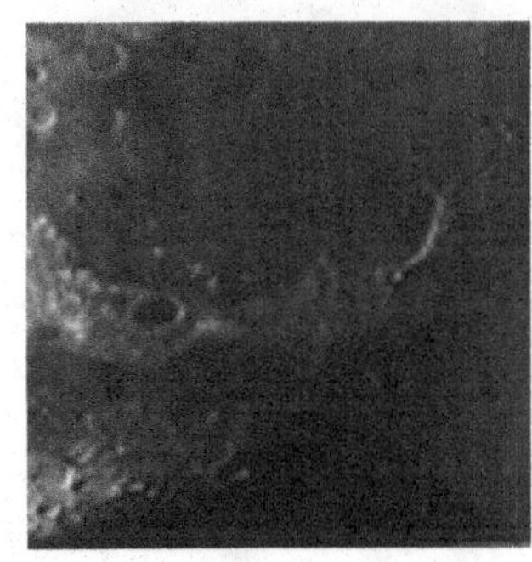

Plate 6d

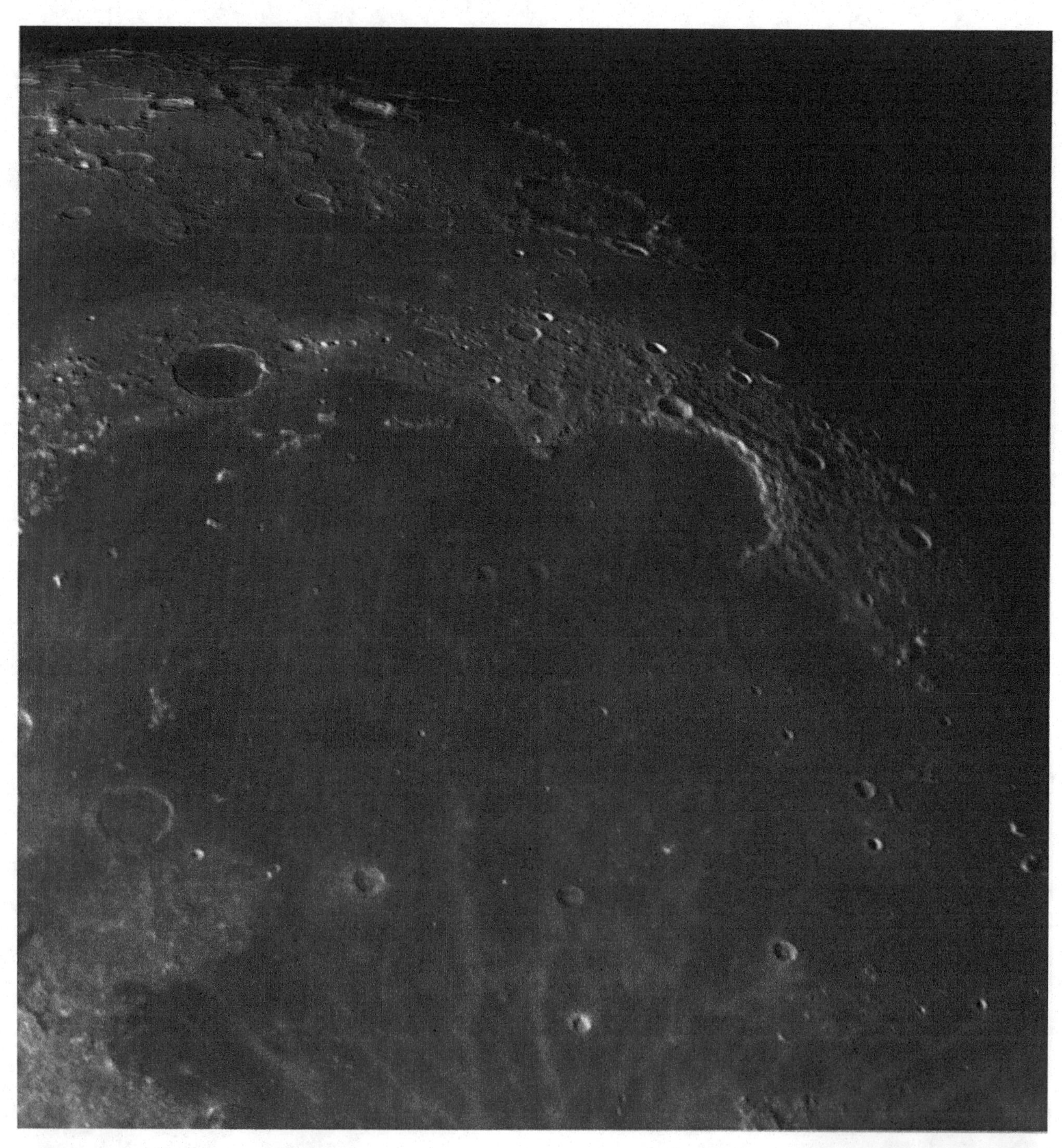

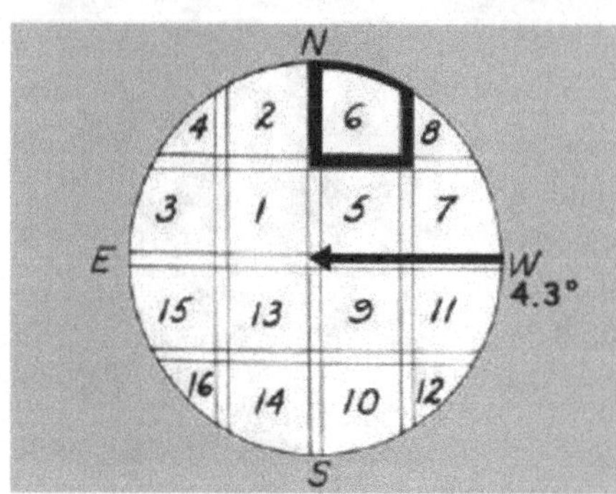

2232 UT Dec 23 1966. Moon's age 11.8 days. Sun's selenographic colongitude = 48.9°. Diameter 64 cm. Compare this with Plate 6c, and note how the different librations cause the shapes of the craters to alter.

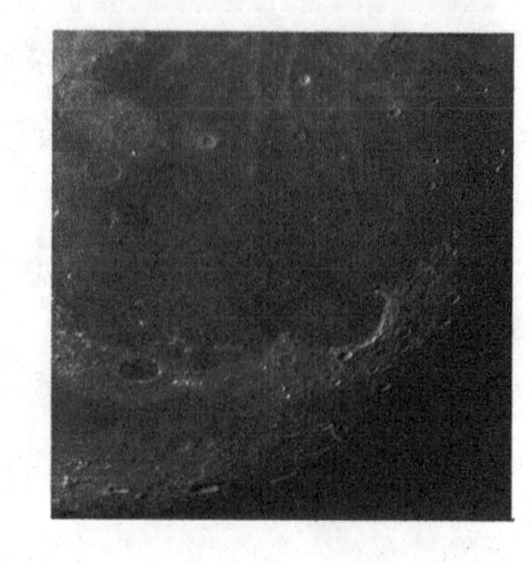

Plate 6e

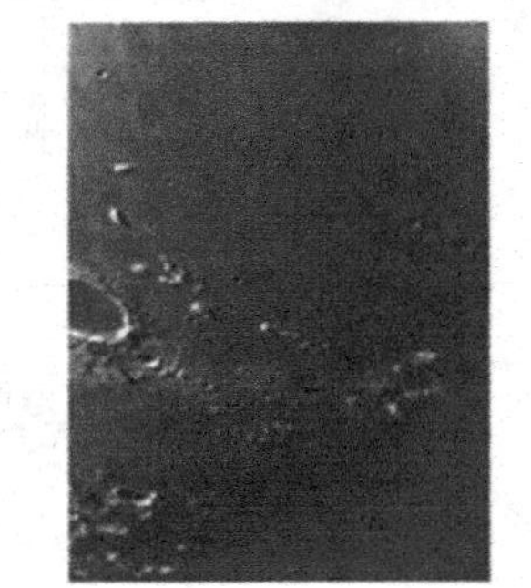

The Montes Recti and surrounding country. 2029 UT Mar 20 1967. Moon's age 9.7 days. Sun's selenographic colongitude = 26.3°. Diameter 91 cm approx. The Montes Recti, which lies in Map 6 d5, is a typical isolated mountain range. Its highest peaks are about 1,800 m high.

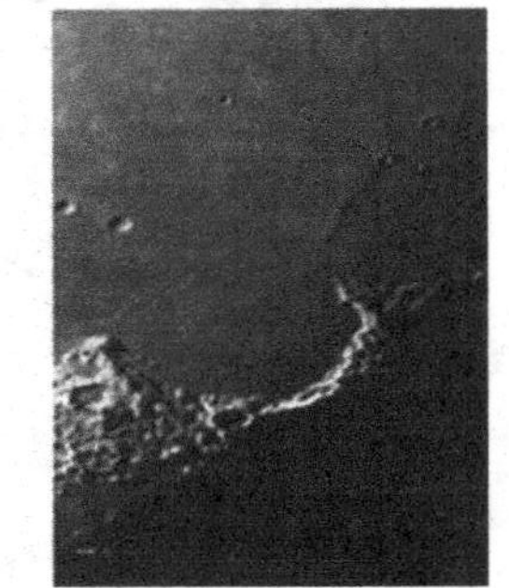

The Sinus Iridum soon after sunrise. 1923 UT Mar 21 1967. Moon's age 10.6 days. Sun's selenographic colongitude = 37.9°. Diameter 91 cm approx. Sinus Iridum lies in Map 6 e4, Cape Heraclides sometimes taking on the appearance of a "Moon Maiden."

Map 7

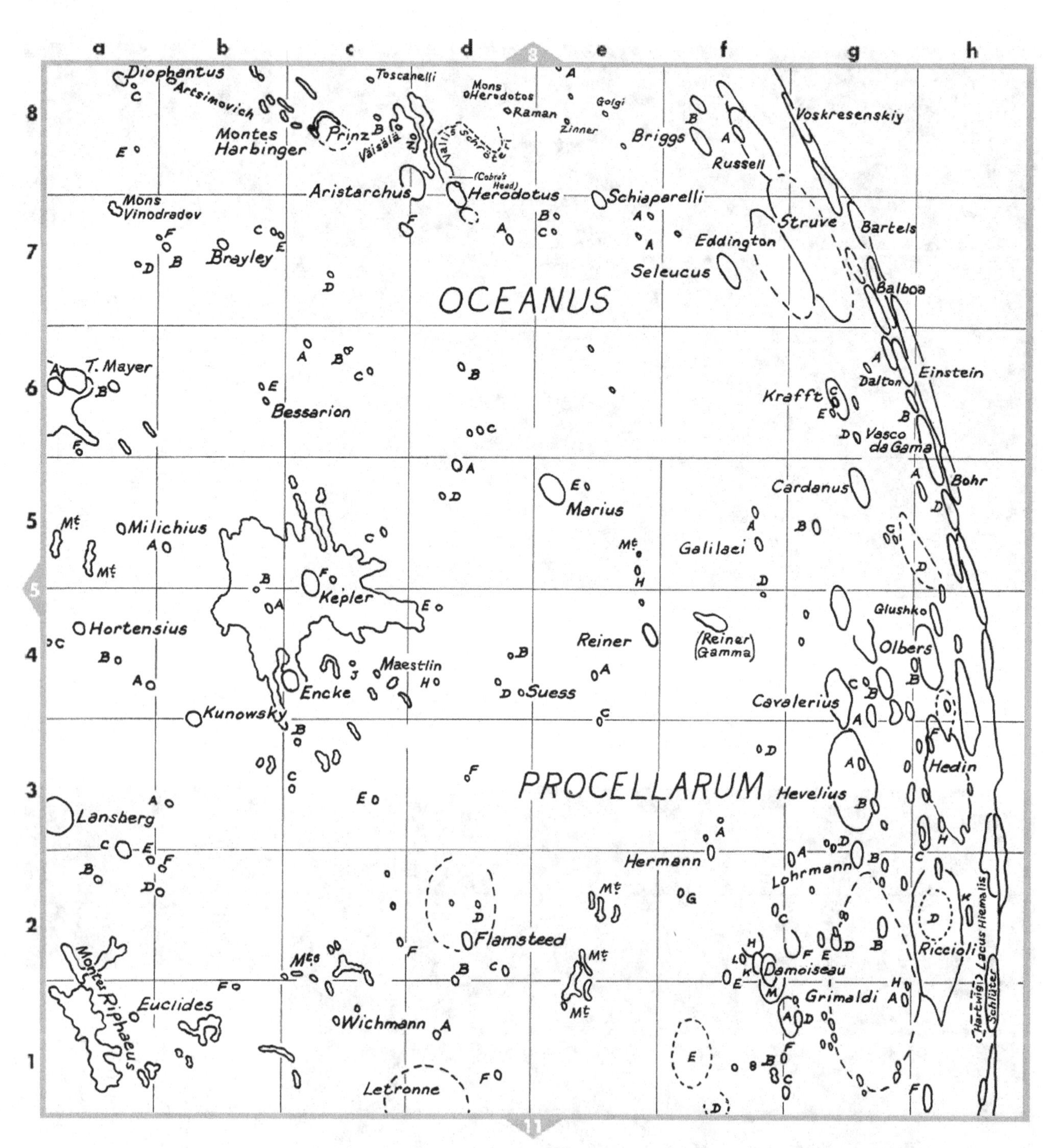

Crater Diameters

Briggs	37 km (f8)
Diophantus	18 km (a8)
Marius	40 km (e5)
Hevelius	117 km (g3)
Euclides	12 km (a1)

This map has been prepared from Plates 7a and 7e. Plates 7b, 7c, and 7d show the same area under different lighting. On Plate 7e there is an insert of the crater Einstein (h6), which is only exposed to view when the libration is extremely favorable.

Plate 7a

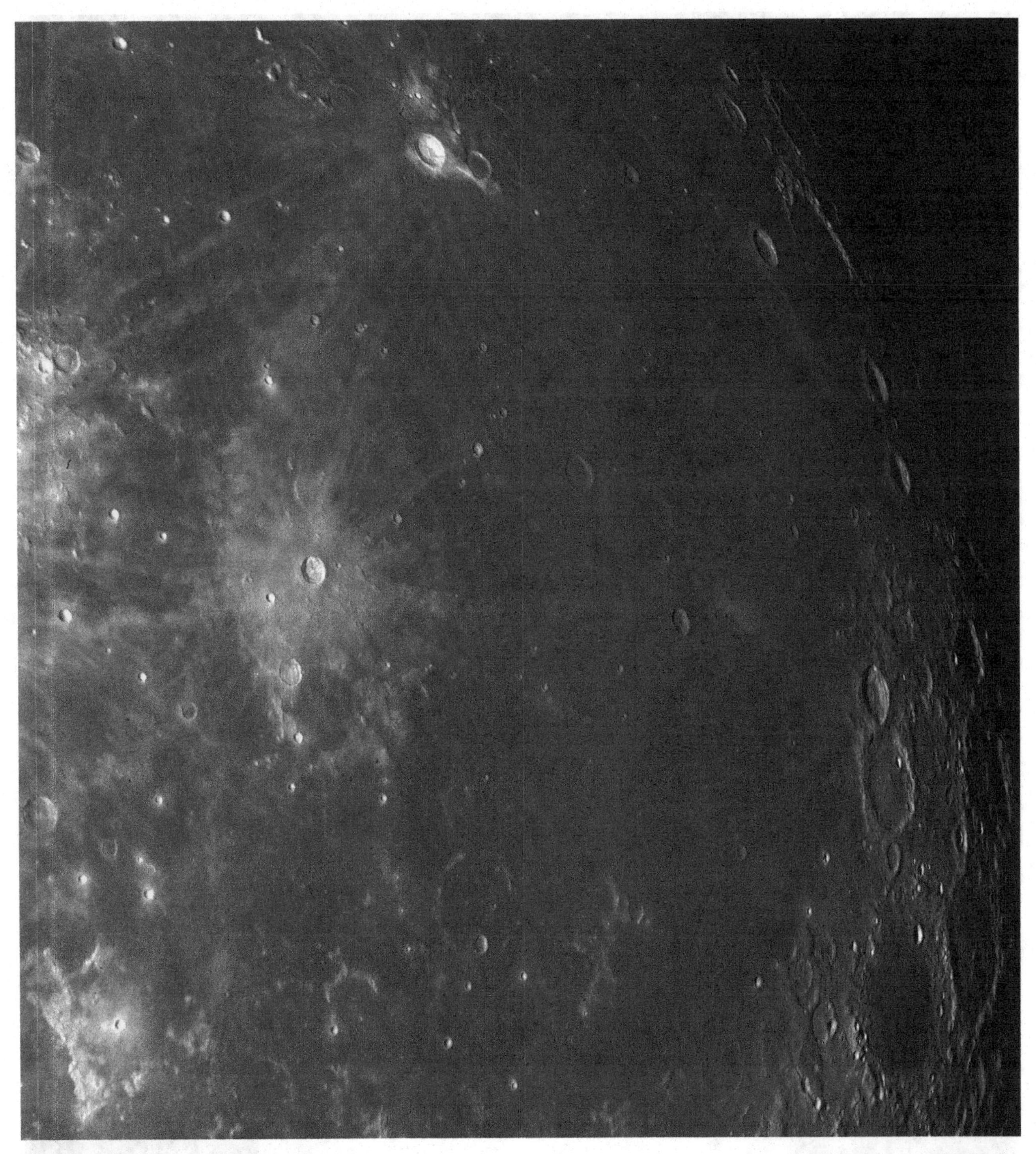

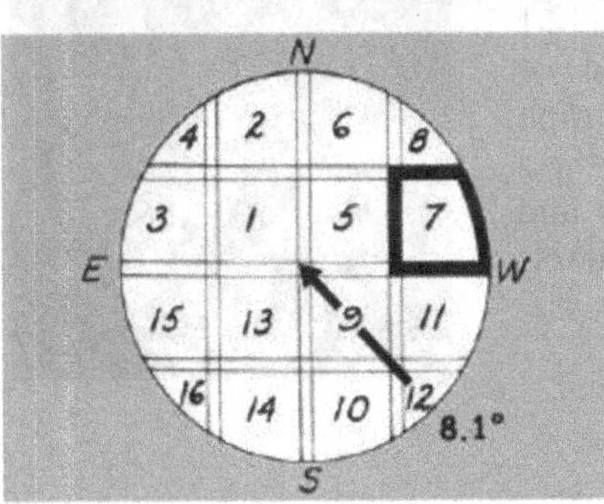

2310 UT Jan 24 1967. Moon's age 14.2 days. Sun's selenographic colongitude = 78.2°. Diameter 64 cm. A comparatively favorable libration here has brought the "limb craters" into view much earlier than usual. Compare the dark floor of Grimaldi (g1) with the extreme brightness of Aristarchus (c8).

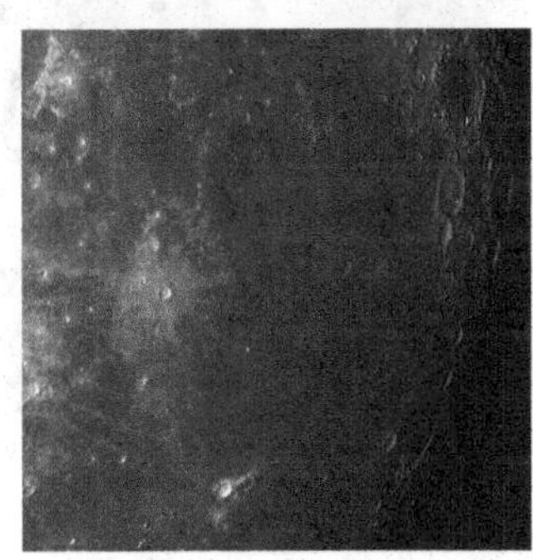

Plate 7b

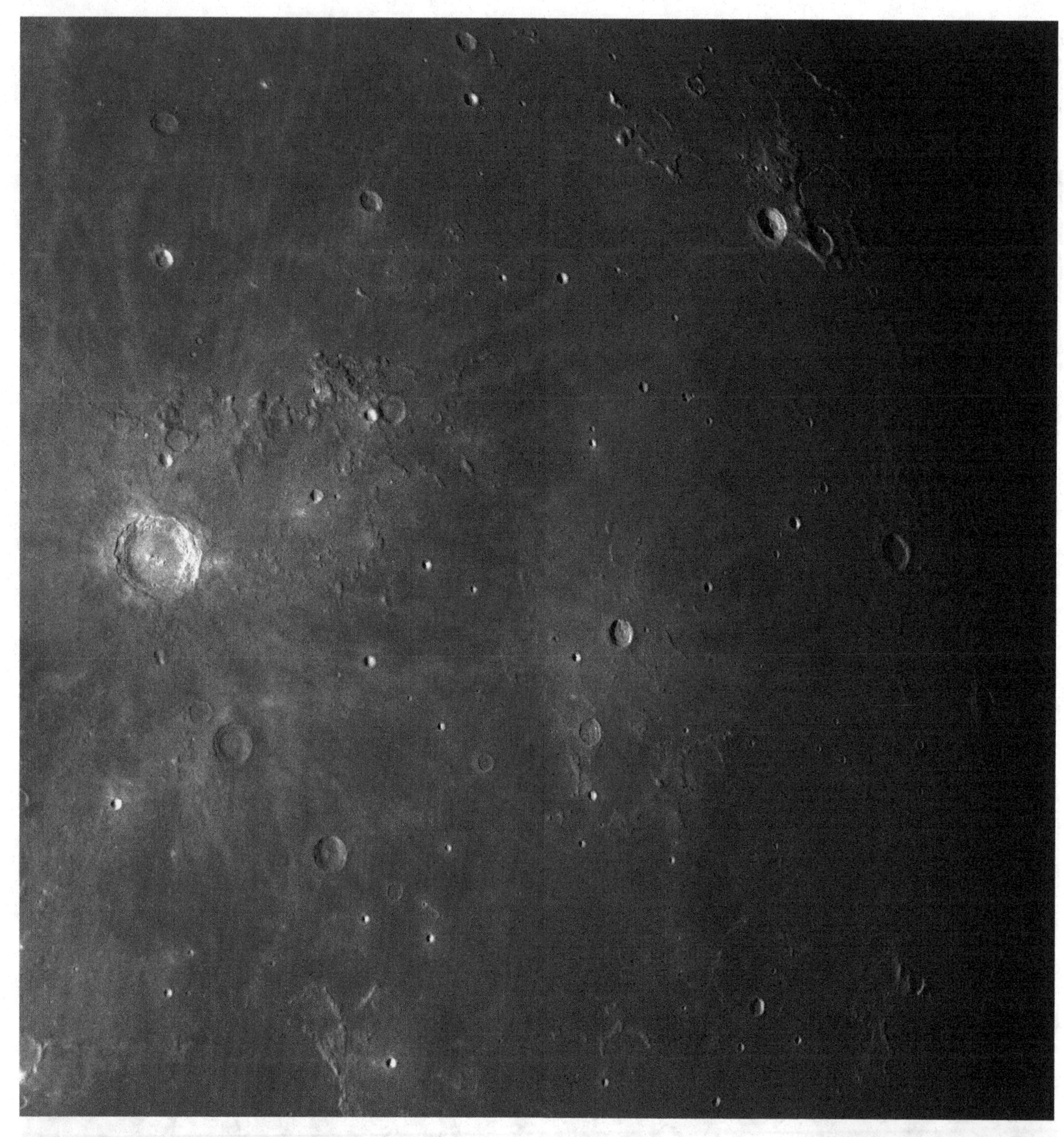

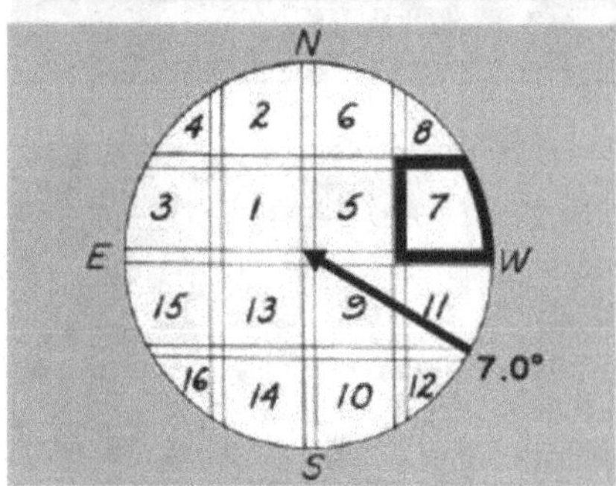

1944 UT Feb 02 1966. Moon's age 12.1 days. Sun's selenographic colongitude = 56.3°. Diameter 64 cm. This photograph extends further to the east than the others in this set. Note Vallis Schröteri (d8) and the low ridges near the terminator.

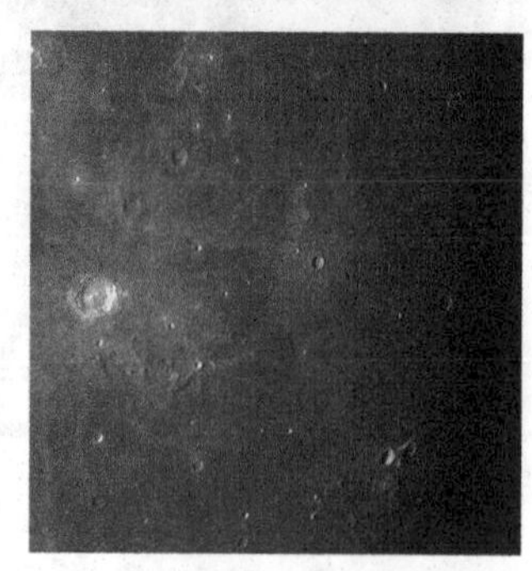

Plate 7c

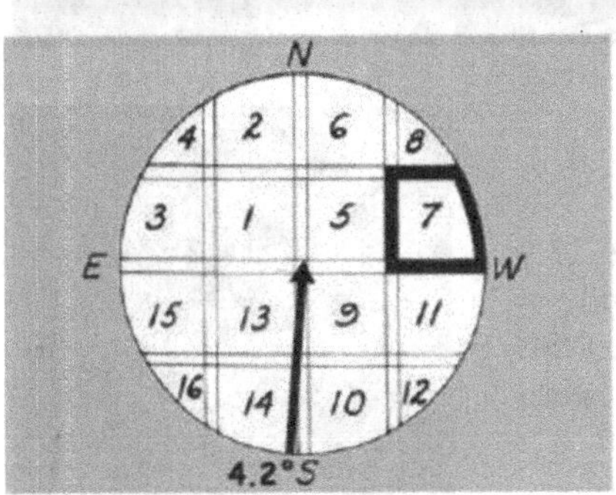

0655 UT Dec 07 1966. Moon's age 24.6 days. Sun's selenographic colongitude = 206.2°. Diameter 64 cm. Compare this with Plate 7a, particularly near the limb.

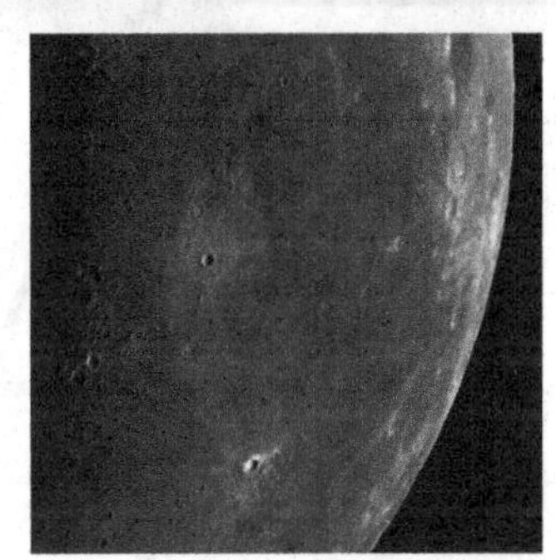

Plate 7d

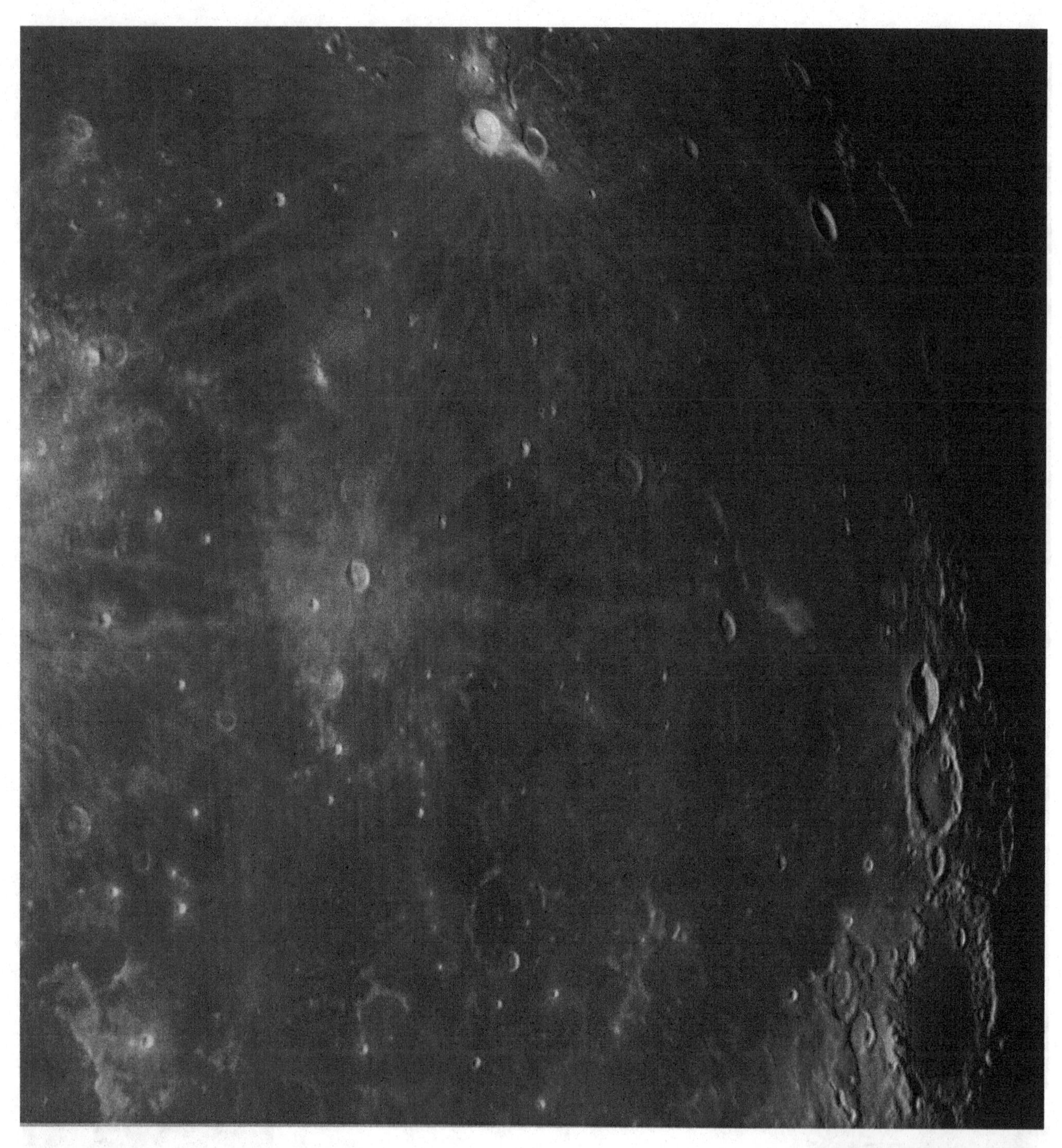

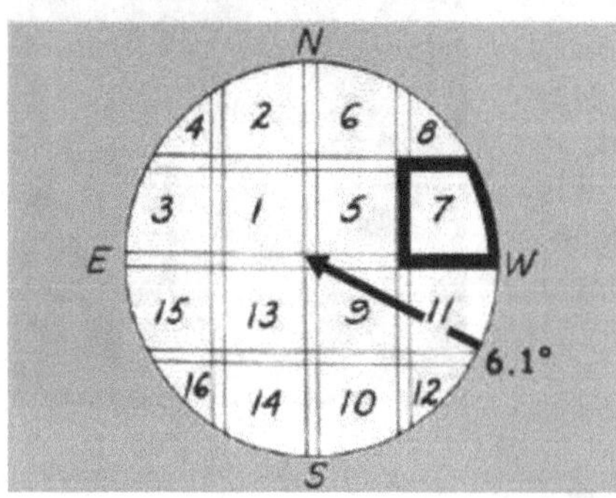

2036 UT Dec 25 1966. Moon's age 13.8 days. Sun's selenographic colongitude = 72.2°. Diameter 64 cm. Note the "St. Andrew's Cross" marking on the western wall of Grimaldi (g1).

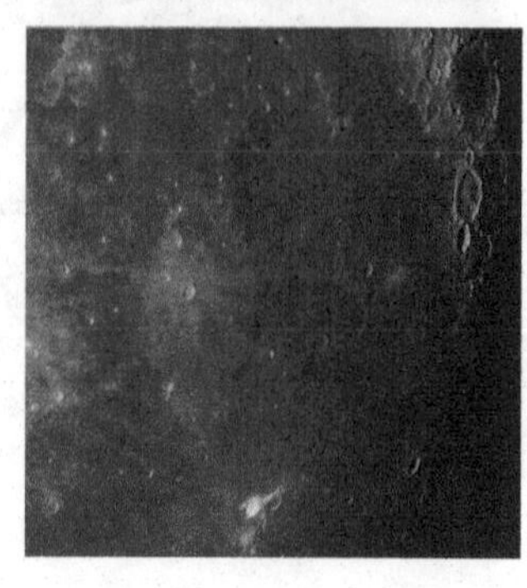

Plate 7e

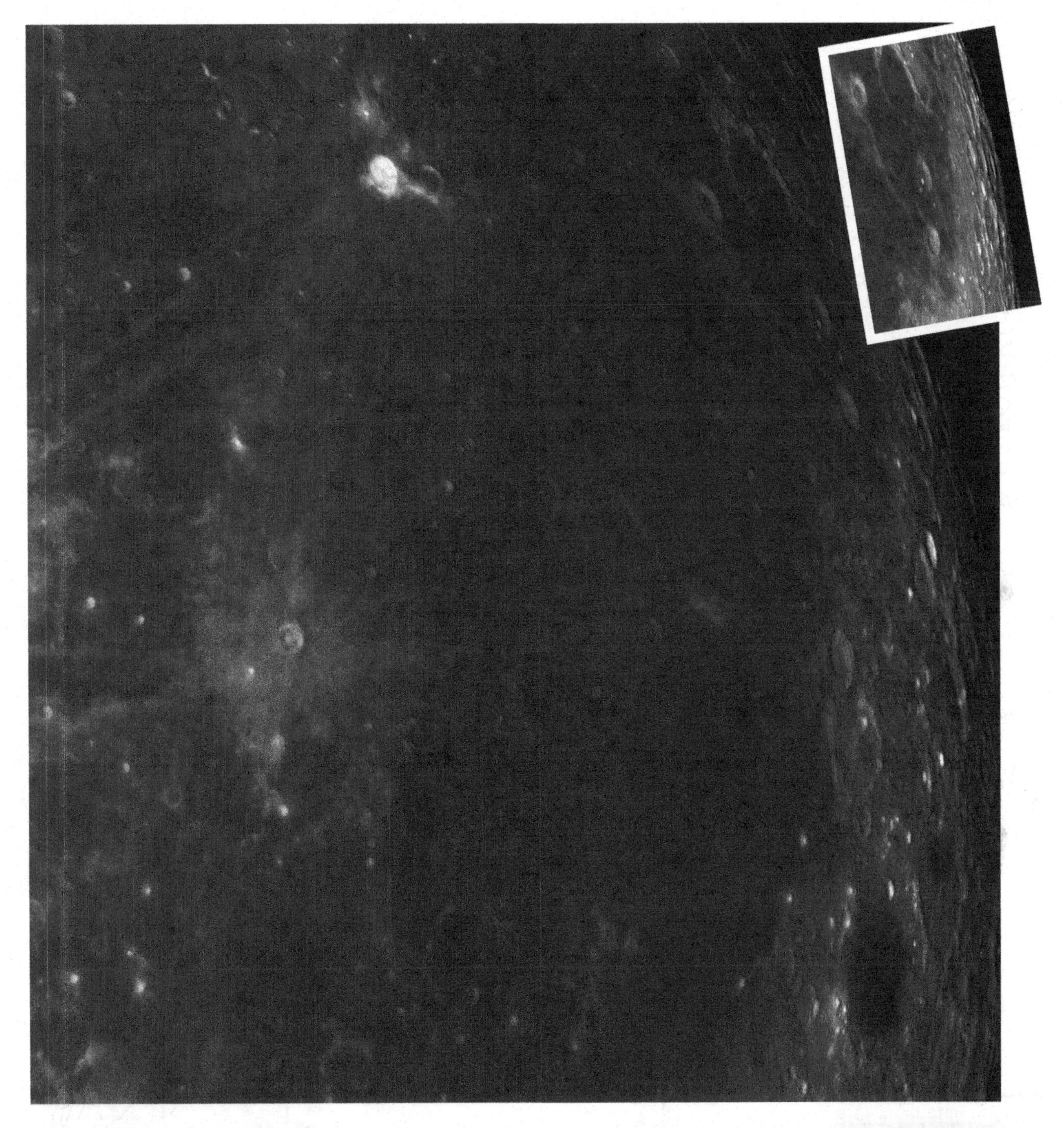

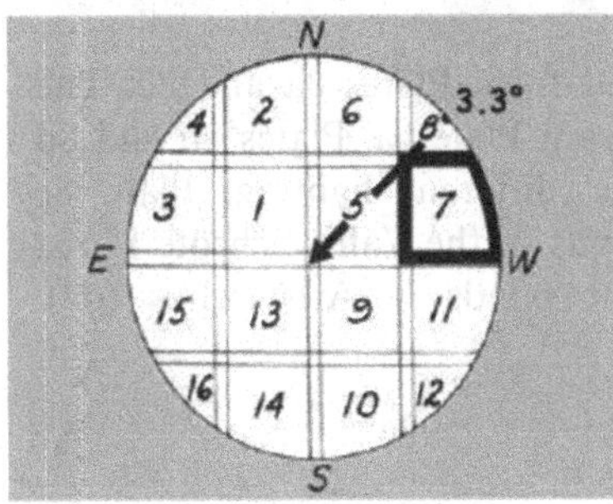

2219 UT Oct 28 1966. Moon's age 14.7 days. Sun's selenographic colongitude = 87.4°. Diameter 64 cm. This was taken 12 hours before Full Moon. Einstein (h6) is just beyond the terminator. Vasco da Gama (h6) is showing plainly. The insert shows Einstein, taken at 2132 UT on Nov 08 1965, with a 6-inch reflector. This crater may be seen clearly on only one or two nights in the average year.

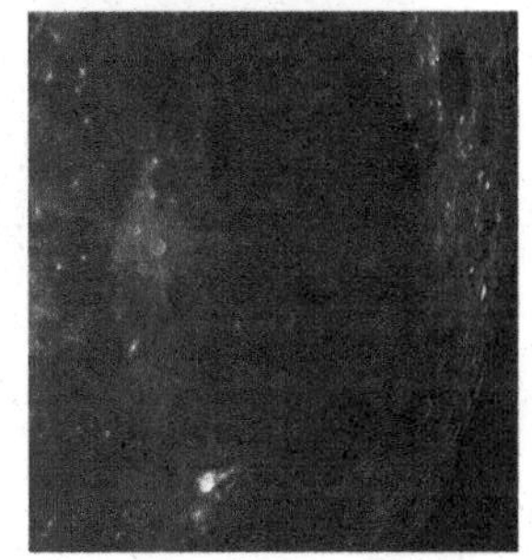

Map 8

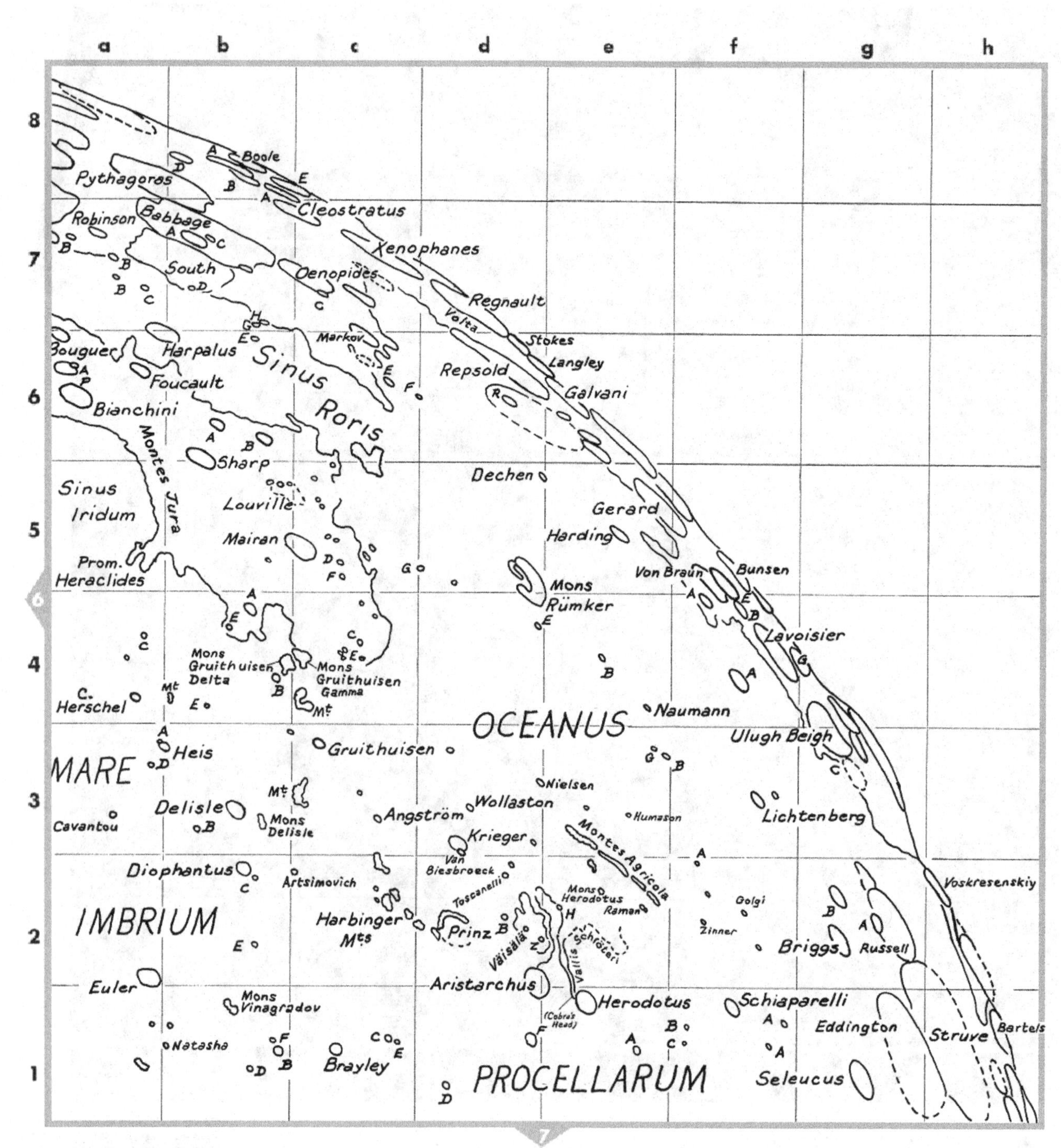

Crater Diameters

Pythagoras	145 km (a8)
Gerard	99 km (e5)
Seleucus	45 km (g1)
Wollaston	10 km (d3)
Euler	26 km (a2)

This map has been prepared from Plates 8a and 8c. The extreme NE corner comes from Plate 6a. Plates 8b and 8d show the same area under different lighting. Plate 8e shows larger-scale photographs of the Vallis Schröteri area (e2), Rümker (d5), and the bands in Aristarchus (d2).

Plate 8a

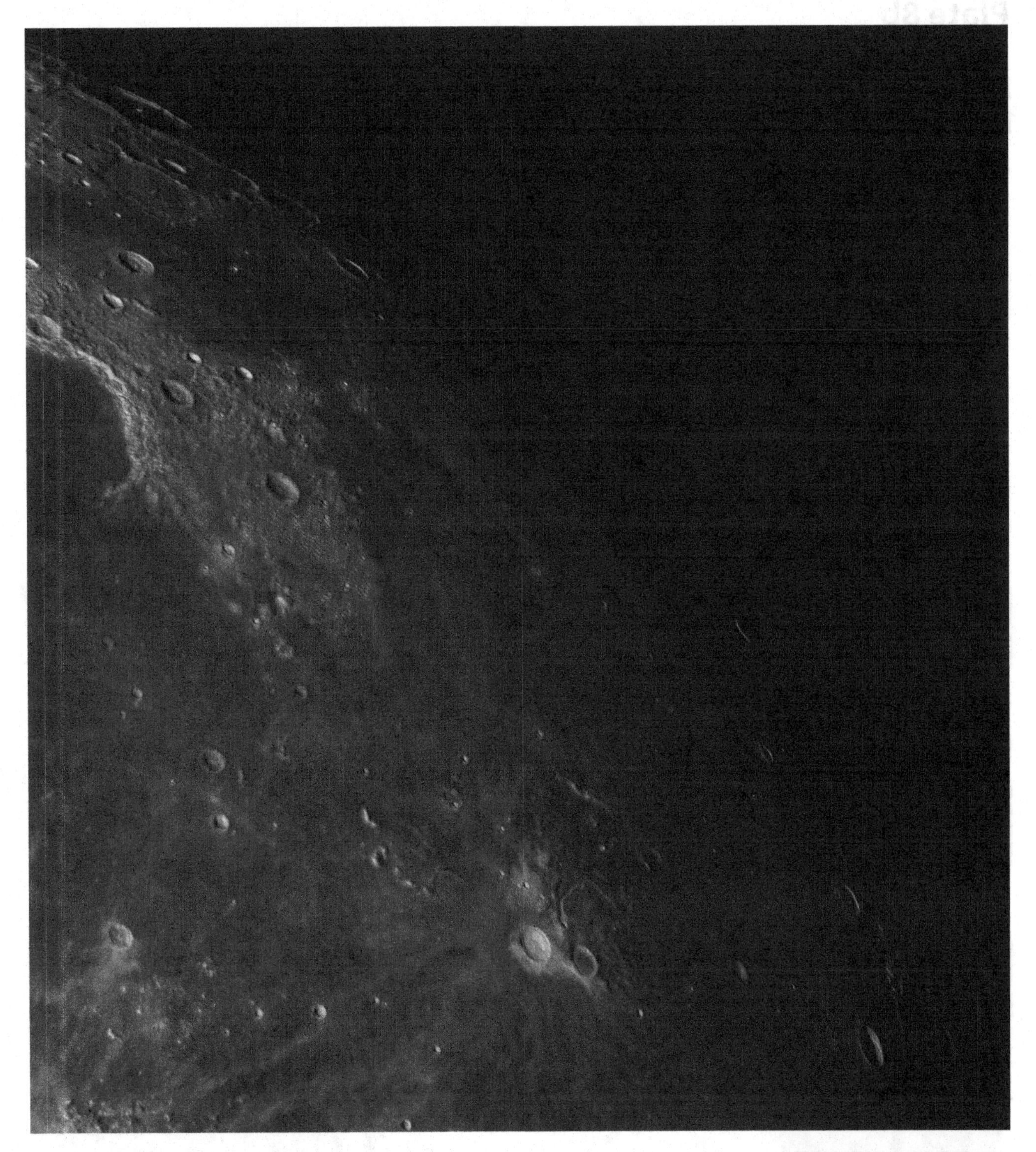

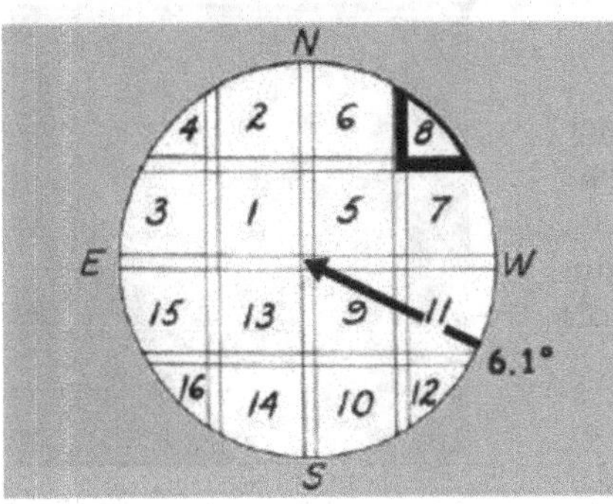

2040 UT Dec 25 1966. Moon's age 13.8 days. Sun's selenographic colongitude = 72.2°. Diameter 64 cm. Pythagoras (a8) is on the terminator near the top. Rümker (d5) looks more like a mound than a crater.

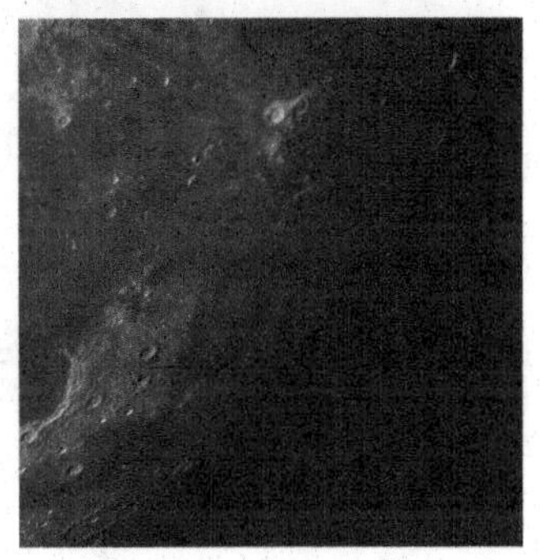

Plate 8b

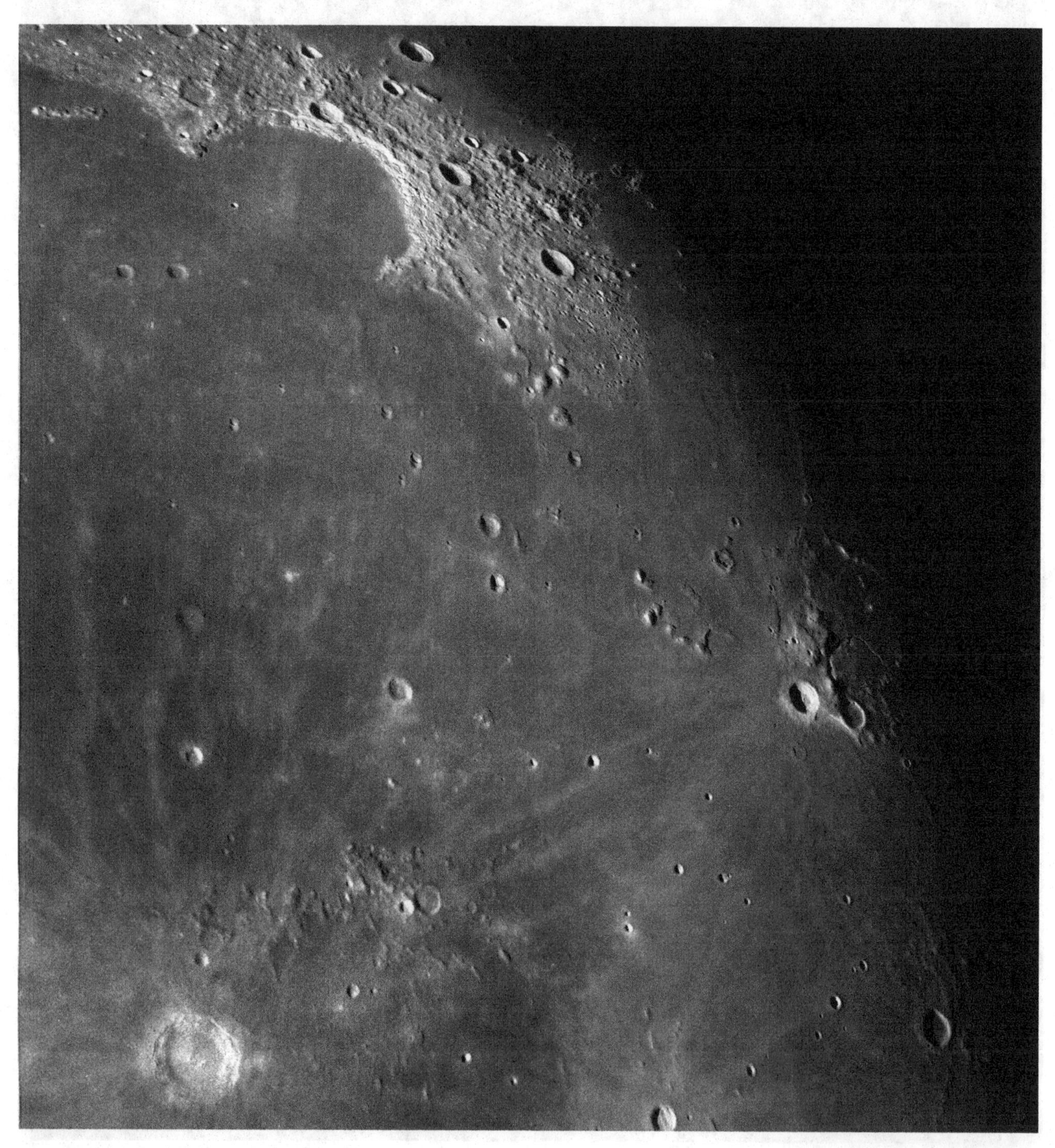

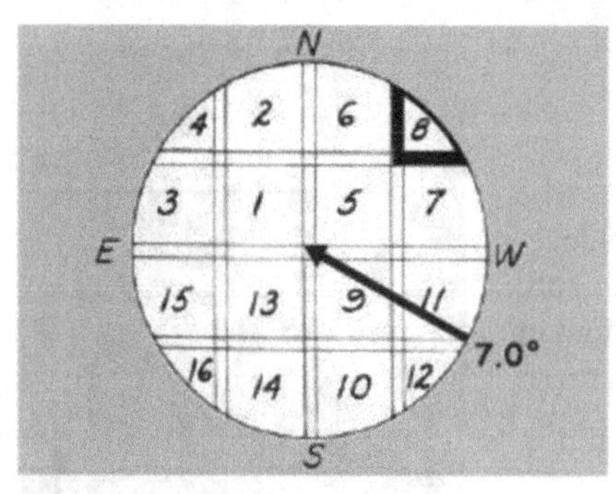

1944 UT Feb 02 1966. Moon's age 12.1 days. Sun's selenographic colongitude = 56.3°. Diameter 64 cm. The Montes Jura (a6) rise to about 6,100 m. The Montes Harbinger (c2) are about 2,400 m high.

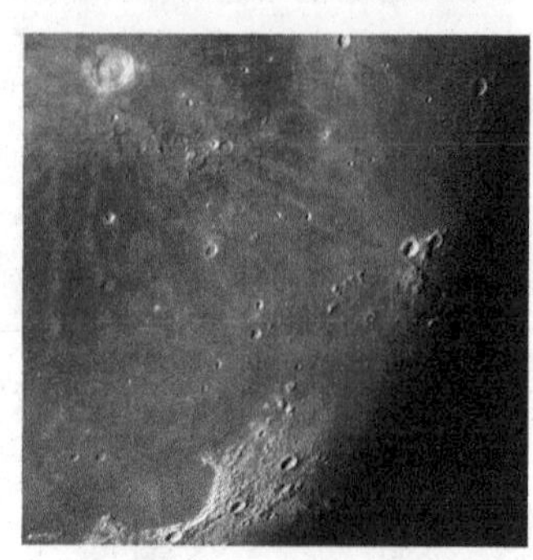

Plate 8c

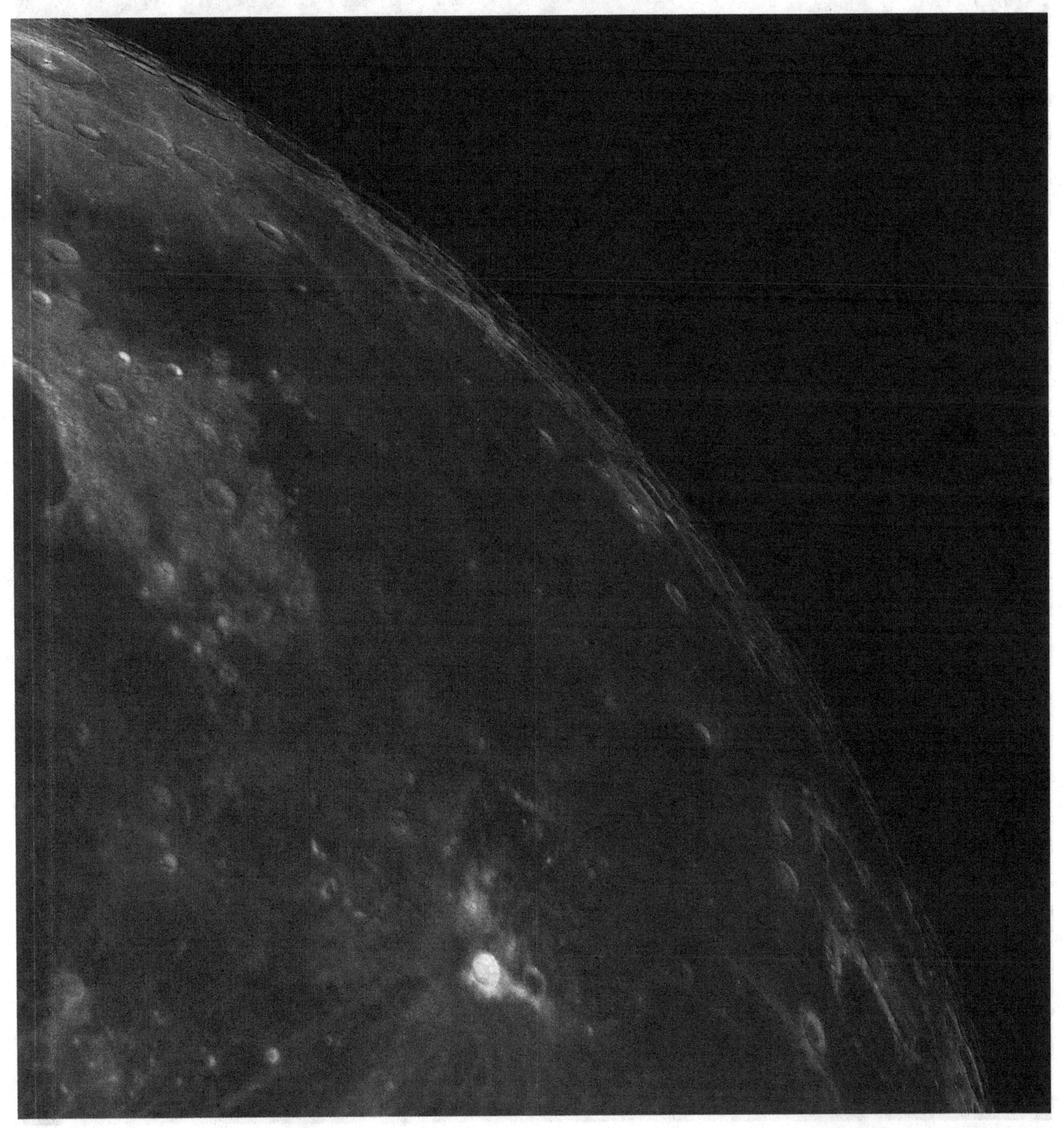

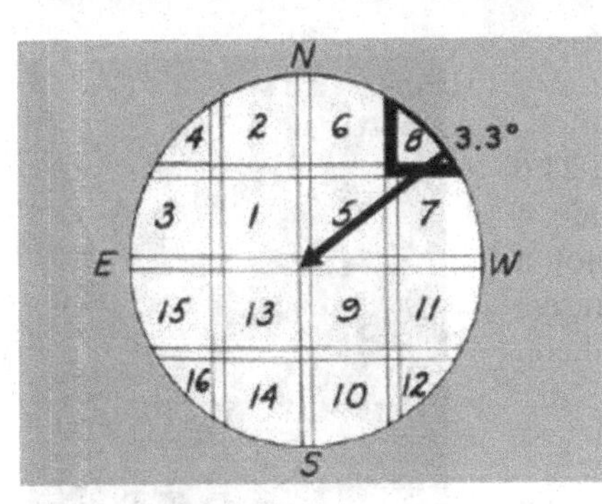

2219 UT Oct 28 1966. Moon's age 14.7 days. Sun's selenographic colongitude = 87.4°. Diameter 64 cm. This was taken 12 hours before Full Moon. The libration is favorable; the craters on the terminator (limb) will not often show up like this.

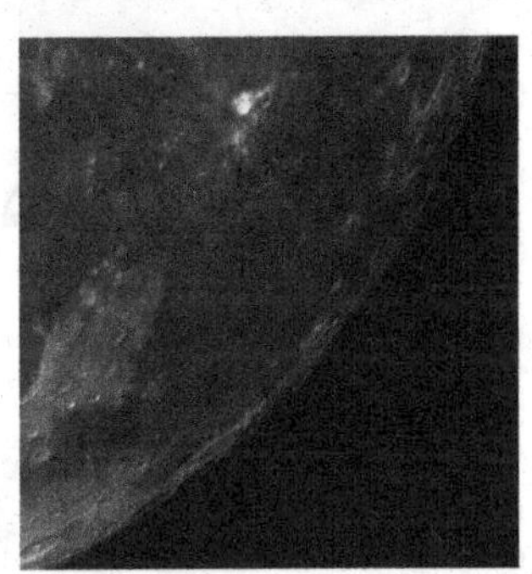

Plate 8d

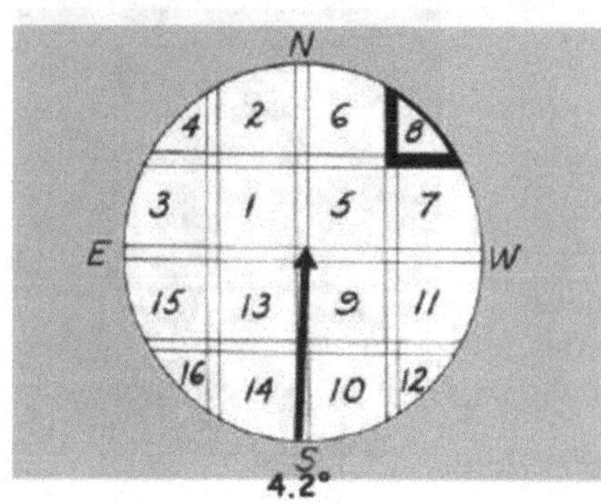

0655 UT Dec 07 1966. Moon's age 24.6 days. Sun's selenographic colongitude = 206.2°. Diameter 64 cm. This is not a favorable libration. Many of the craters shown on Plate 8c are out of sight here, beyond the limb.

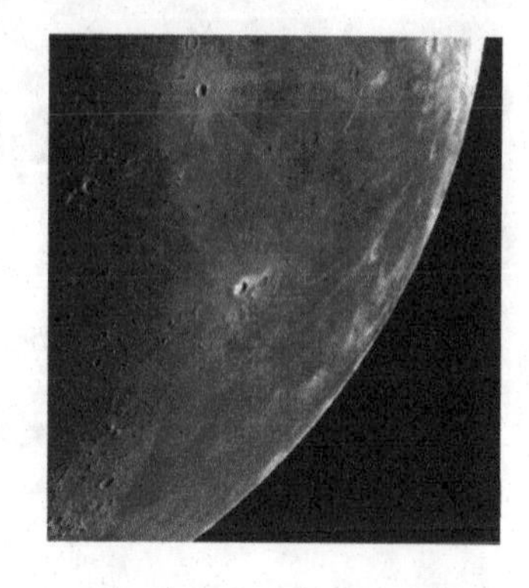

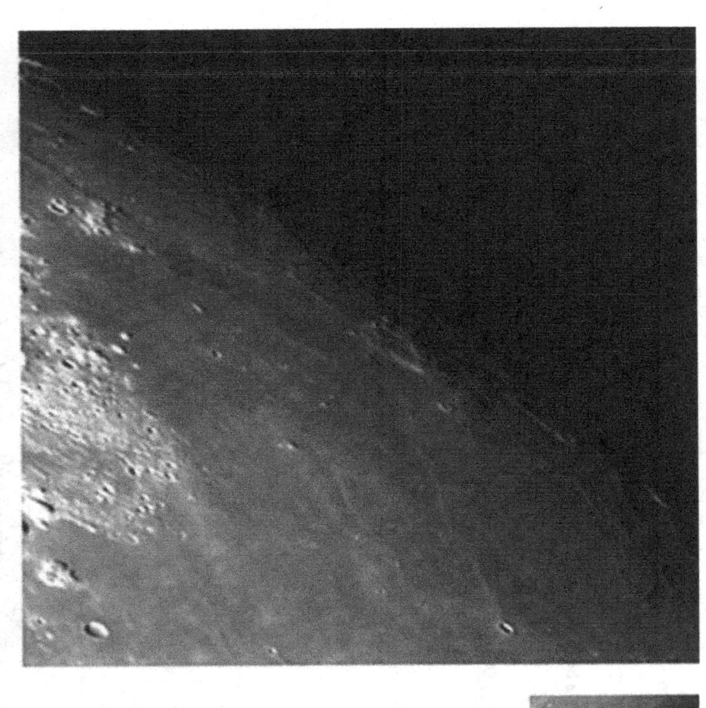

Vallis Schröteri, Aristarchus, and Herodotus. 2033 UT Apr 21 1967. Moon's age 12.0 days. Sun's selenographic colongitude = 56.4°. Diameter 94 cm approx. Vallis Schröteri lies on Map 8 e2.

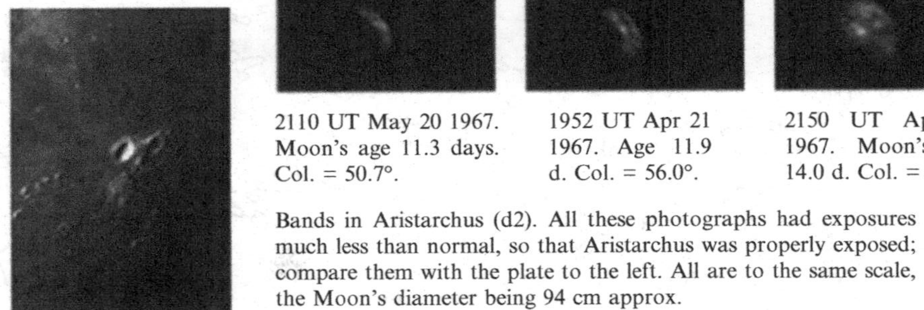

Mons Rümker. 1944 UT Mar 23 1967. Moon's age 12.6 days. Sun's selenographic colongitude = 62.4°. Diameter 89 cm approx. Mons Rümker lies in Map 8 d5.

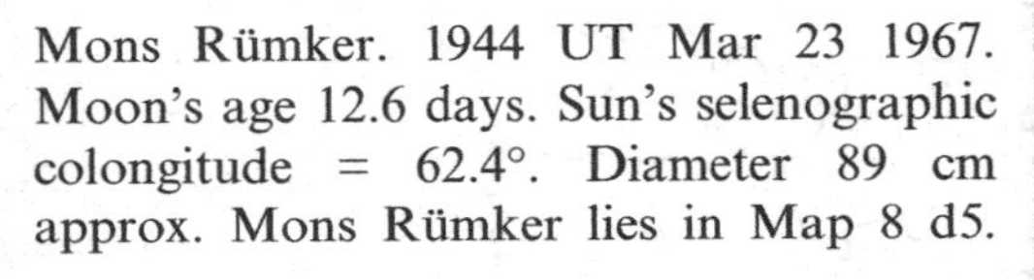

2110 UT May 20 1967. Moon's age 11.3 days. Col. = 50.7°.

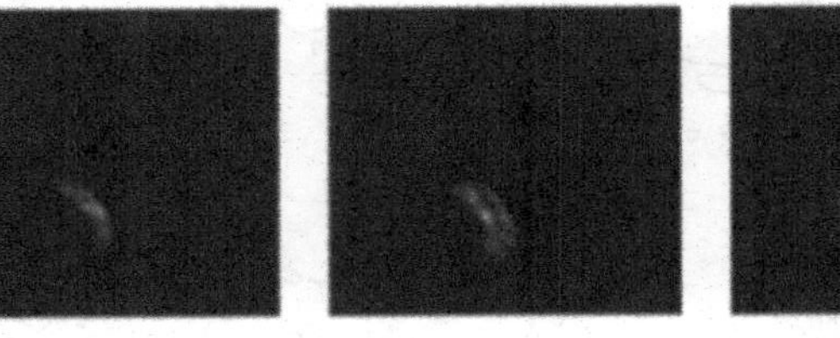

1952 UT Apr 21 1967. Age 11.9 d. Col. = 56.0°.

2150 UT Apr 23 1967. Moon's age 14.0 d. Col. = 81.4°.

Bands in Aristarchus (d2). All these photographs had exposures much less than normal, so that Aristarchus was properly exposed; compare them with the plate to the left. All are to the same scale, the Moon's diameter being 94 cm approx.

Plate 8e

Map 9

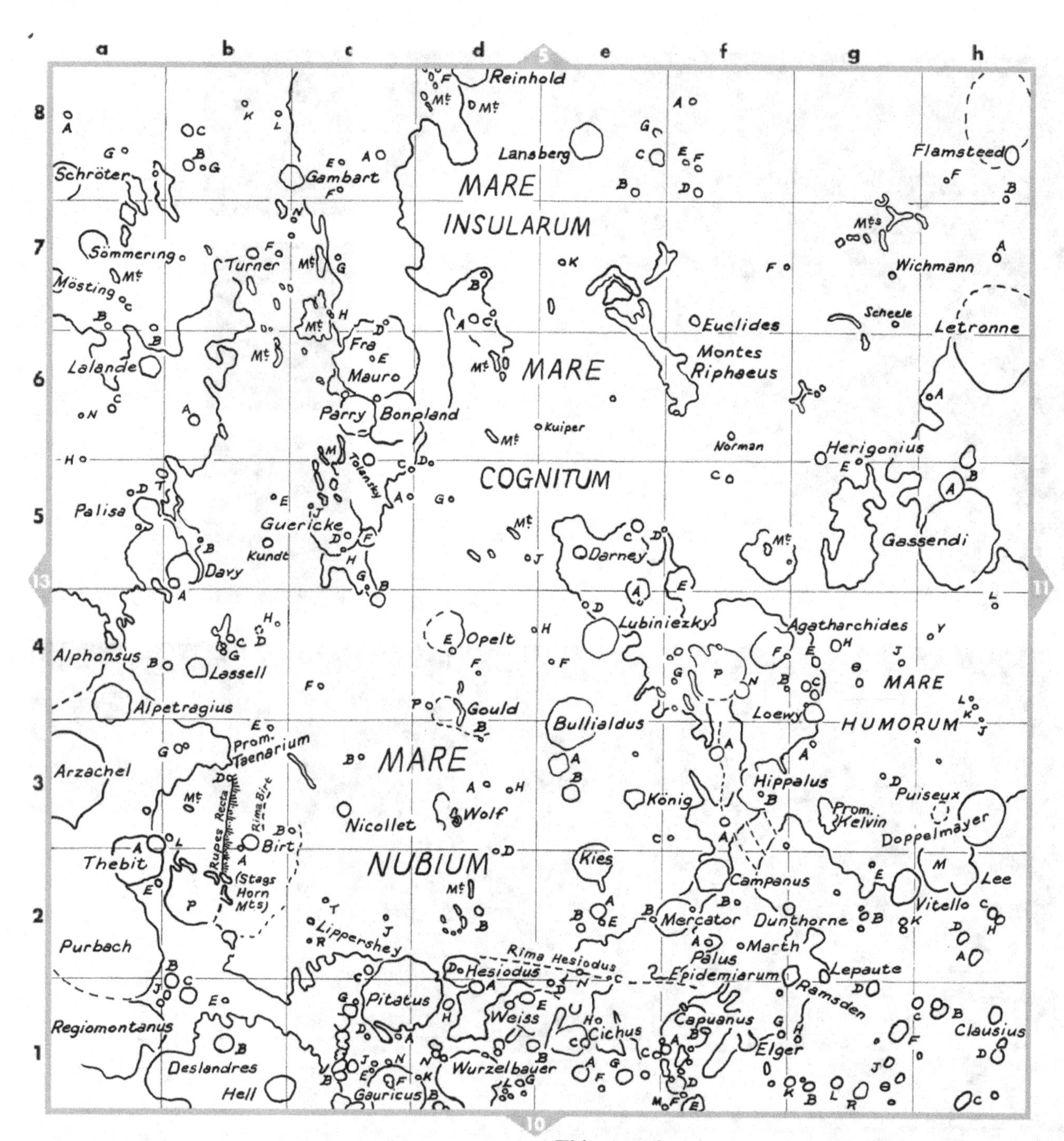

Crater Diameters

Flamsteed	19 km (h8)
Gambart	25 km (c8)
Bullialdus	61 km (e3)
Lepaute	16 km (g2)
Hell	33 km (b1)

This map has been prepared from Plate 9a. Plates 9b, 9c, 9d, and 9e show the same area under different lighting. Plate 9f shows larger scale photographs of the Rupes Recta area (b3). Plate 9g shows larger scale photographs of the area around Bullialdus and Kies (e2), and of the Mare Humorum clefts and ridges (f3).

Plate 9a

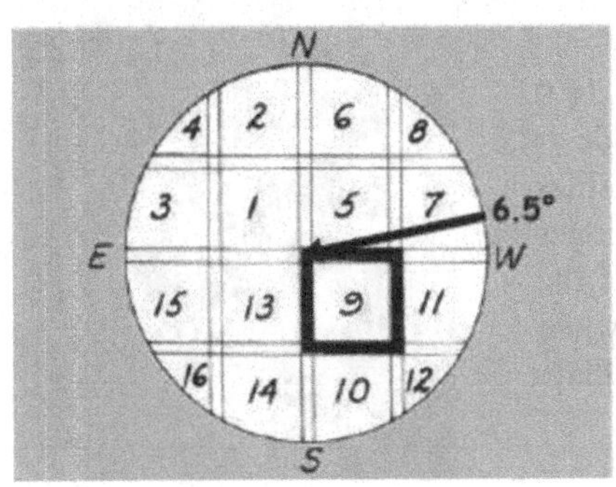

0317 UT Aug 09 1966. Moon's age 22.0 days. Sun's selenographic colongitude = 181.6°. Diameter 64 cm. Note the Rupes Recta or Straight Wall (b3); it is about 110 km long and 250 to 300 m high and slopes up towards the west. The Montes Riphaeus (f6) are about 900 m high.

Plate 9b

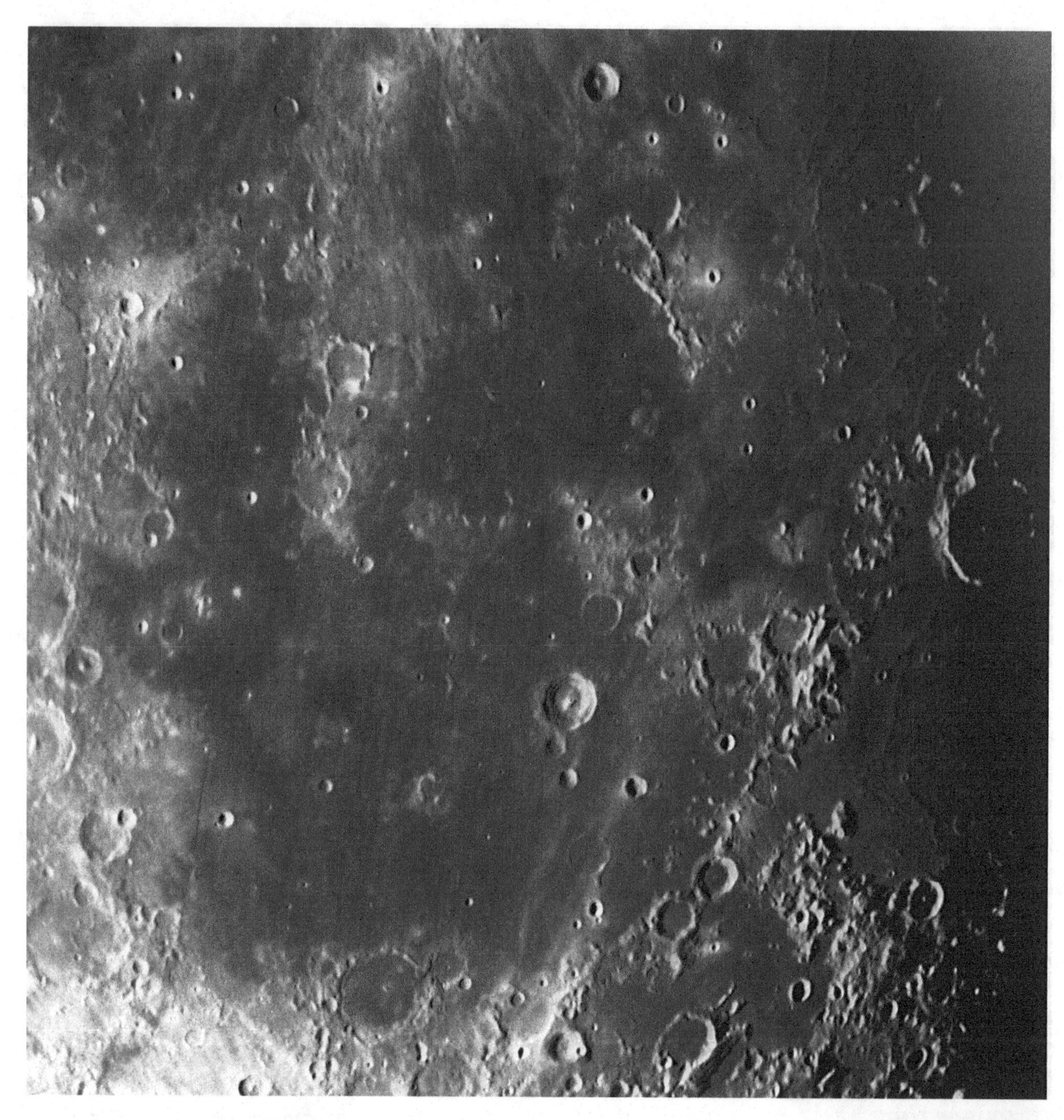

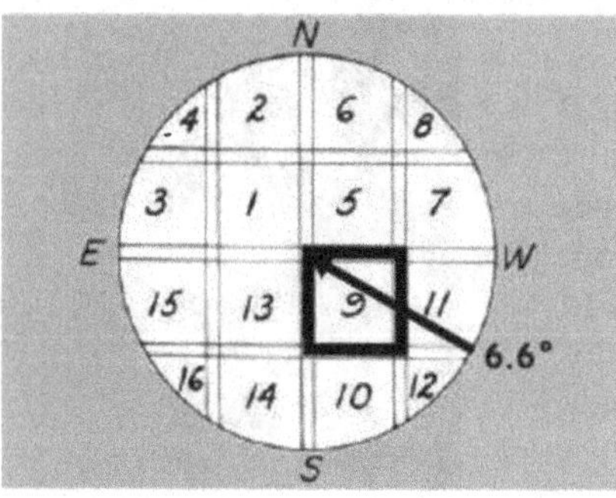

1758 UT Jan 21 1967. Moon's age 11.0 days. Sun's selenographic colongitude = 39.2°. Diameter 64 cm. Note the rilles NW from Campanus (f2) and the ridges running down the right-hand side of the photograph. These are only visible when the lighting is very oblique. (See Plate 9e also.)

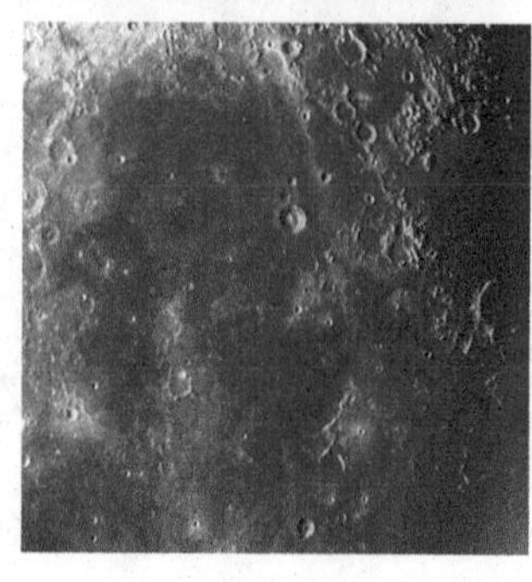

Plate 9c

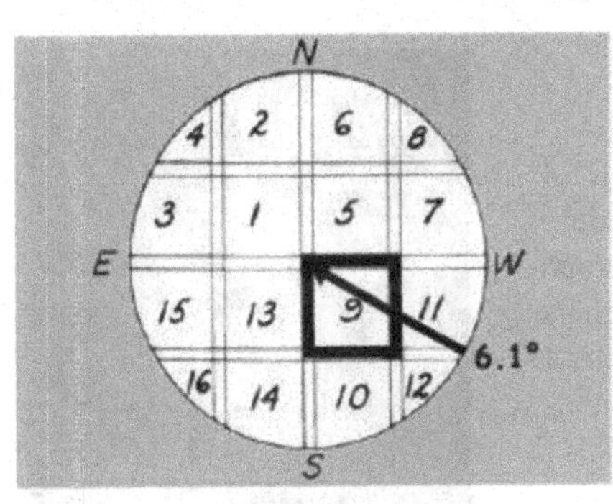

2135 UT Dec 25 1966. Moon's age 13.8 days. Sun's selenographic colongitude = 72.7°. Diameter 64 cm. Compare this with Plates 9b and 9a, which show almost the same area under morning and afternoon lighting.

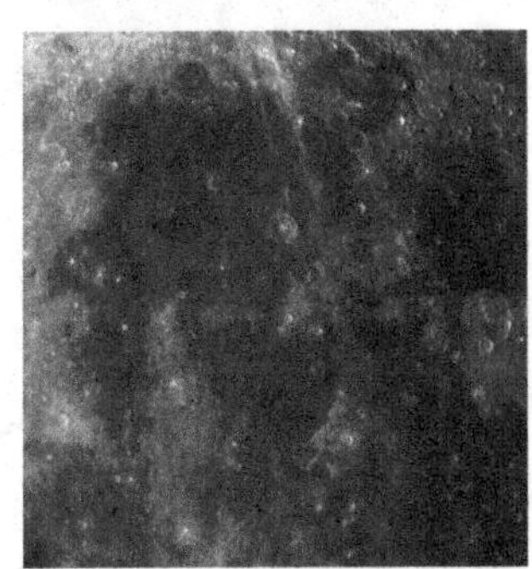

Plate 9d

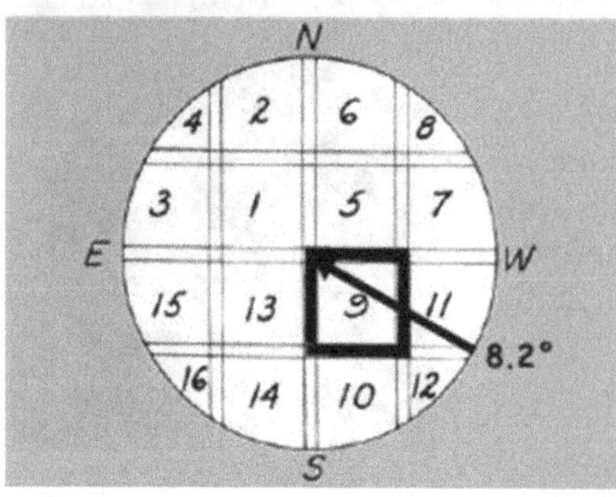

0516 UT Oct 06 1966. Moon's age 21.4 days. Sun's Selenographic Colongitude = 170.5°. Diameter 64 cm. This photograph extends further to the East than the others in this group and overlaps into sections 1 and 13.

Plate 9e

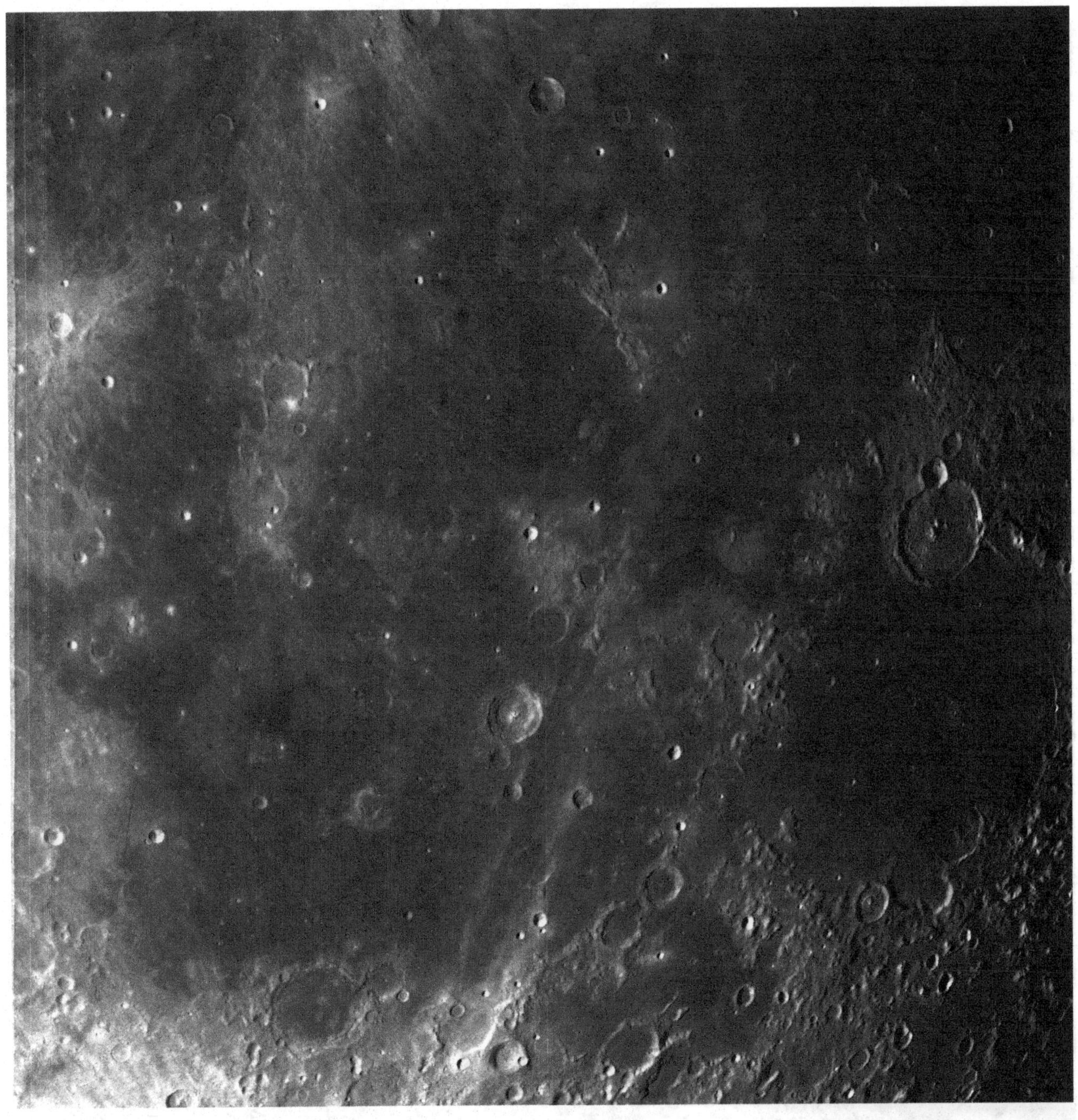

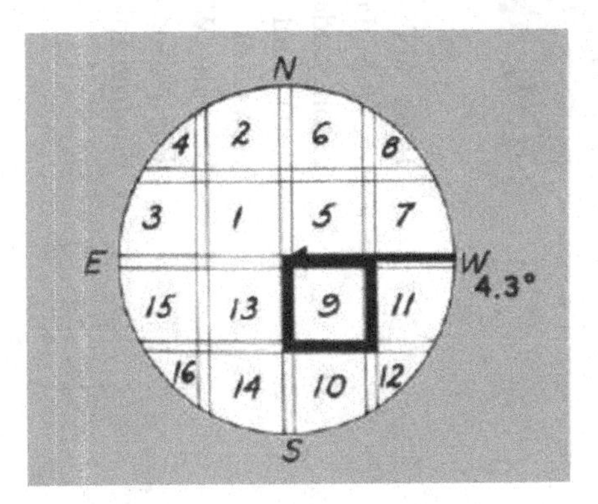

2236 UT Dec 23 1966. Moon's age 11.8 days. Sun's selenographic colongitude = 48.9°. Diameter 64 cm. The Moon here is less than one day older than it is in Plate 9b and yet the rilles and dorsa shown in the latter have virtually disappeared.

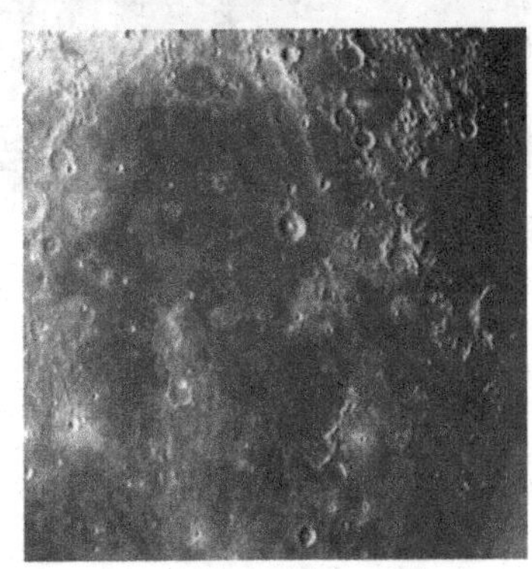

Plate 9f

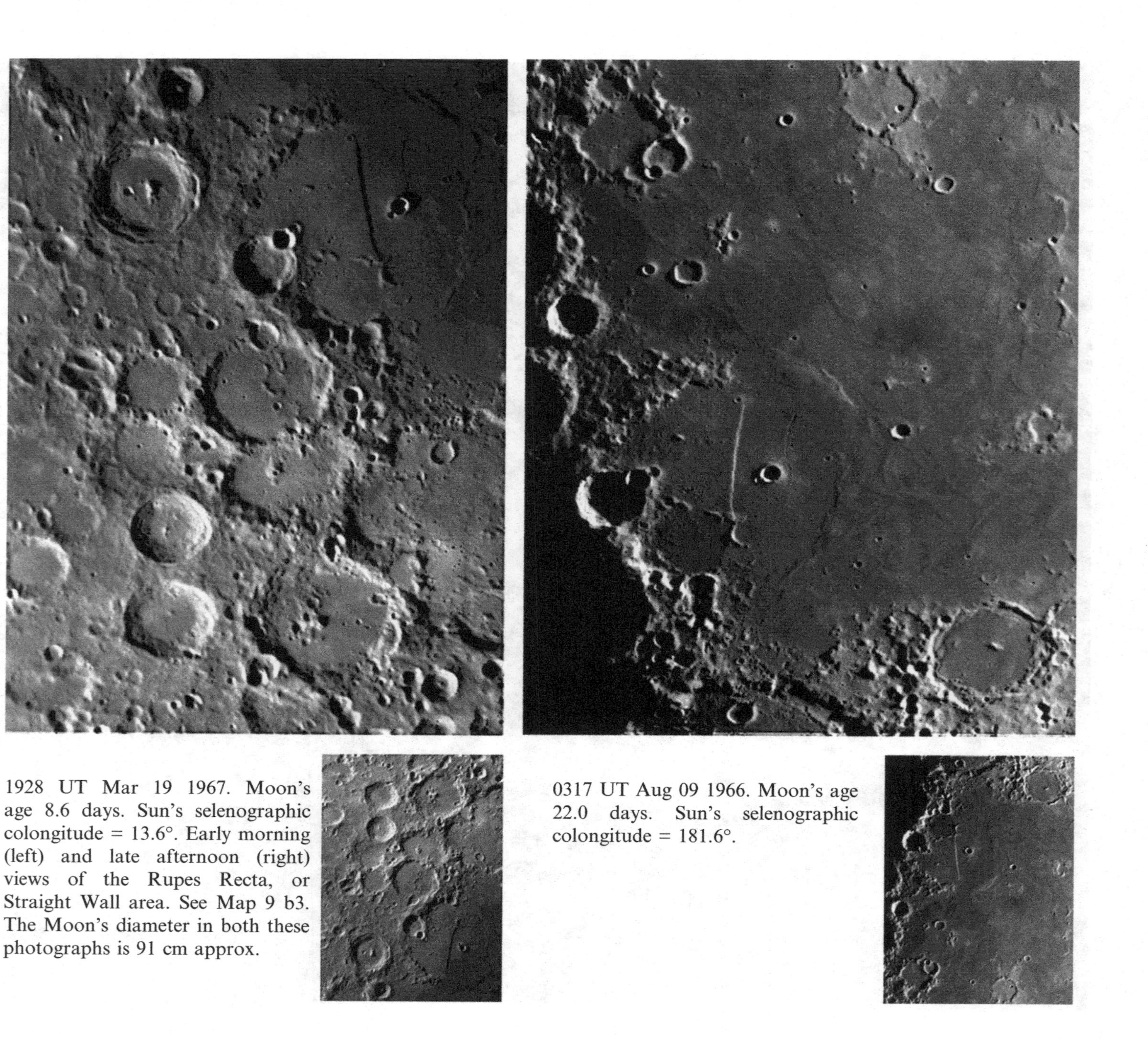

1928 UT Mar 19 1967. Moon's age 8.6 days. Sun's selenographic colongitude = 13.6°. Early morning (left) and late afternoon (right) views of the Rupes Recta, or Straight Wall area. See Map 9 b3. The Moon's diameter in both these photographs is 91 cm approx.

0317 UT Aug 09 1966. Moon's age 22.0 days. Sun's selenographic colongitude = 181.6°.

Plate 9g

Bullialdus, Kies and the Kies dome, and the Hesiodus Cleft. 2056 UT Mar 20 1967. Moon's age 9.7 days. Sun's selenographic colongitude = 26.5°. Diameter 91 cm approx. The Kies dome lies just under one diameter west from Kies, which is shown in Map 9 e2.

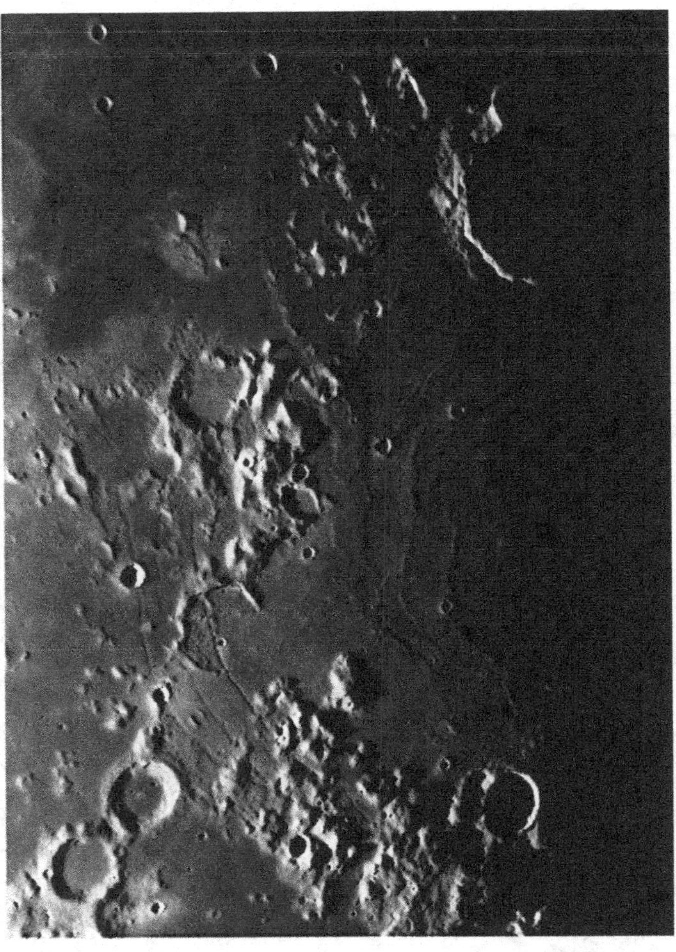

Rilles and dorsa on the eastern shores of the Mare Humorum. 1932 UT Mar 21 1967. Moon's age 10.6 days. Sun's selenographic colongitude = 37.9°. Diameter 91 cm approx. Campanus and Mercator (top left) lie in Map 9 f2.

Map 10

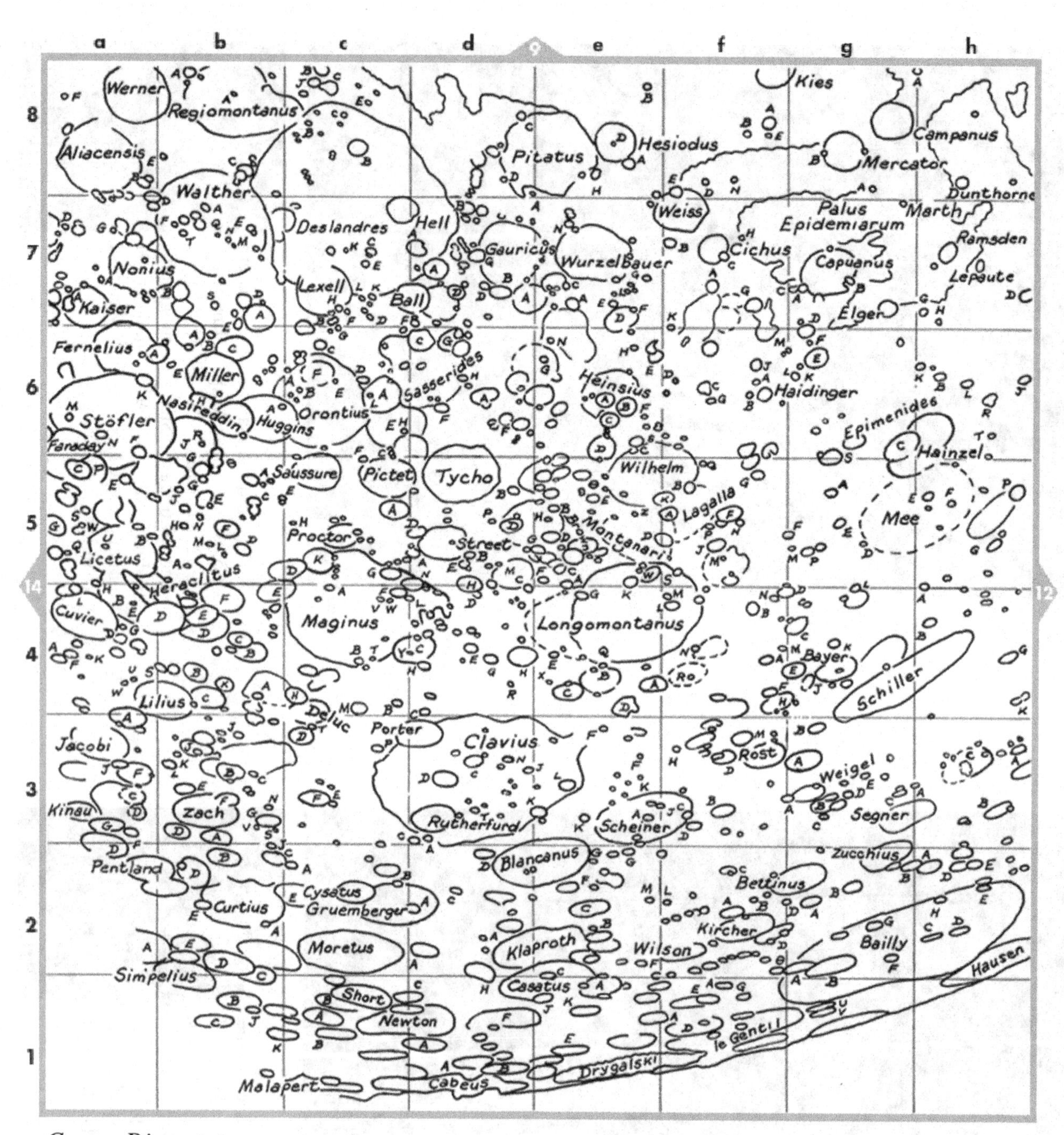

Crater Diameters

Mercator	46 km (g8)
Werner	71 km (a8)
Tycho	86 km (d5)
Bettinus	72 km (g2)
Moretus	114 km (c2)

This map has been prepared from Plate 10a. Plates 10b, 10c, and 10d show the same area under different lighting. Plate 10e shows larger scale photographs of the area between Clavius (d3) and Bailly (g2).

Plate 10a

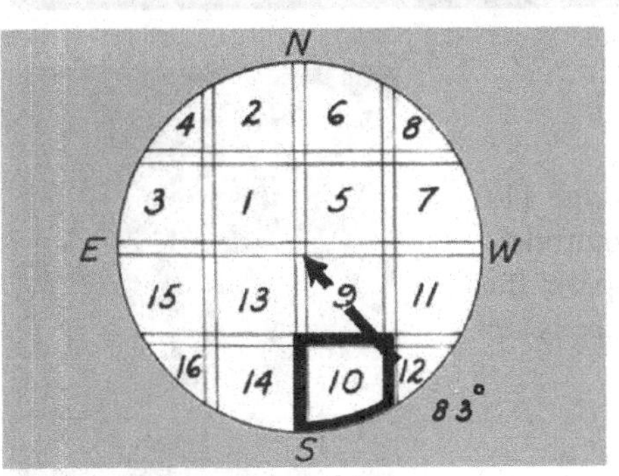

0609 UT Nov 04 1966. Moon's age 21.0 days. Sun's Selenographic Colongitude = 164.3°. Diameter 64 cm. This is a favourable libration for this area.

Plate 10b

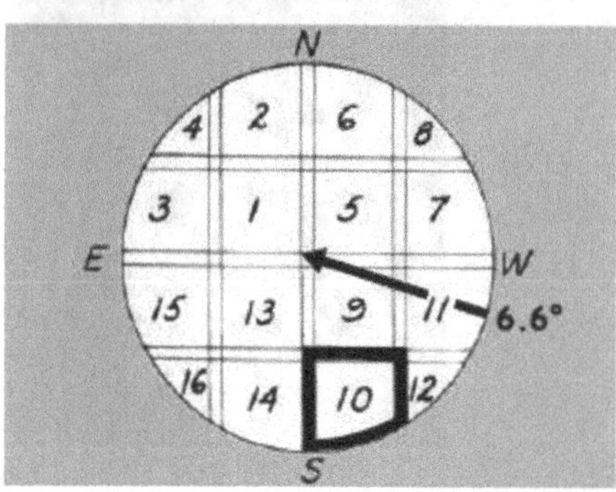

1758 UT Jan 21 1967. Moon's age 11.0 days. Sun's selenographic colongitude = 39.2°. Diameter 64 cm. Compare this with Plate 10a; the libration here is not so favorable.

Plate 10c

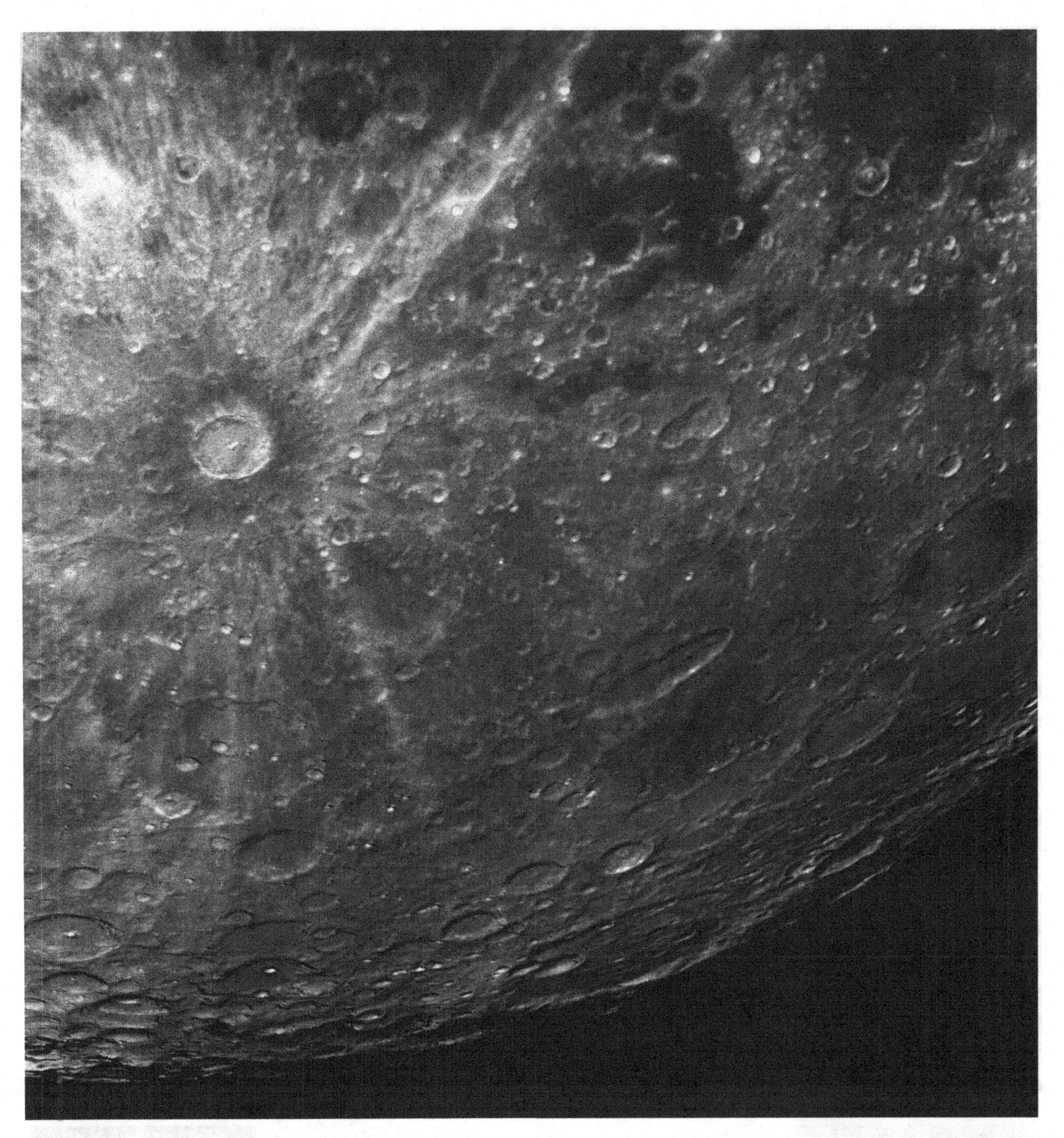

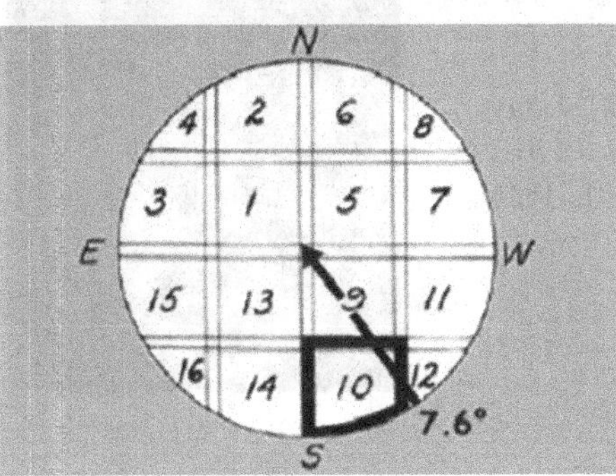

2104 UT Feb 23 1967. Moon's age 14.5 days. Diameter 64 cm. Sun's Selenographic Colongitude = 82.0°. This was taken about 20 hours before Full Moon. Compare it with Plate 10d, which shows almost the same area.

Plate 10d

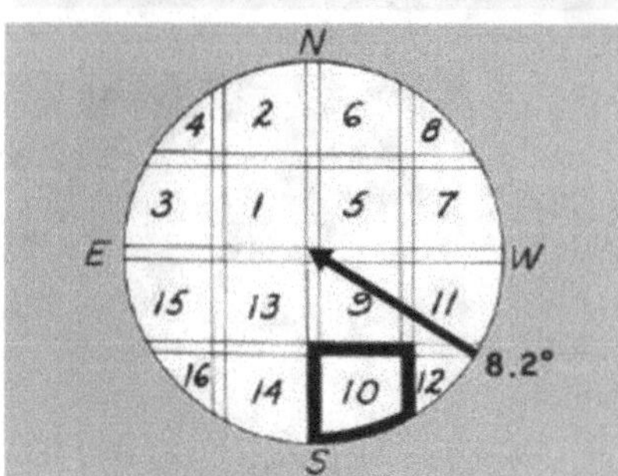

0516 UT Oct 06 1966. Moon's age 21.4 days. Sun's selenographic colongitude = 170.5°. Diameter 64 cm. Compare this with Plates 10a and 10c.

Plate 10e

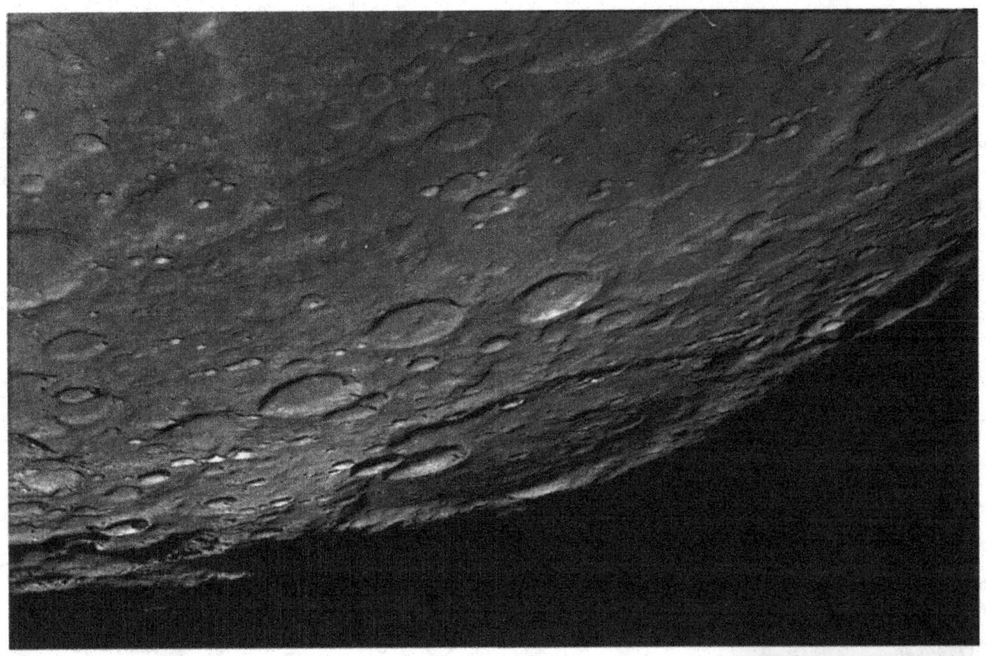

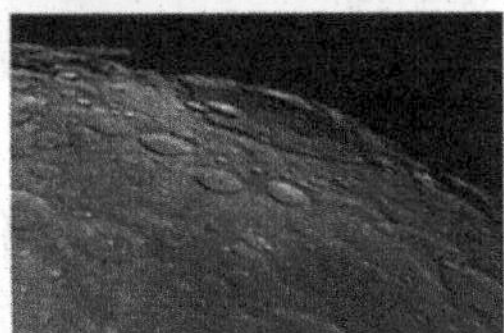

Bailly. 2259 UT Mar 05 1966. Moon's age 13.5 days. Selenographic col. 75.1°. Diameter 91 cm approx. Bailly lies in Map 10 g2.

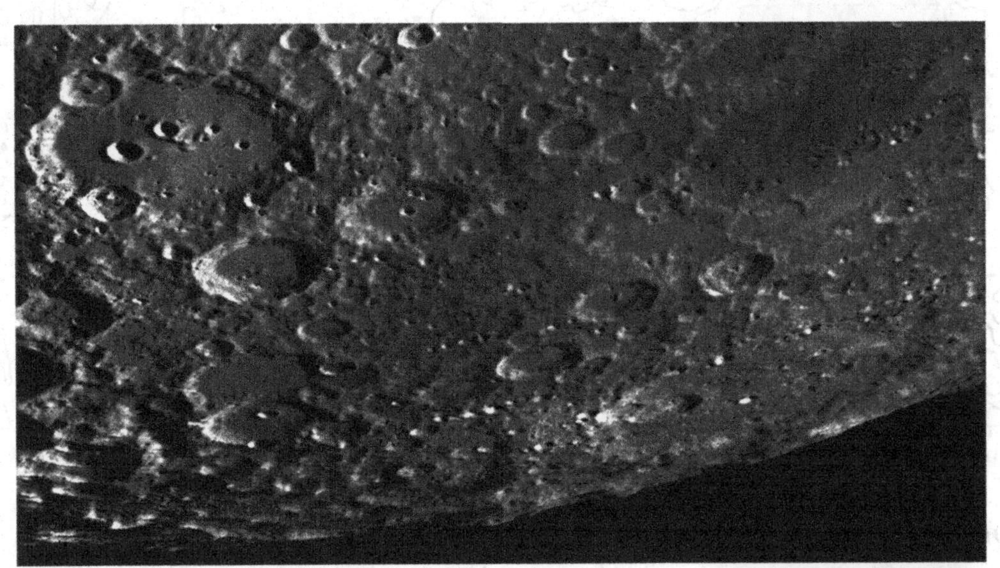

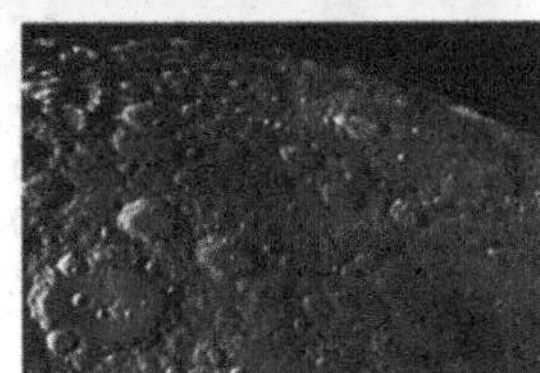

Clavius to Bailly. 0516 UT Oct 06 1966. Moon's age 21.4 days. Sun's selenographic colongitude = 170.5°. Diameter 91 cm approx. This is an enlargement of part of Plate 10d. Compare the Bailly area here with the same region in the top photograph.

Map 11

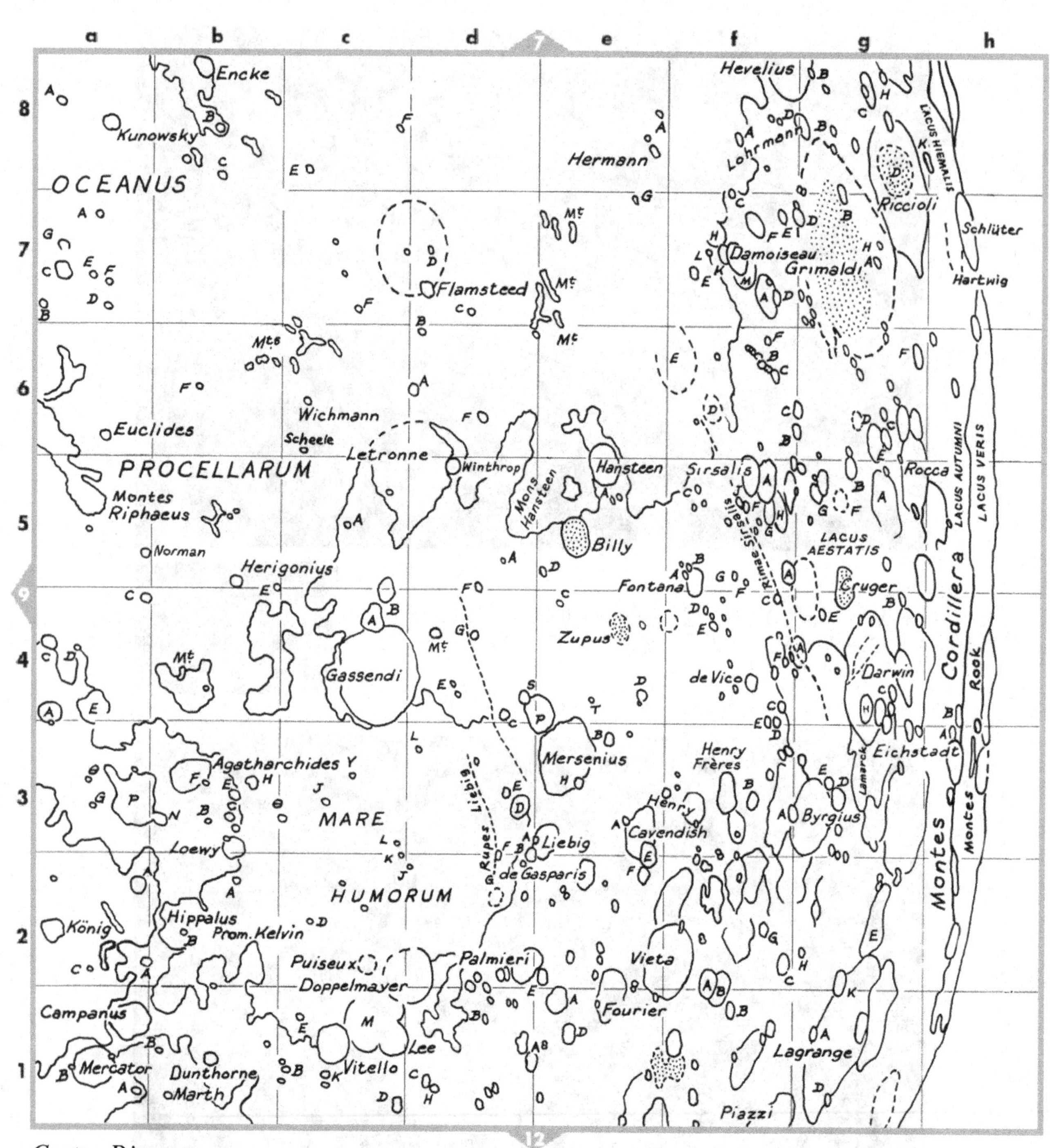

Crater Diameters

Damoiseau	37 km (f7)
Kunowsky	18 km (a8)
Billy	46 km (e5)
Vieta	89 km (f2)
Dunthorne	15 km (b1)

This map has been prepared from Plates 11a and 11c. Plates 11b and 11d show the same area under different lighting. Plate 11e shows larger scale photographs of Darwin, Rimae Sirsalis, and Grimaldi (g4 to g7) and of the Mare Humorum and Gassendi (c3).

Plate 11a

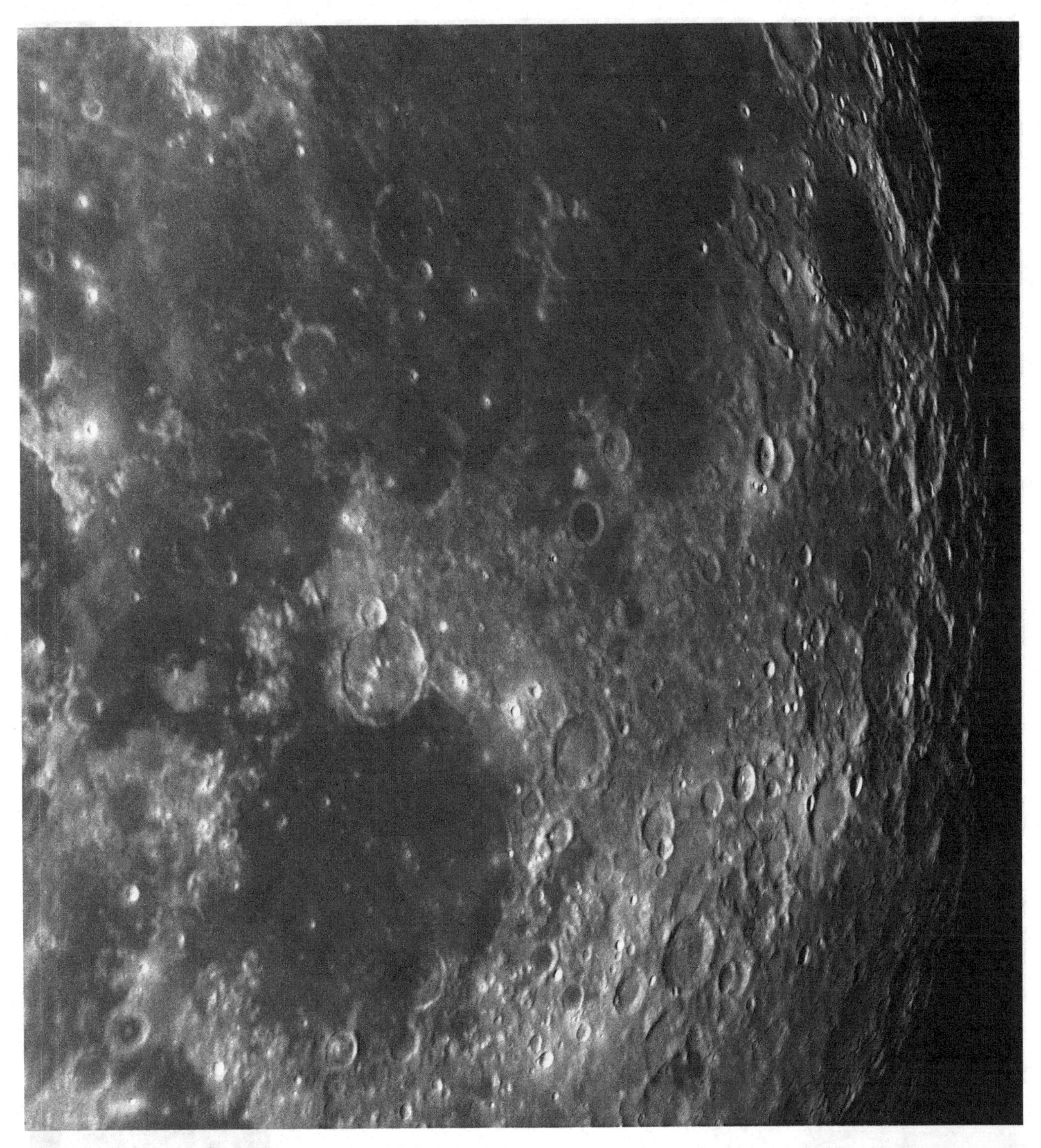

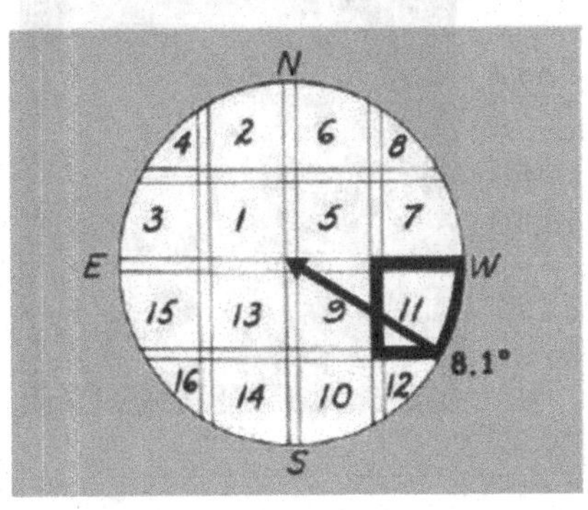

2310 UT Jan 24 1967. Moon's age 14.2 days. Sun's selenographic colongitude = 78.2°. Diameter 64 cm. This is a fairly favorable libration for this area. Byrgius A (f3) may be seen as a "ray center" on Plates 11b and 11c.

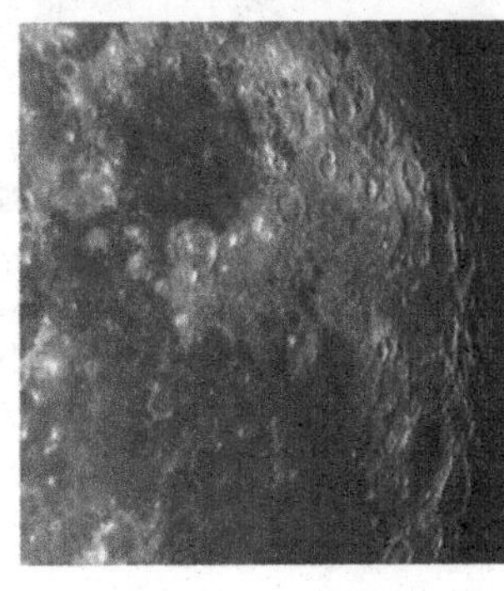

Plate 11b

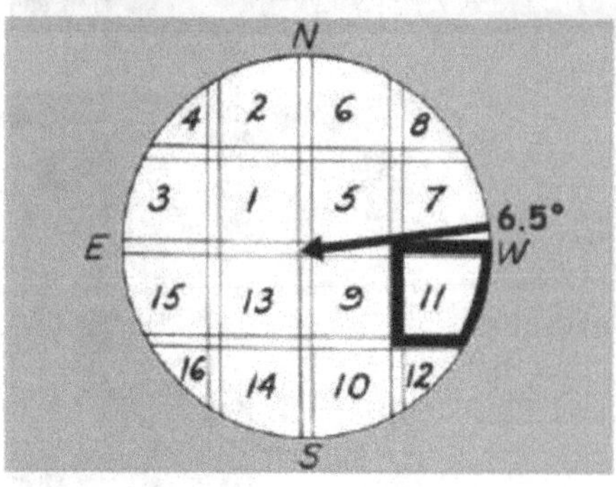

0317 UT Aug 09 1966. Moon's age 22.0 days. Sun's selenographic colongitude = 181.6°. Diameter 64 cm. Compare this with Plate 11a. The "ray center" (top right) is Byrgius A (f3).

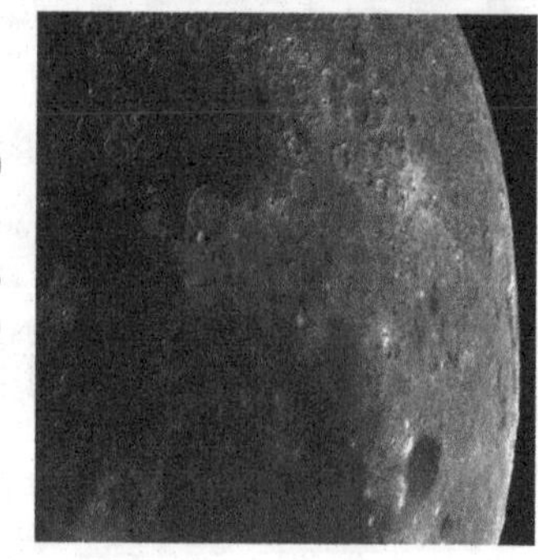

Plate 11c

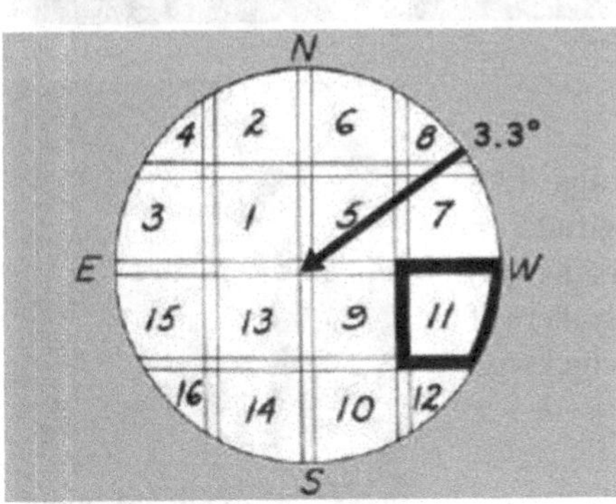

2229 UT Oct 28 1966. Moon's age 14.7 days. Sun's selenographic colongitude = 87.5°. Diameter 64 cm. This was taken about 12 hours before Full Moon. Compare the detail on the limb here with that on Plate 11b. Although the libration in Plate 11b is better than it is here, there is virtually nothing to be seen on the limb, since the lighting is not suitable.

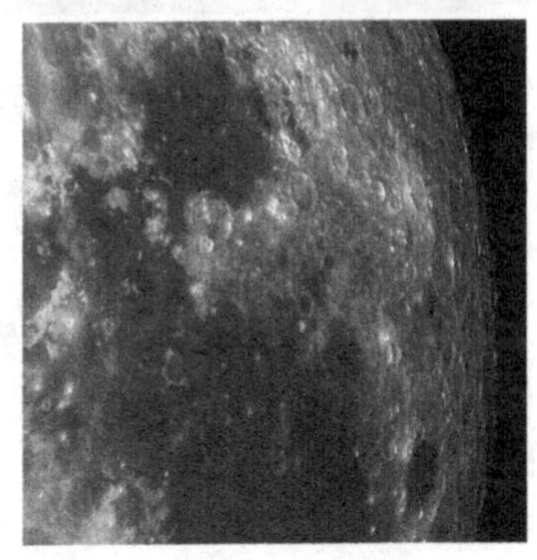

Plate 11d

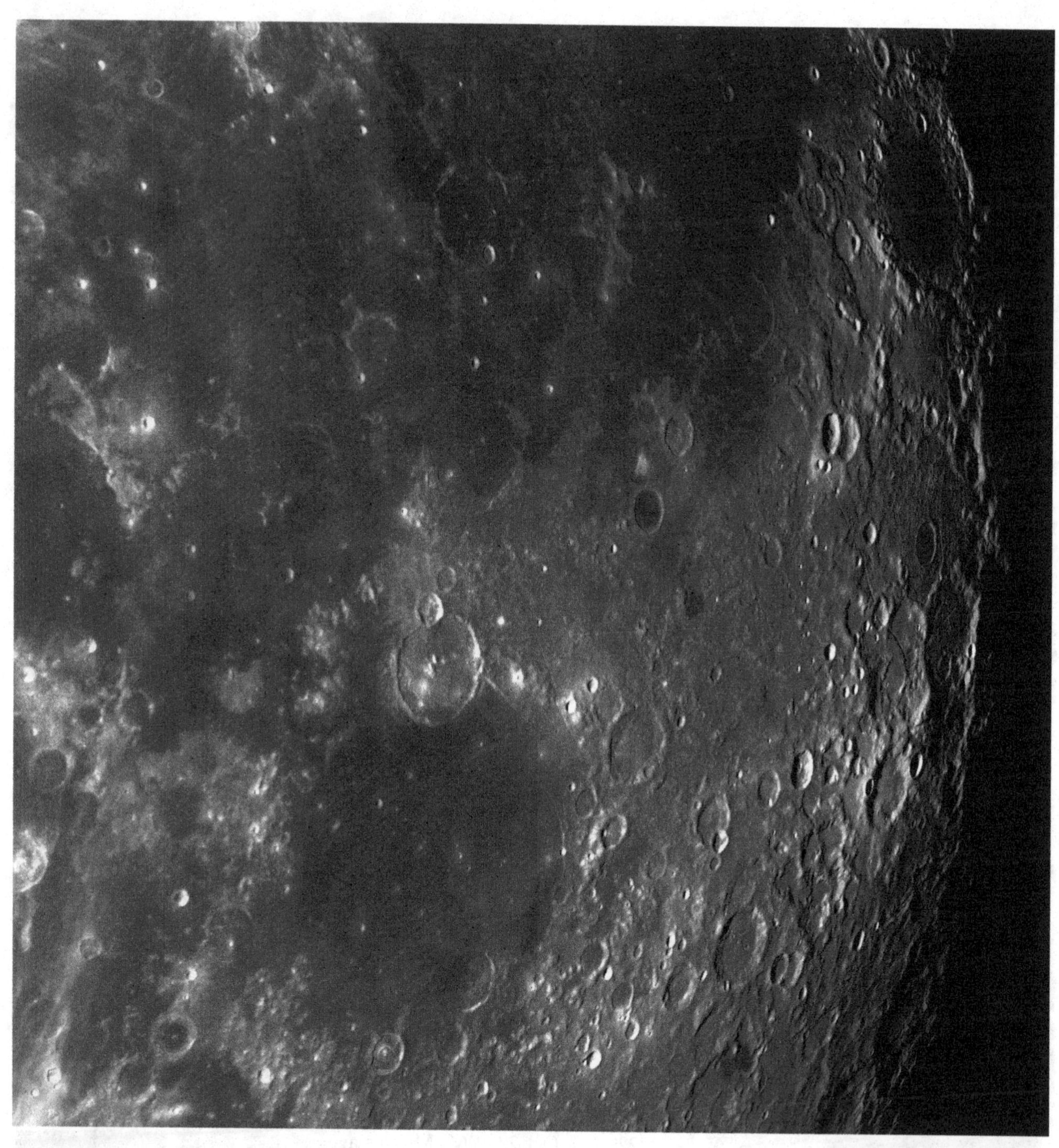

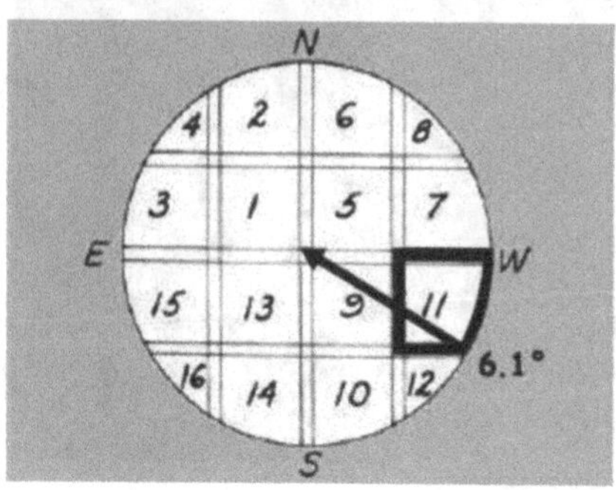

2036 UT Dec 25 1966. Moon's age 13.8 days. Sun's selenographic colongitude = 72.2°. Diameter 64 cm. Rimae Sirsalis (f5) will probably show up more clearly here if the reader turns the page 90° in a clockwise direction.

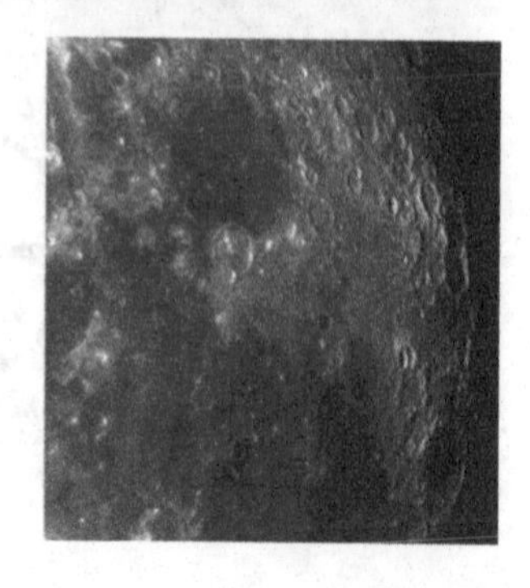

Plate 11e

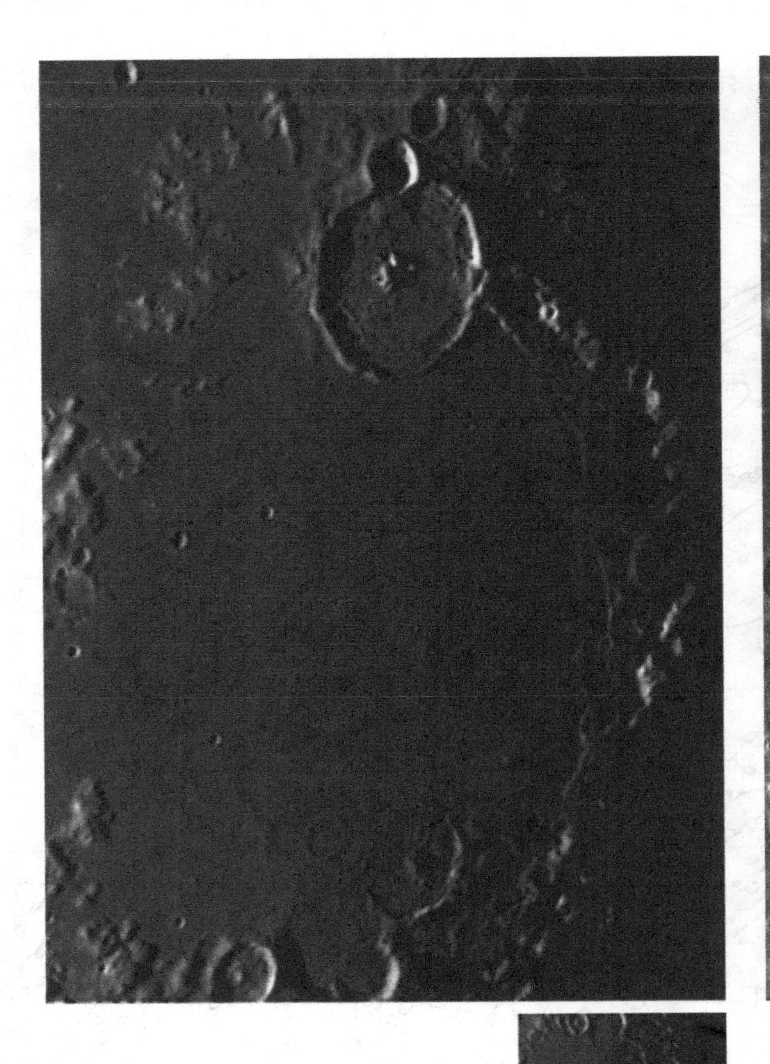

The Mare Humorum and Gassendi (c4). 2106 UT Apr 20 1967. Moons's age = 10.9 days. Col. = 44.5°. Diameter 94 cm approx.

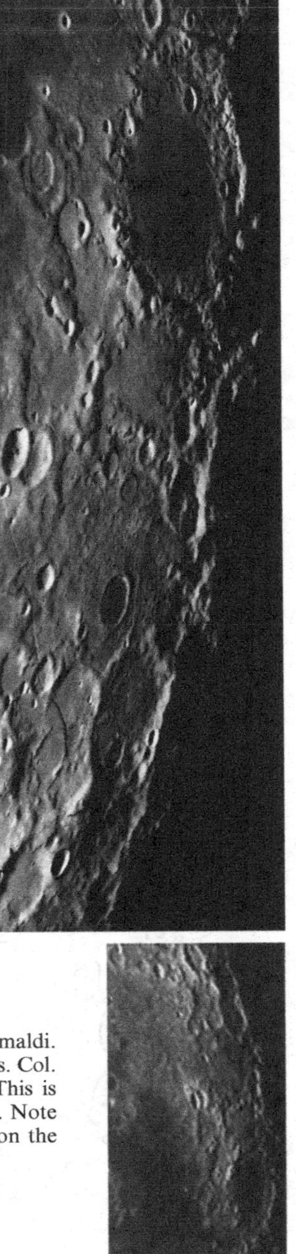

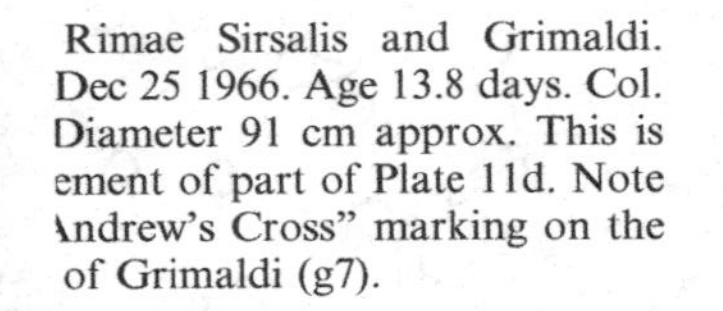

Rimae Sirsalis and Grimaldi. Dec 25 1966. Age 13.8 days. Col. Diameter 91 cm approx. This is ement of part of Plate 11d. Note \ndrew's Cross" marking on the of Grimaldi (g7).

Map 12

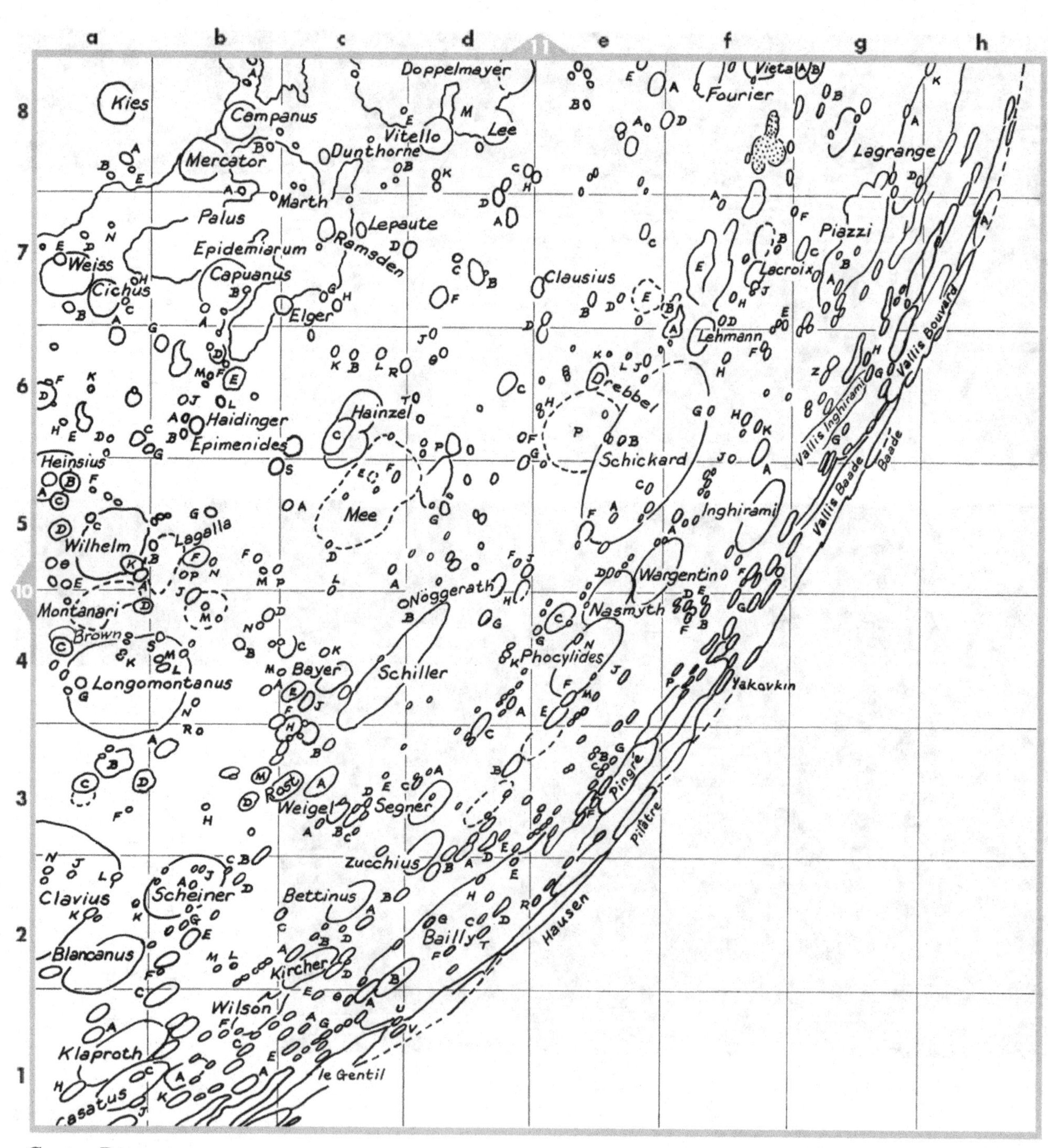

Crater Diameters

Lacroix	36 km (f7)
Vitello	43 km (d8)
Schickard	208 km (e6)
Zucchius	63 km (d3)
Casatus	103 km (a1)

This map has been prepared from Plate 12a. Plates 12b, 12c, and 12d show the same area under different lighting. Plate 12e shows larger scale photographs of Bailly (d2) and the area around Wargentin (e5).

Plate 12a

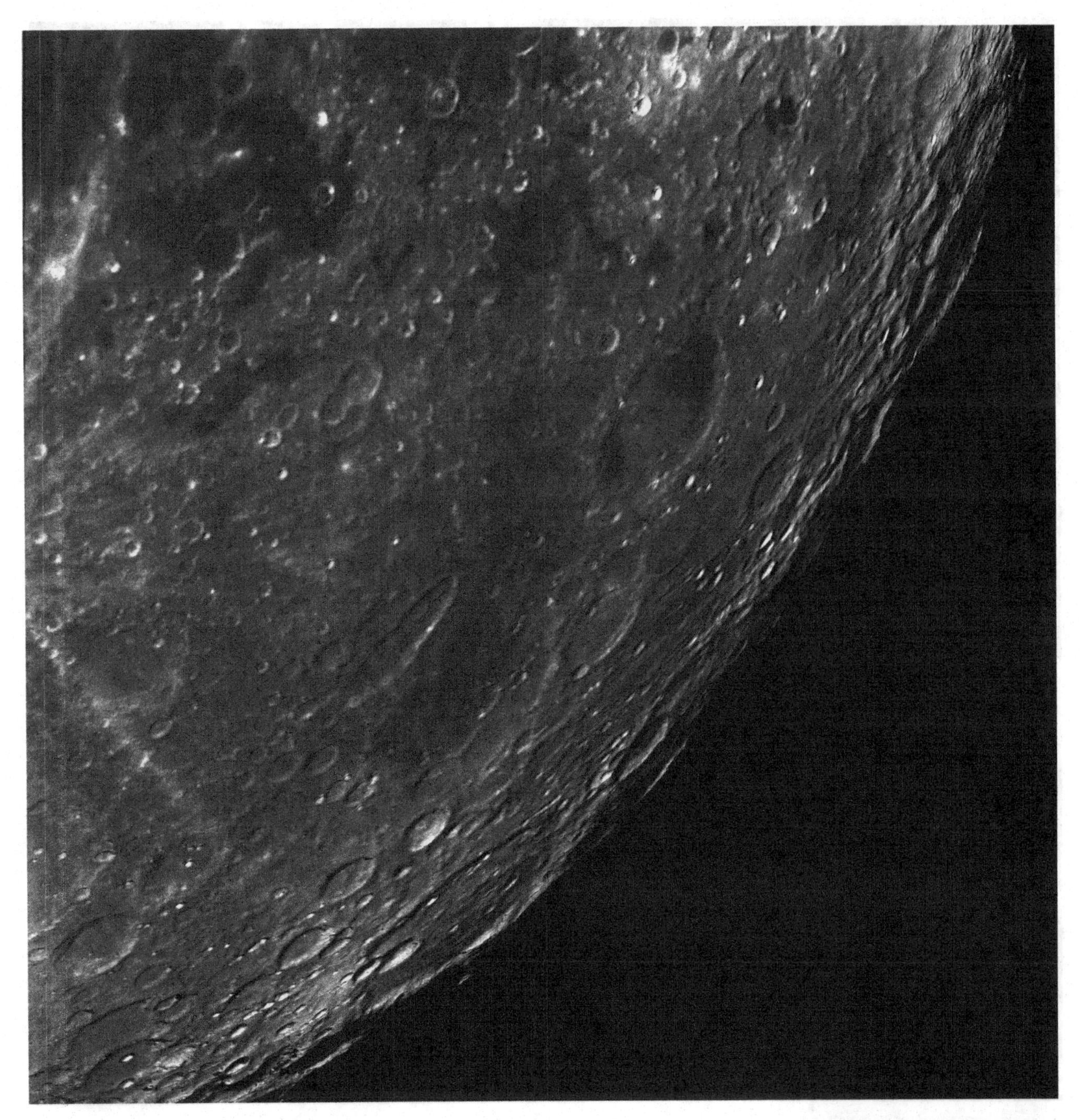

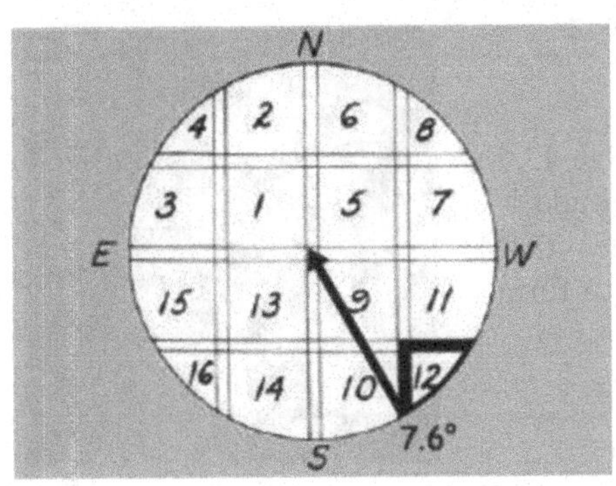

2104 UT Feb 23 1967. Moon's age 14.5 days. Sun's selenographic colongitude = 82.0°. Diameter 64 cm. This is a good libration for this area. Bailly (d2) will not often be seen as well as this.

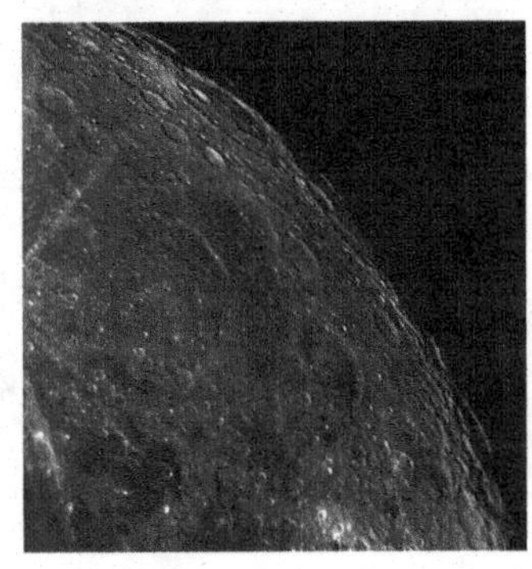

Plate 12b

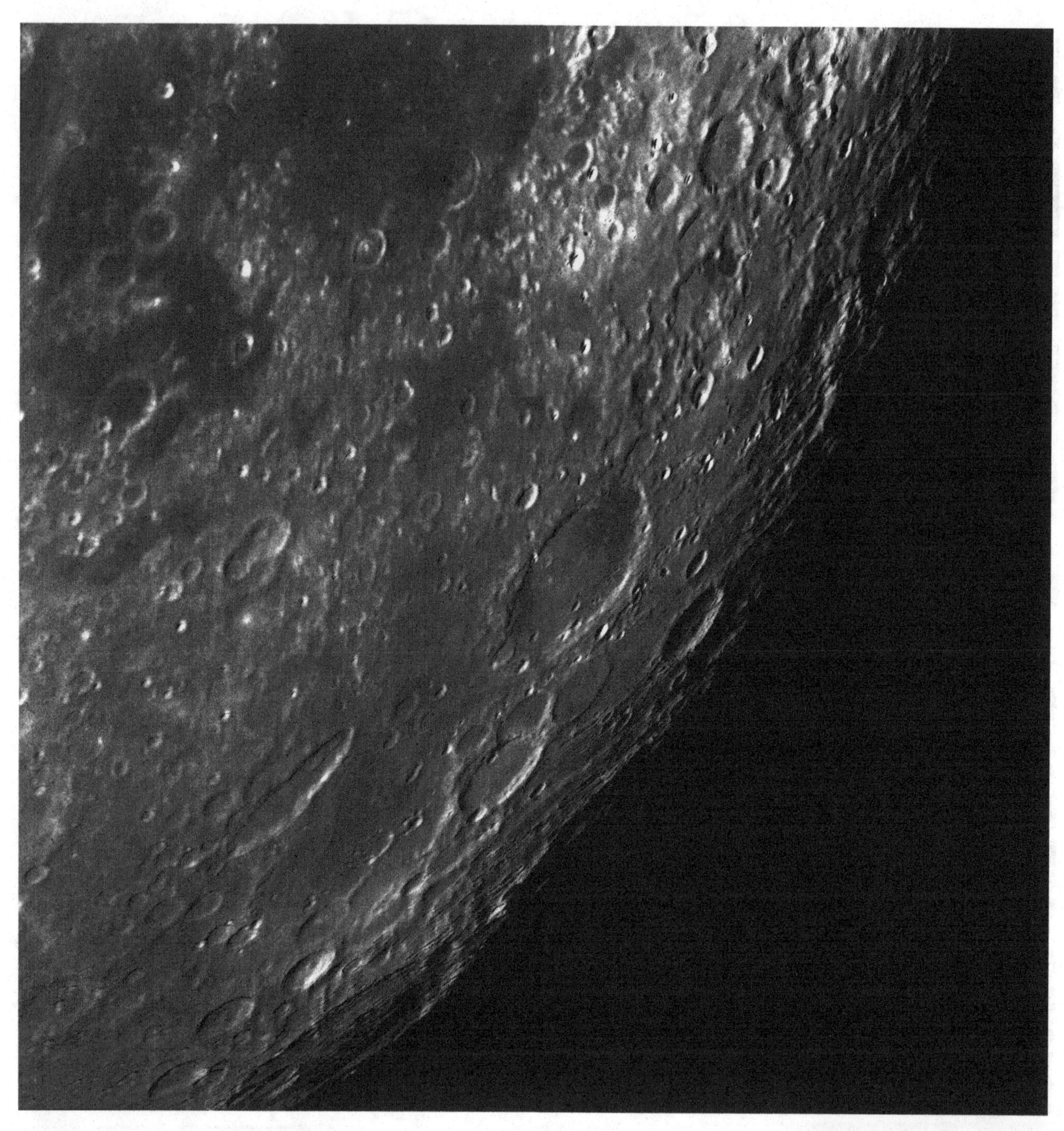

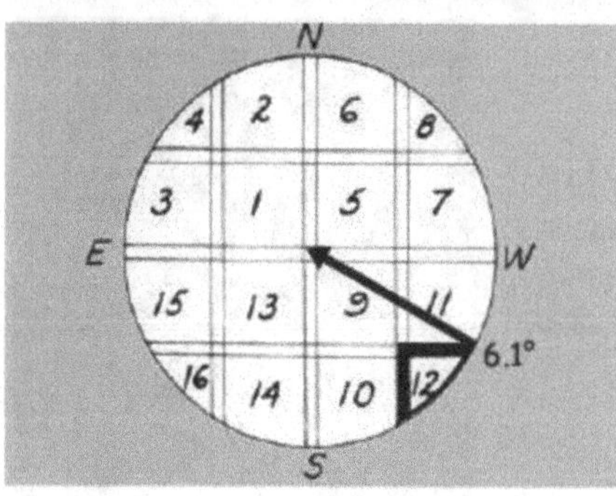

2036 UT Dec 25 1966. Moon's age 13.8 days. Sun's selenographic colongitude = 72.2°. Diameter 64 cm. Note that Wargentin (e5) has been filled in to form a plateau crater.

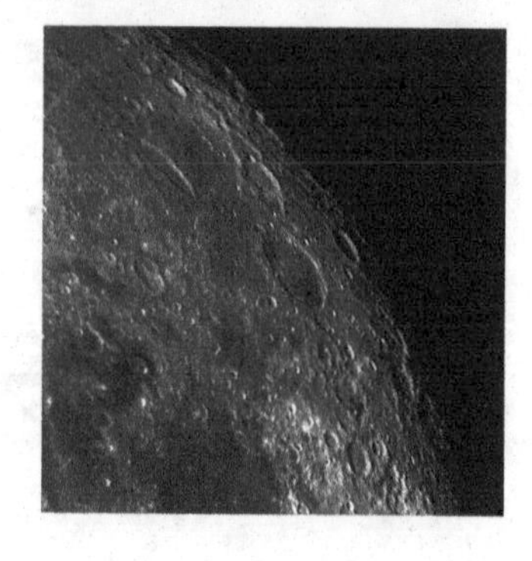

Plate 12c

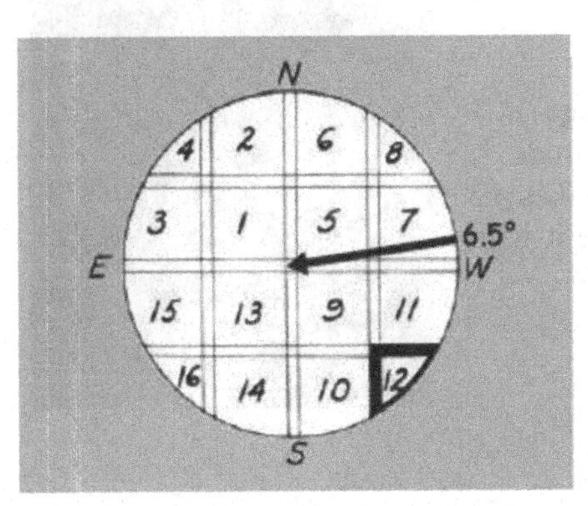

0317 UT Aug 09 1966. Moon's age 22.0 days. Sun's selenographic colongitude = 181.6°. Diameter 64 cm. Compare this with Plate 12a, particularly near Bailly (d2). Inset: Kircher (c2) to Schiller (c4). 1958 UT Apr 21 1967. Moon's age 11.9 days. Sun's selenographic colongitude = 56.1°. Diameter 94 cm approx.

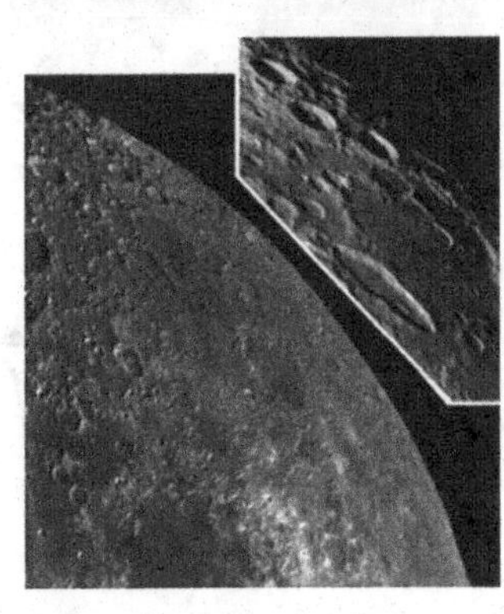

Plate 12d

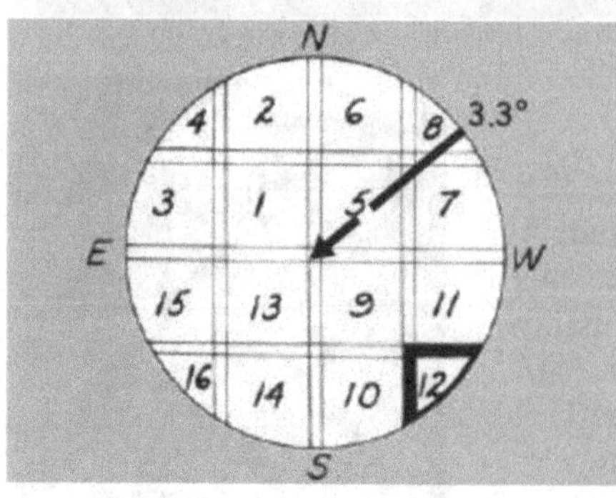

2229 UT Oct 28 1966. Moon's age 14.7 days. Sun's selenographic colongitude = 87.5°. Diameter 64 cm. This was taken 12 hours before Full Moon. Compare it with Plates 12a and 12c.

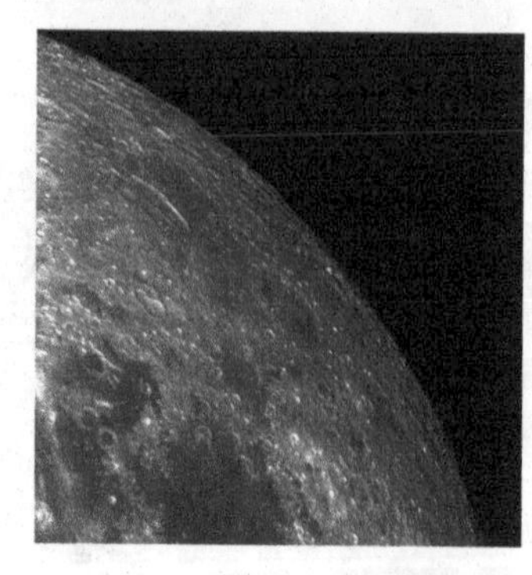

Plate 12e

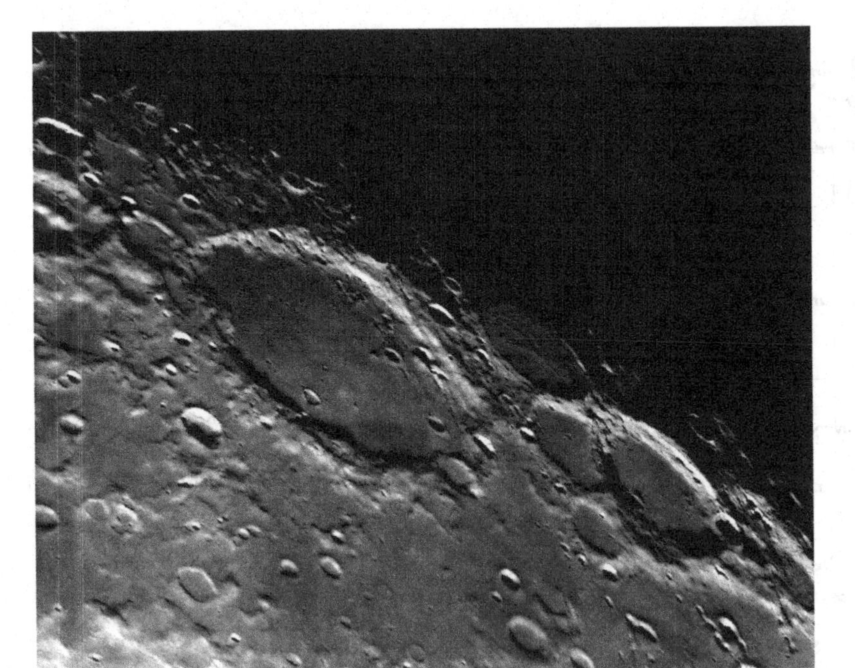

Phocylides, Wargentin and Schickard (e4, e5 and e6). 1938 UT Mar 23 1967.Moon's age 12.6 days. Sun's selenographic col. = 62.3°. Diameter 91 cm approx. These craters are shown under similar lighting on Plate 12b.

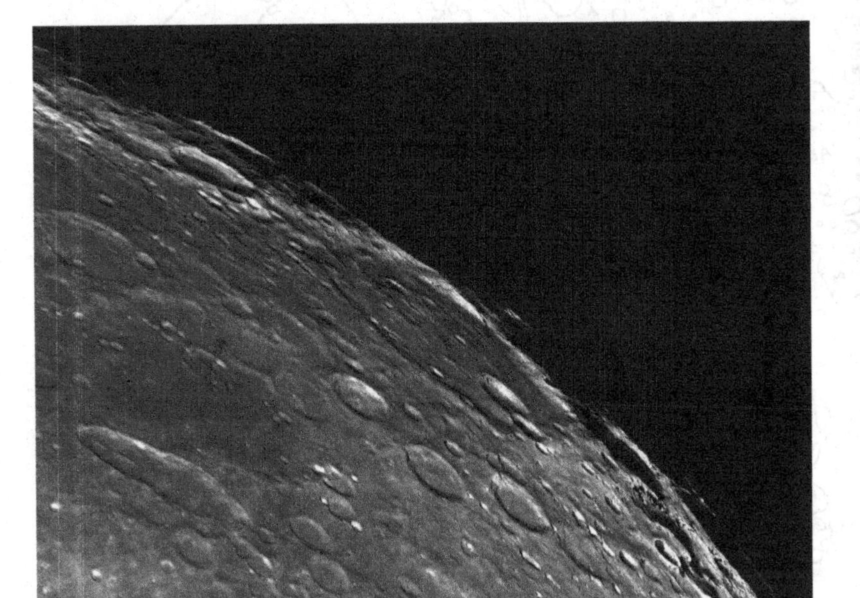

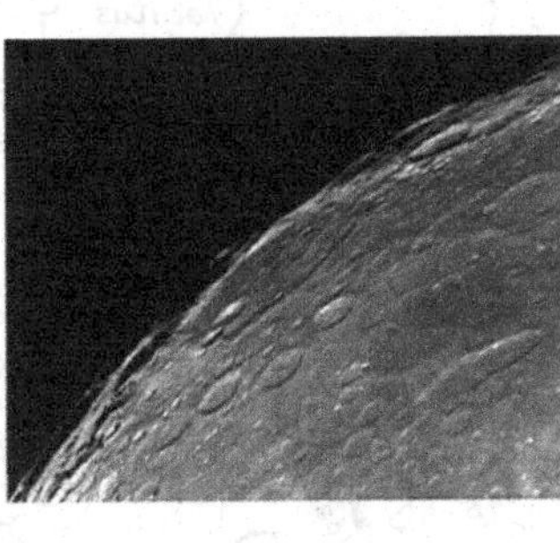

Bailly (d2). 2104 UT Feb 23 1967. Age 14.5 days. Col. = 82.0°. Diameter 98 cm approx. This is an enlargement of part of Plate 12a.

Map 13

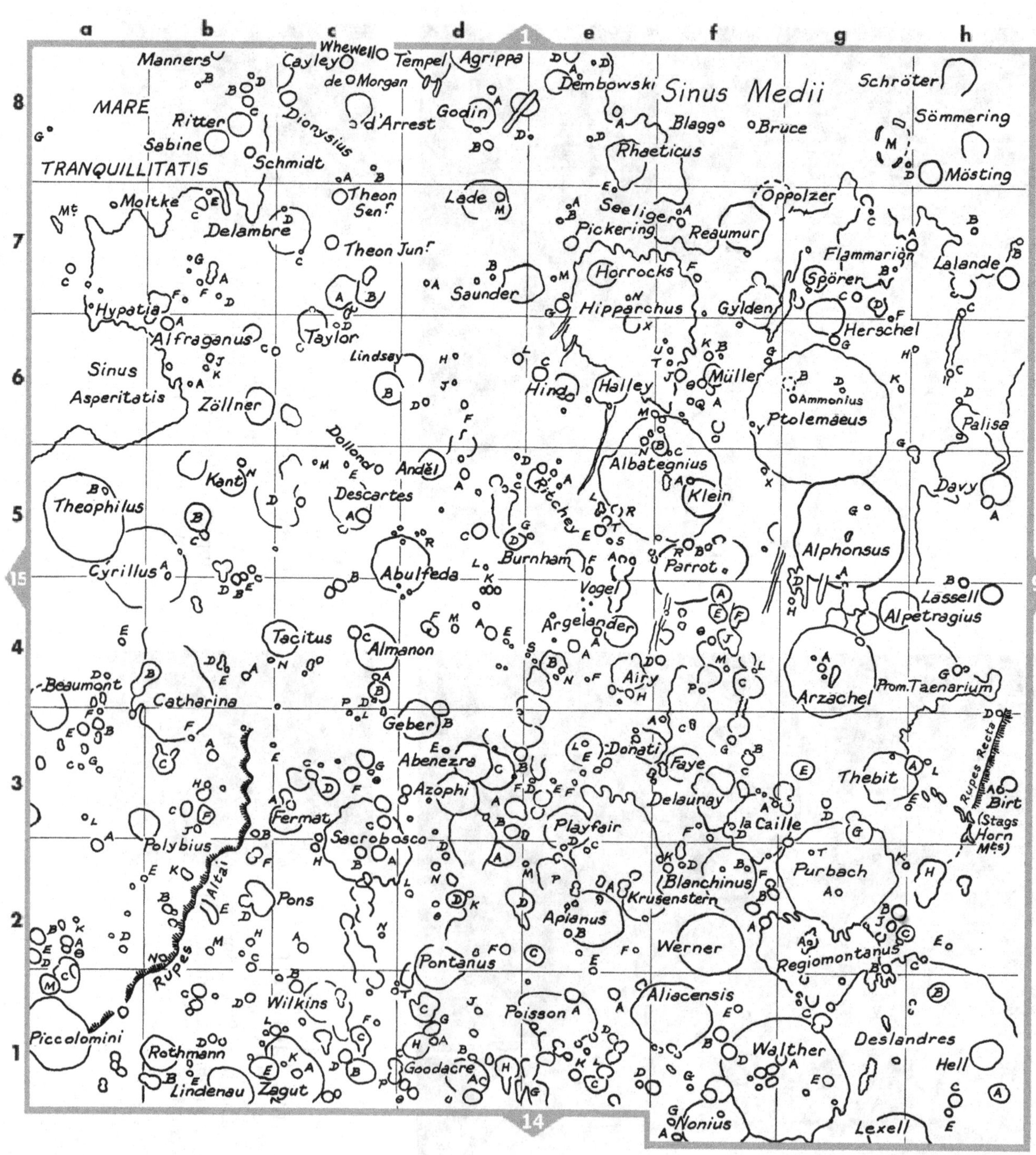

Crater Diameters

Mösting	25 km (h8)
Sabine	30 km (b8)
Argelander	34 km (e4)
Werner	71 km (f2)
Rothmann	43 km (b1)

This map has been prepared from Plate 13a. Plates 13b, 13c and 13d show the same area under different lighting. Plate 13e shows larger scale photographs of Abenezra (d3) and the area around Walther (g1). Plate 13f shows the area from Ptolemaeus (g6) to Hell (h1) on a larger scale. Plate 13g shows the Ptolemaeus–Alphonsus–Arzachel chain on a larger scale.

Plate 13a

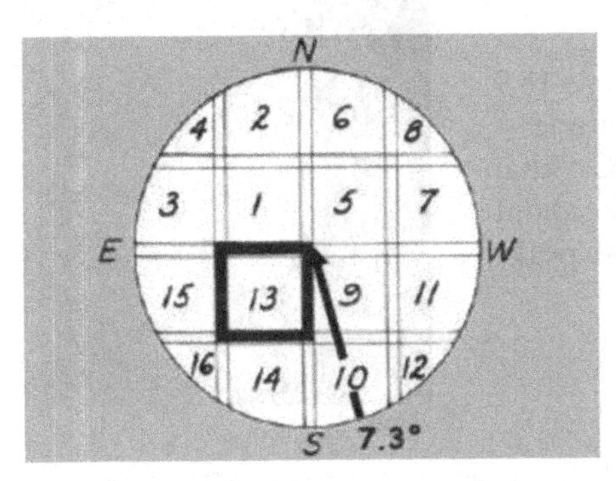

2046 UT Apr 28 1966. Moon's age 7.9 days. Sun's selenographic colongitude = 12.3°. Diameter 64 cm. Note the small crater Regiomontanus A (g2), which lies on the summit of a mountain.

Plate 13b

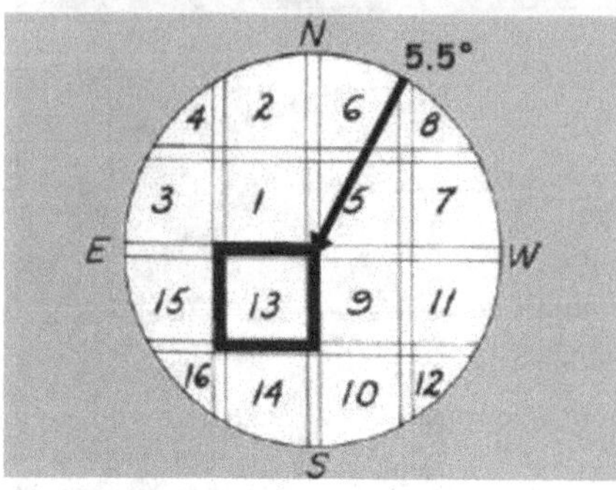

0215 UT Aug 06 1966. Moon's age 18.9 days. Sun's selenographic colongitude = 144.4°. Diameter 64 cm. The Rupes Altai (b2) form an escarpment that is about 1,800 m high generally; individual peaks may rise as much again.

Plate 13c

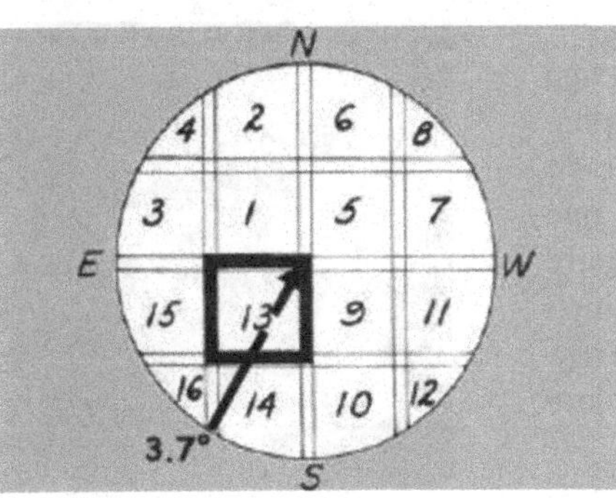

2334 UT Feb 26 1967. Moon's age 17.6 days. Sun's selenographic colongitude = 119.7°. Diameter 64 cm. This photograph extends further to the south and east than the other members of this group. At the bottom it overlaps into the areas covered by Maps 14 and 16.

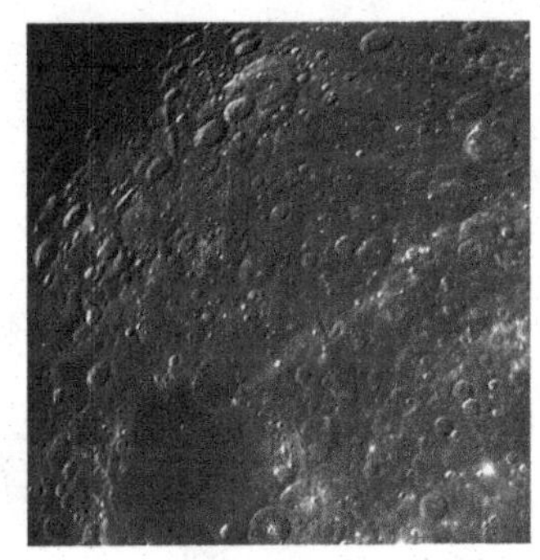

Plate 13d

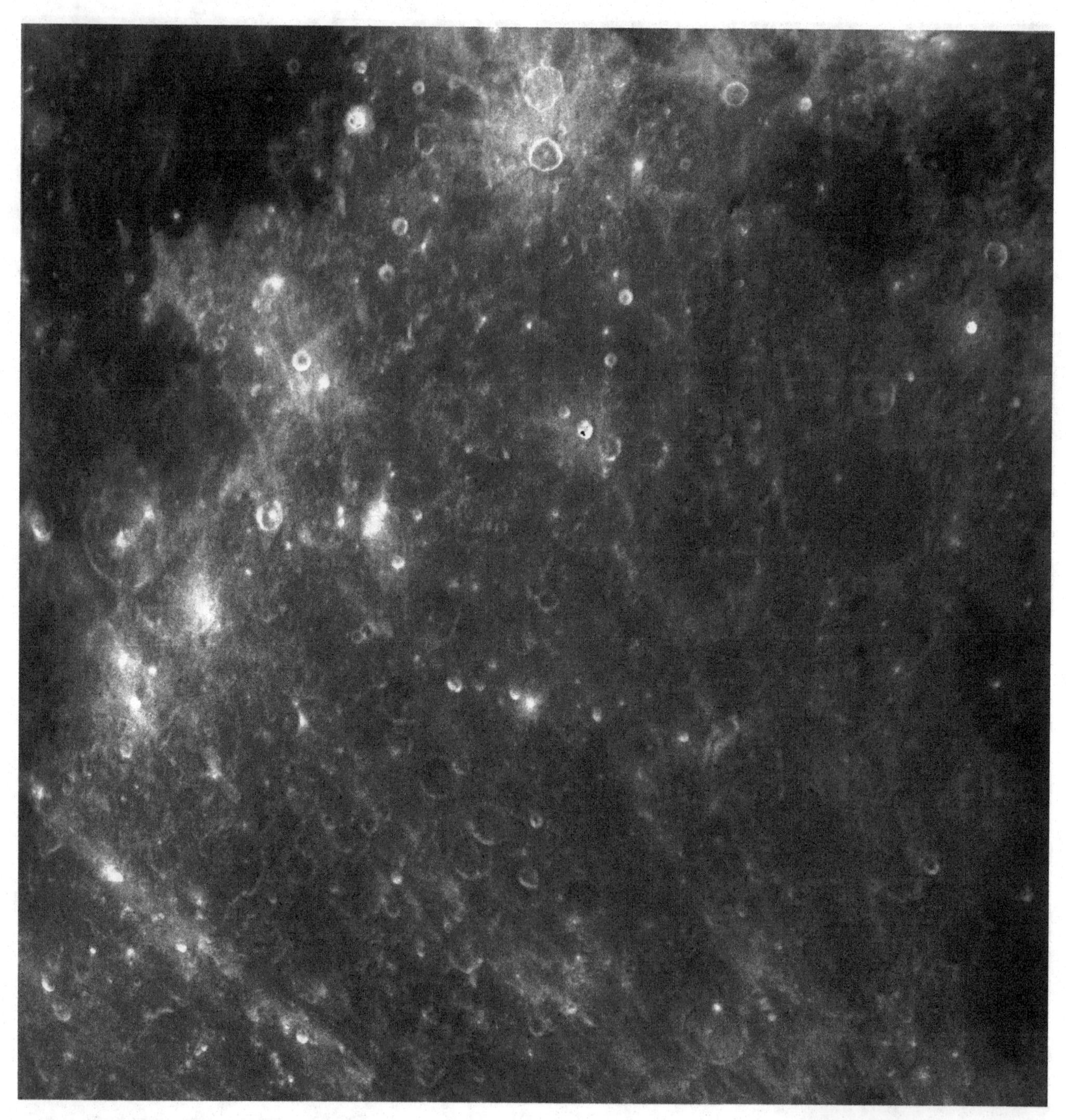

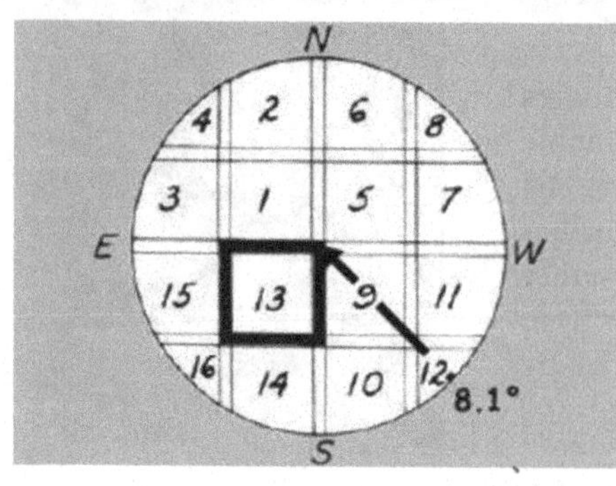

2314 UT Jan 24 1967. Moon's age 14.2 days. Sun's selenographic colongitude = 78.2°. Diameter 64 cm. This was taken just over one day before Full Moon. Compare it with Plates 13a and 13b, which show almost exactly the same area. Almost all the bright rays radiate from Tycho (Map 10 d5).

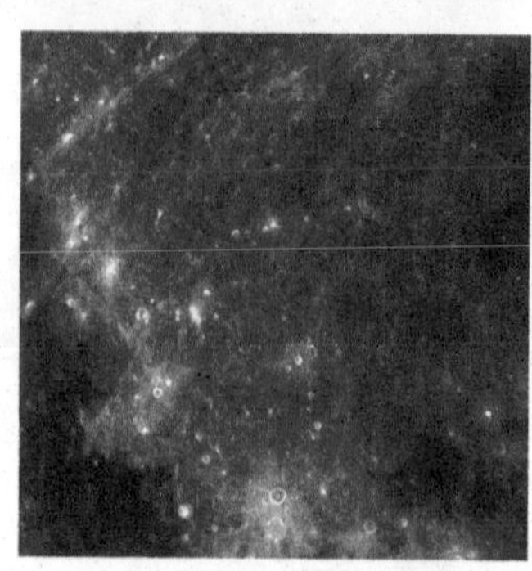

Plate 13e

The area around Walther (g1). 1950 UT Apr 17 1967. Moon's age 7.9 days. Sun's selenographic colongitude = 7.3°. Diameter 94 cm approx. Note the mountaintop crater Regiomontanus A (g2) and the dark bands in Stöfler (top left).

The area around Abenezra (d3). 2008 UT May 16 1967. Moon's age 7.2 days. Sun's selenographic colongitude = 1.4°. Diameter 94 cm approx. Note the radial dark bands in Abenezra C.

Plate 13f

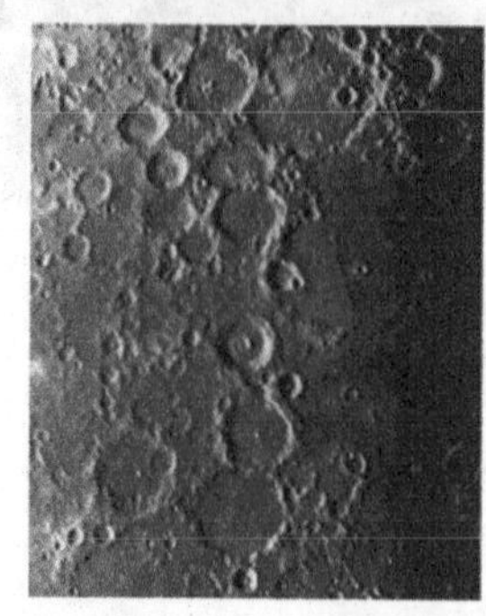

Ptolemaeus (g6) to Hell (h1). 1959 UT Mar 19 1967. Moon's age 8.6 days. Sun's selenographic colongitude = 13.8°. Diameter 91 cm.

Plate 13g

Ptolemaeus, Alphonsus, and Arzachel (g6, g5 and g4) soon after sunrise. 1951 UT Apr 17 1967. Moon's age 7.9 days. Sun's selenographic colongitude = 7.3°. Diameter 91 cm approx. The diameter of Albategnius C (f5) is 6 km.

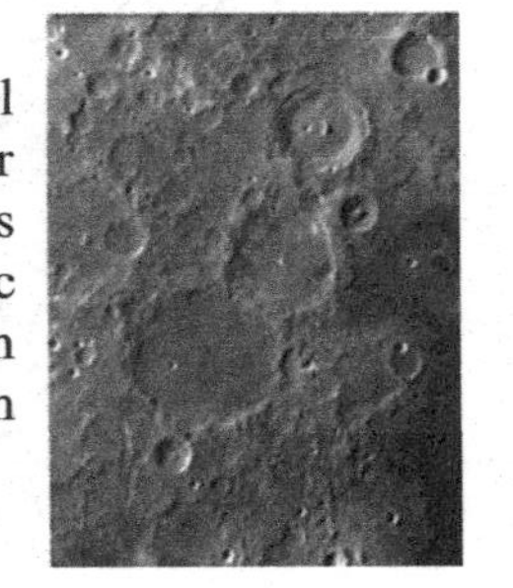

Ptolemaeus, Alphonsus, and Arzachel (g6, g5 and g4) about two days after sunrise. 2050 UT Mar 20 1967. Moon's age 9.7 days. Sun's selenographic colongitude = 26.4°. Diameter 91 cm approx. Note the dark patches in Alphonsus (g5).

Map 14

Crater Diameters

Walther	134 km (f7)
Stiborius	44 km (a7)
Cuvier	77 km (e4)
Clavius	225 km (g3)
Manzinus	98 km (d2)

This map has been prepared from Plate 14a. Plates 14b, 14c, and 14d show the same area under different lighting. Plate 14e shows larger scale photographs of the area around Maurolycus (d6) and Stöfler (e6), and Clavius (g3).

Plate 14a

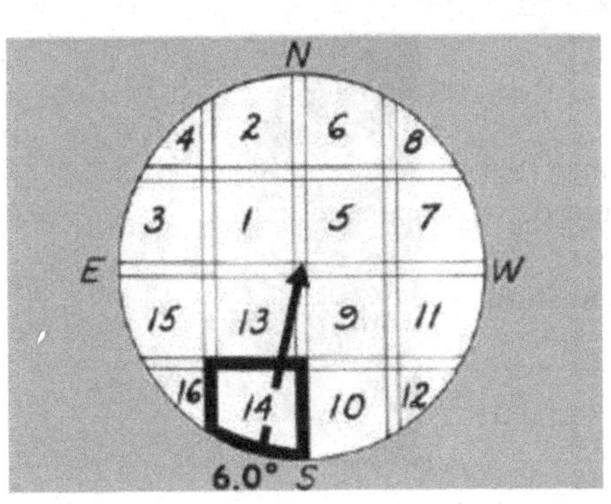

2122 UT May 28 1966. Moon's age 8.4 days. Sun's selenographic colongitude = 19.0°. Diameter 64 cm. This is quite a good libration for the area at the bottom of the photograph.

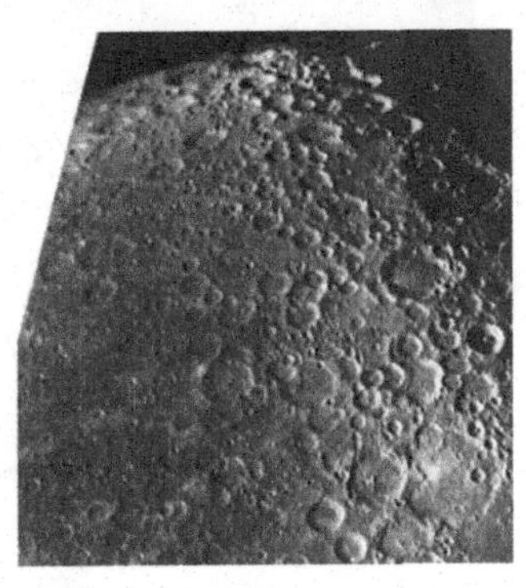

Plate 14b

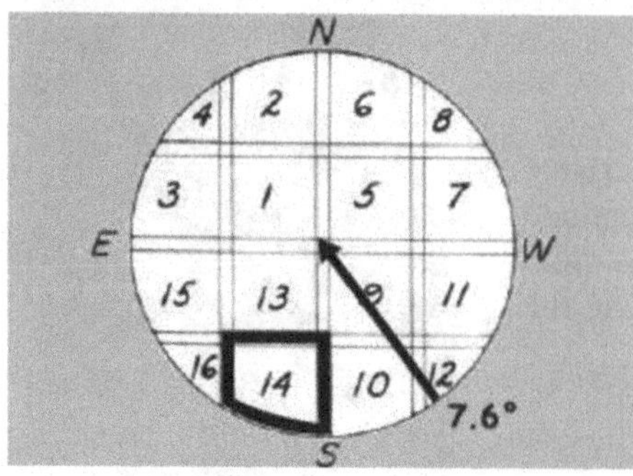

2336 UT Dec 01 1966. Moon's age 19.4 days. Sun's selenographic colongitude = 141.7°. Diameter 64 cm. Compare this with Plates 14a and 14c, which show almost exactly the same area.

Plate 14c

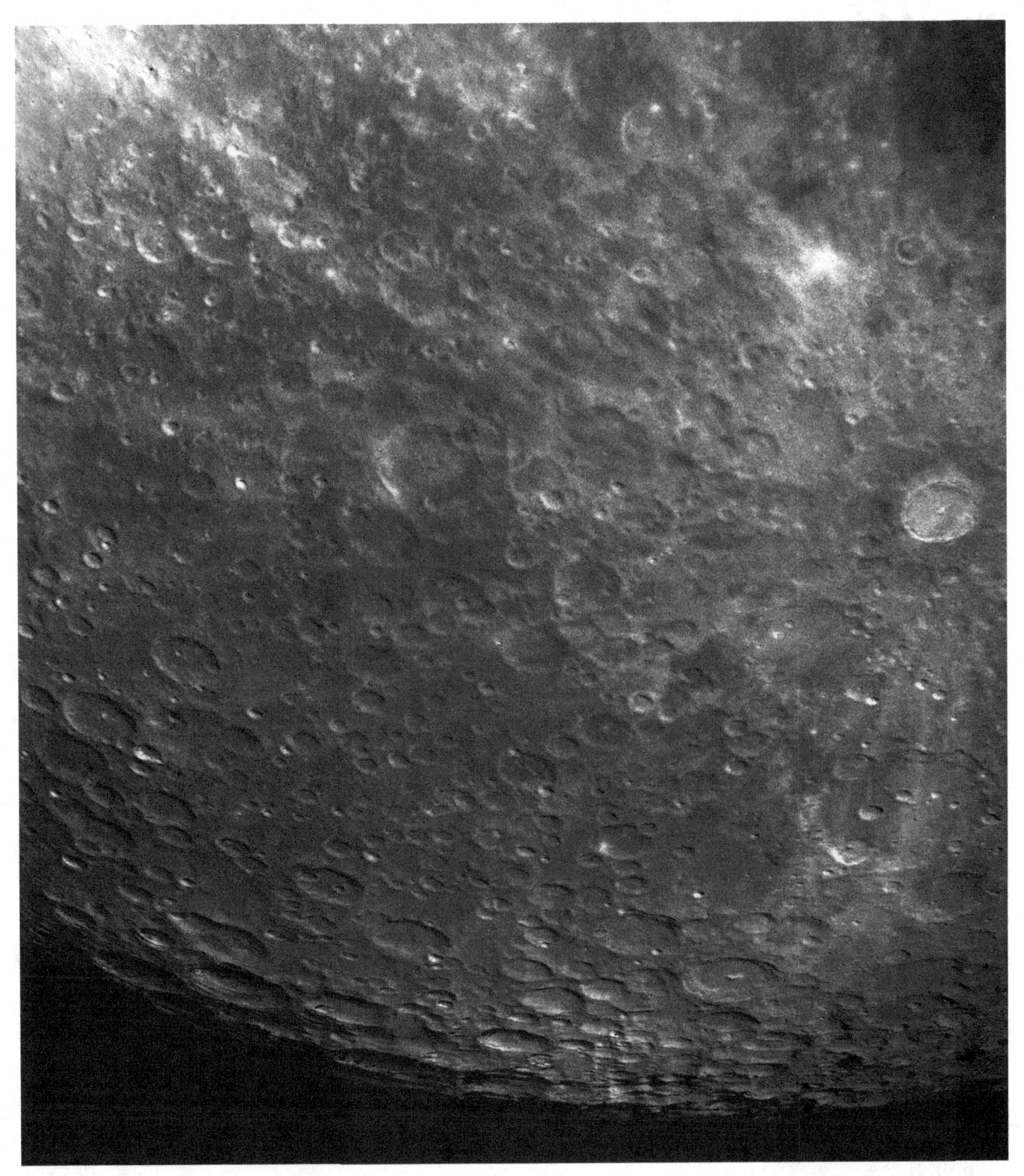

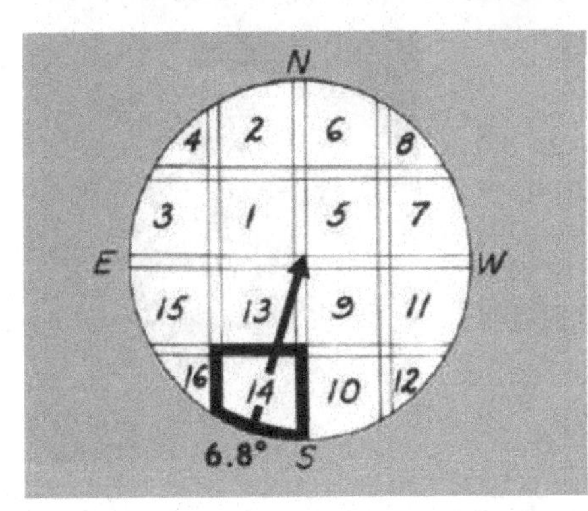

2326 UT Feb 06 1966. Moon's age 16.3 days. Sun's selenographic colongitude = 106.7°. Diameter 64 cm. This was taken over just one day after Full Moon, and the rays from Tycho are very prominent. Compare this with Plates 14a and 14b, which show almost exactly the same area.

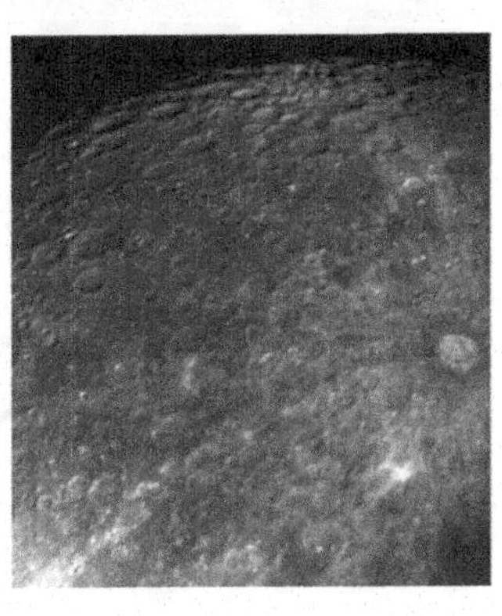

Plate 14d

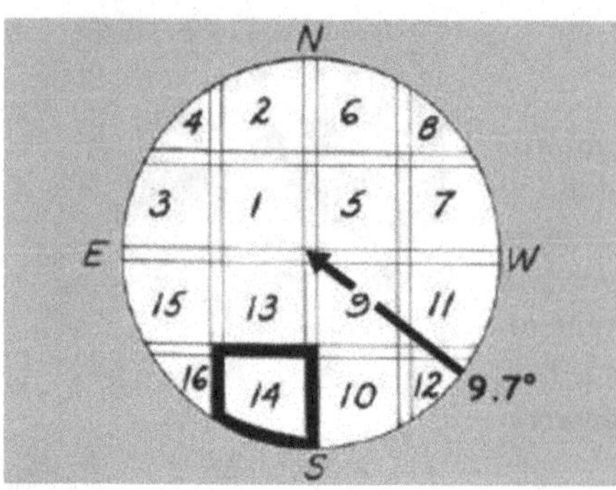

1959 UT Mar 19 1967. Moon's age 8.6 days. Sun's selenographic colongitude = 13.8°. Diameter 64 cm. Compare this with Plate 14a. The Moon's age is very nearly the same in both cases, but the librations are quite different.

Plate 14e

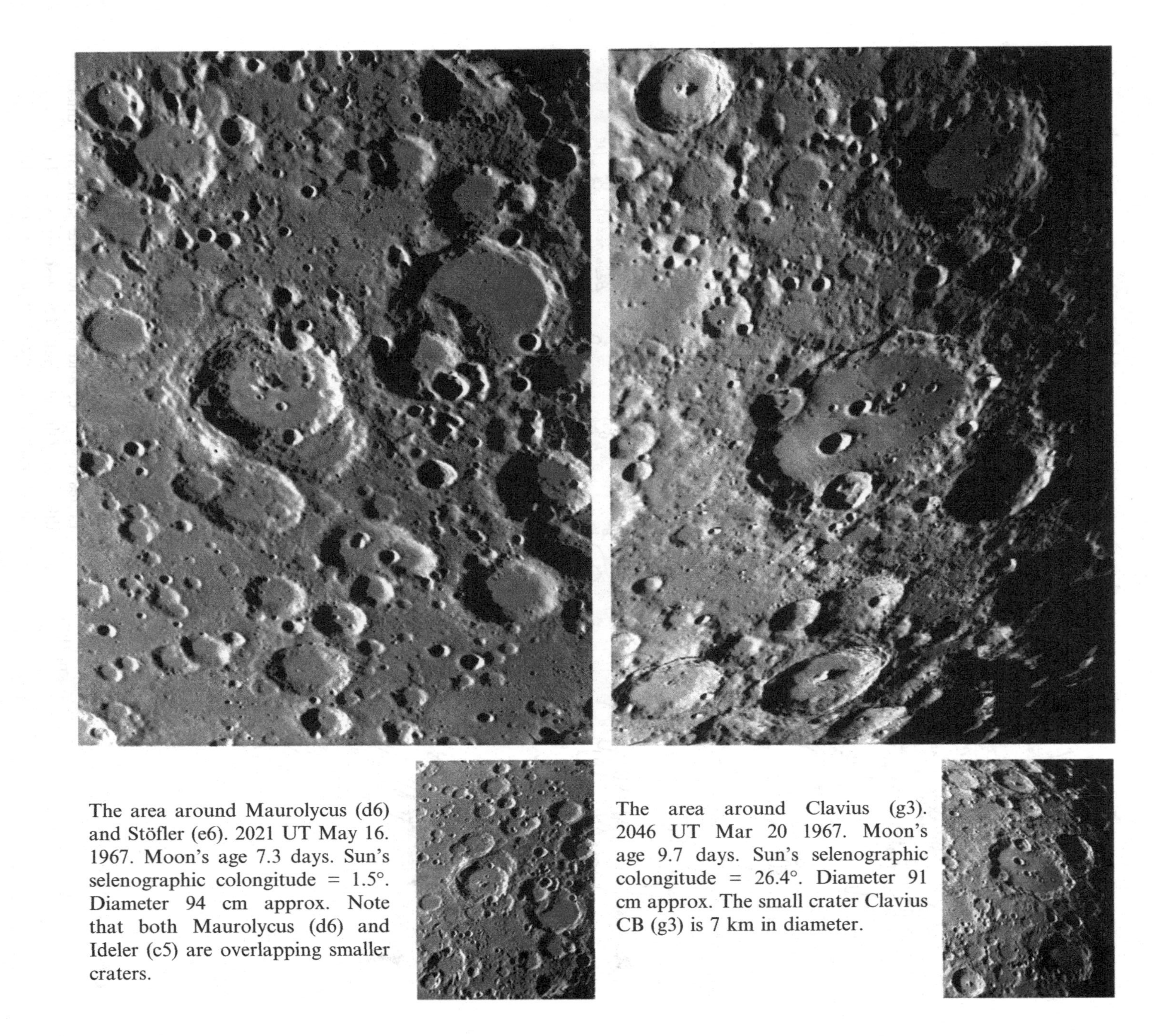

The area around Maurolycus (d6) and Stöfler (e6). 2021 UT May 16. 1967. Moon's age 7.3 days. Sun's selenographic colongitude = 1.5°. Diameter 94 cm approx. Note that both Maurolycus (d6) and Ideler (c5) are overlapping smaller craters.

The area around Clavius (g3). 2046 UT Mar 20 1967. Moon's age 9.7 days. Sun's selenographic colongitude = 26.4°. Diameter 91 cm approx. The small crater Clavius CB (g3) is 7 km in diameter.

Map 15

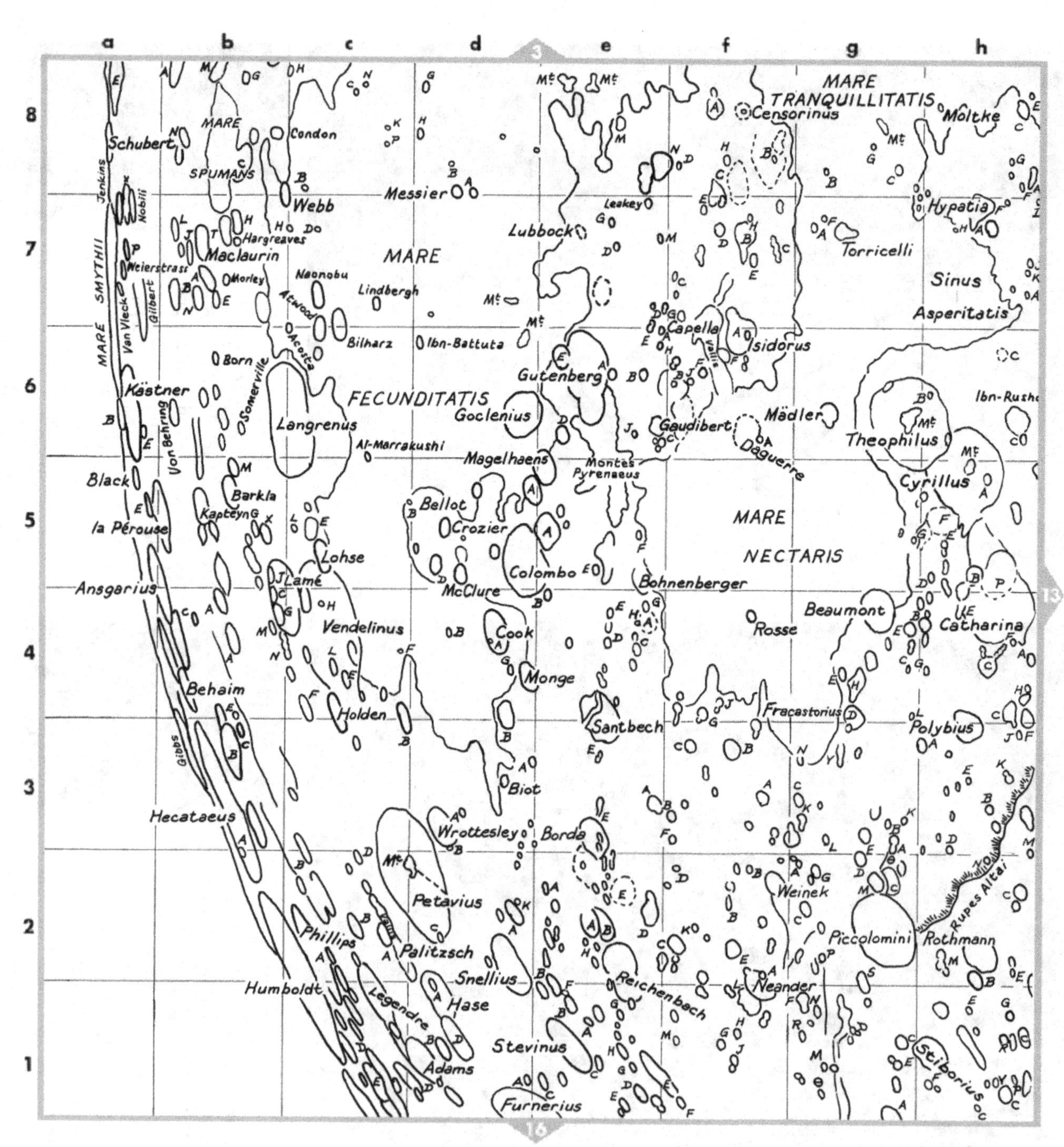

Crater Diameters

Moltke	6 km (h8)
Webb	22 km (b8)
Theophilus	99 km (g6)
Piccolomini	88 km (g2)
Stevinus	72 km (e1)

This map has been prepared from Plates 15a and 15d. Plates 15b, 15c, and 15e show the same area under different lighting. Plate 15e also shows a larger scale photograph of the area around Humboldt (c2).

Plate 15a

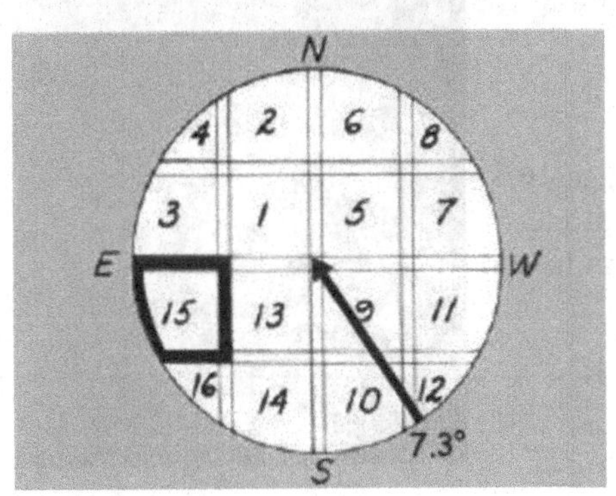

2152 UT Apr 27 1966. Moon's age 7.1 days. Sun's selenographic colongitude = 0.7°. Diameter 64 cm. Compare this with the other members of this group, which show the detail near the limb much more clearly.

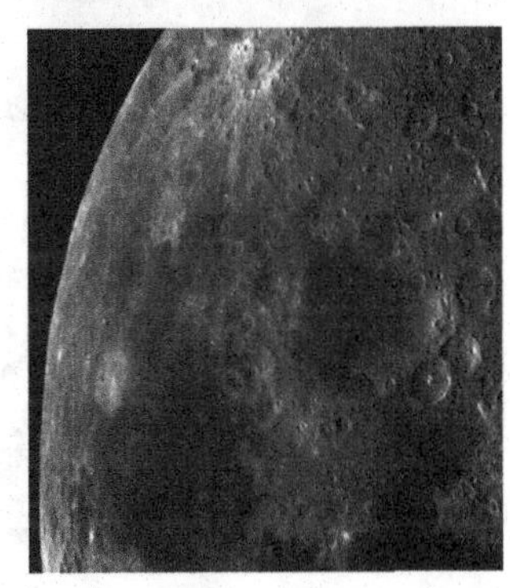

Plate 15b

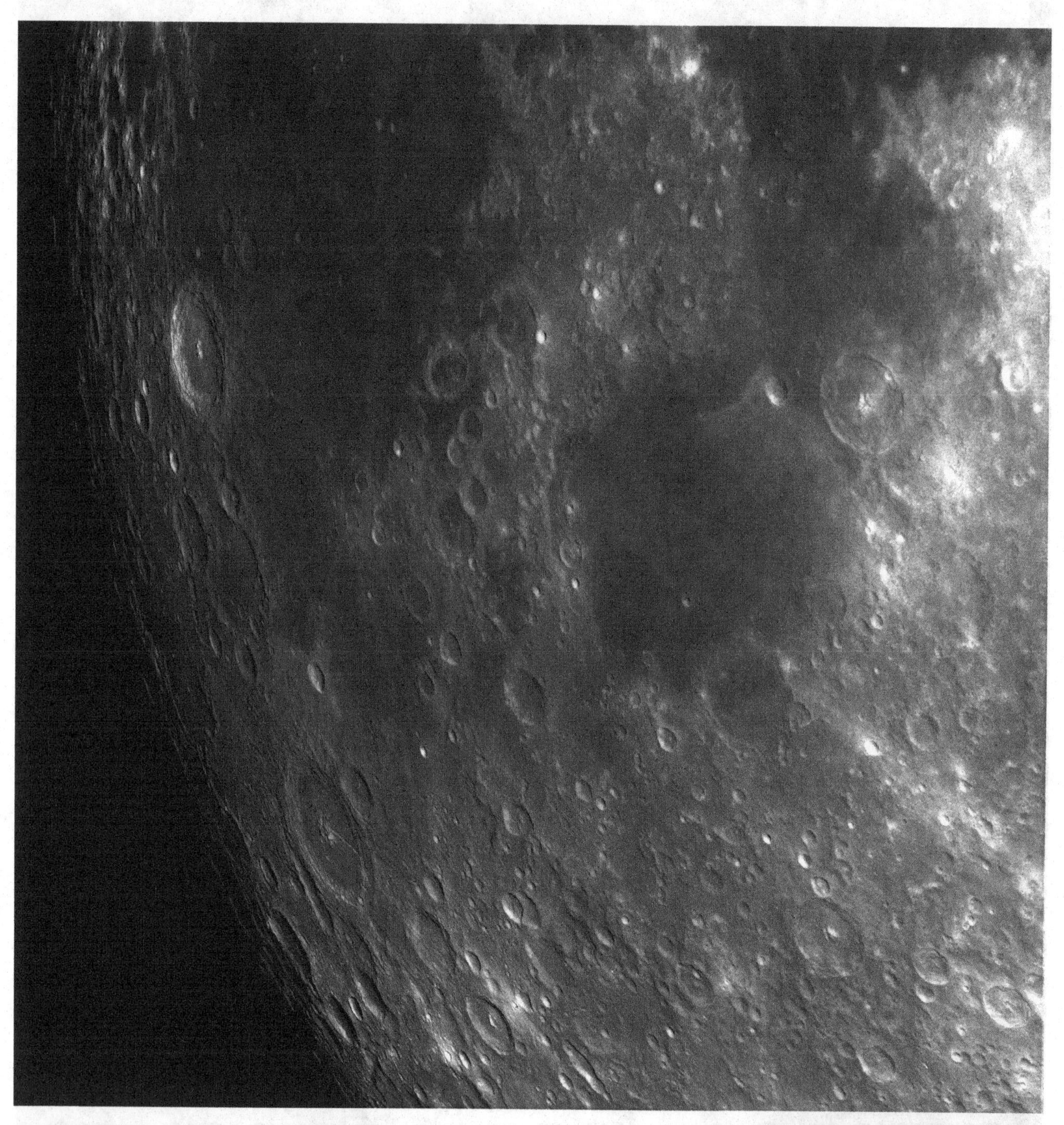

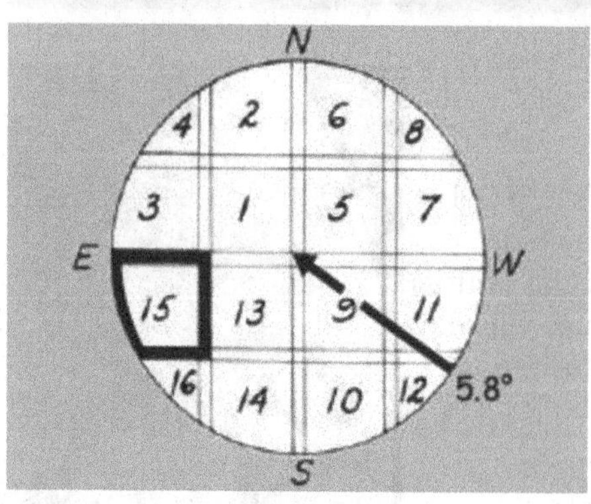

2125 UT Nov 28 1966. Moon's age 16.3 days. Sun's selenographic colongitude = 104.2°. Diameter 64 cm. This was taken about 19 hours after Full Moon.

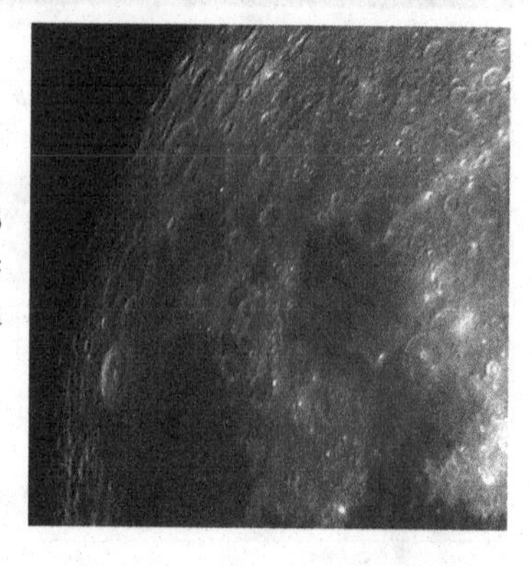

Plate 15c

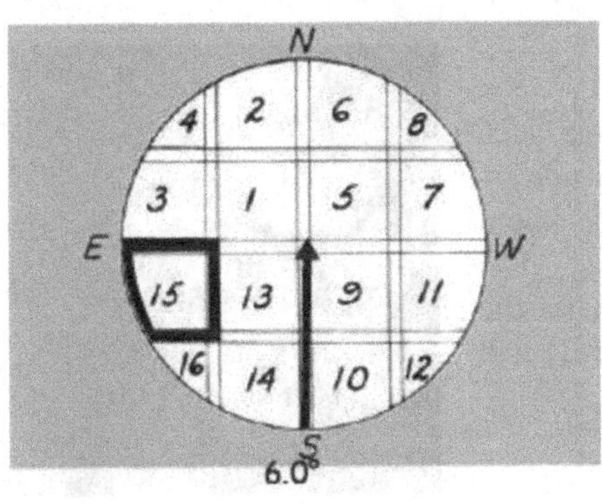

2228 UT Jan 08 1966. Moon's age 17.0 days. Sun's selenographic colongitude = 113.6°. Diameter 64 cm. Petavius (d2) and Langrenus (c6) are very prominent. Compare this with Plate 15b, which shows almost exactly the same area.

Plate 15d

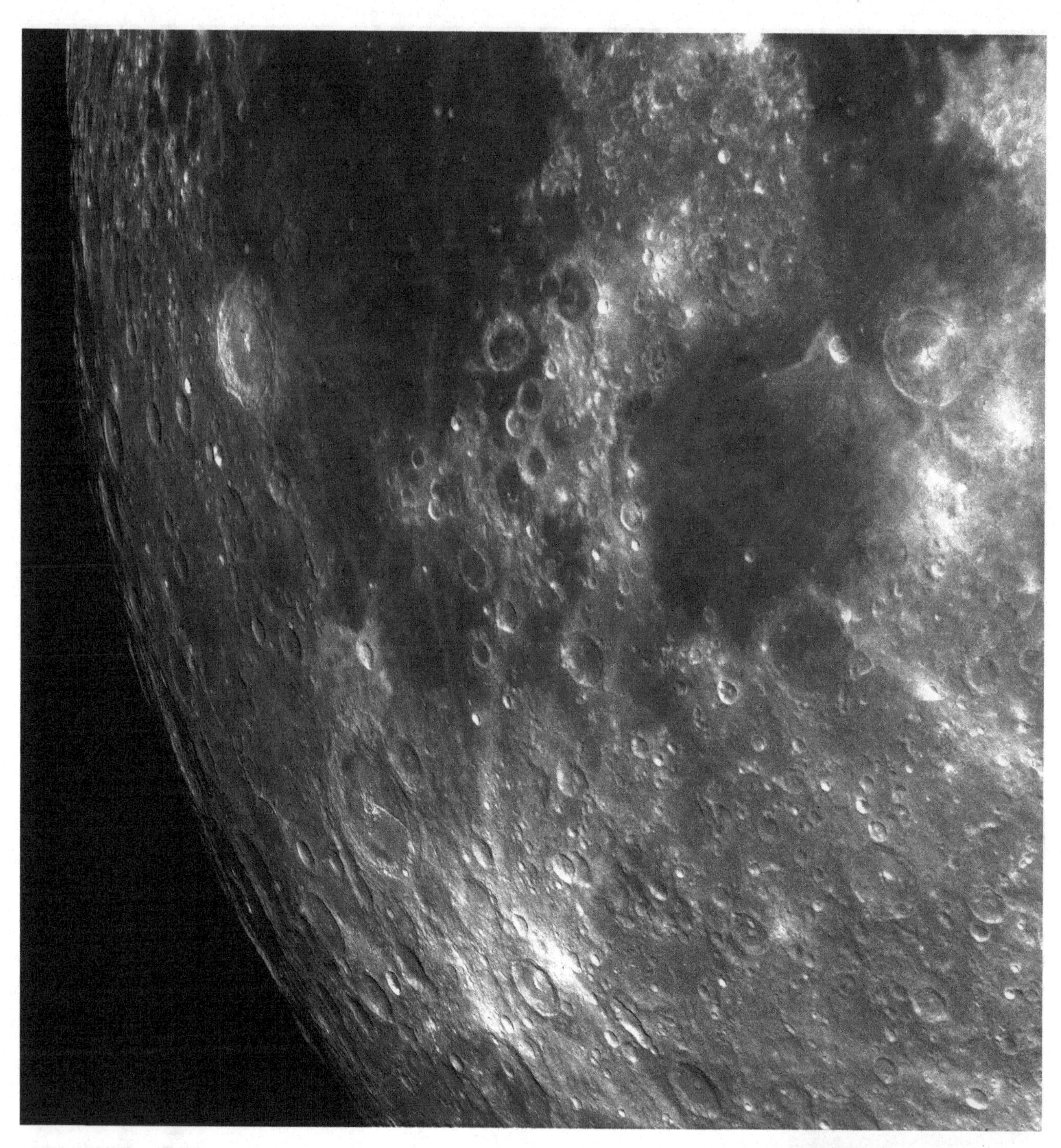

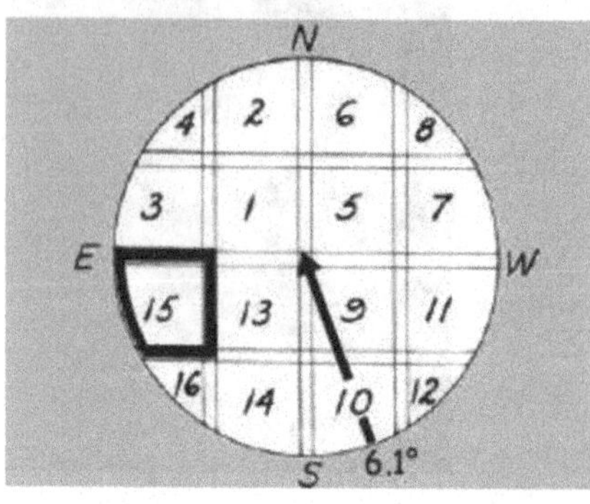

2300 UT Feb 24 1967. Moon's age 15.6 days. Sun's selenographic colongitude = 95.1°. Diameter 64 cm. This was taken about 5 hours after Full Moon. The libration is not favorable, but even so the craters on the limb near Humboldt (c1) will not often be seen like this.

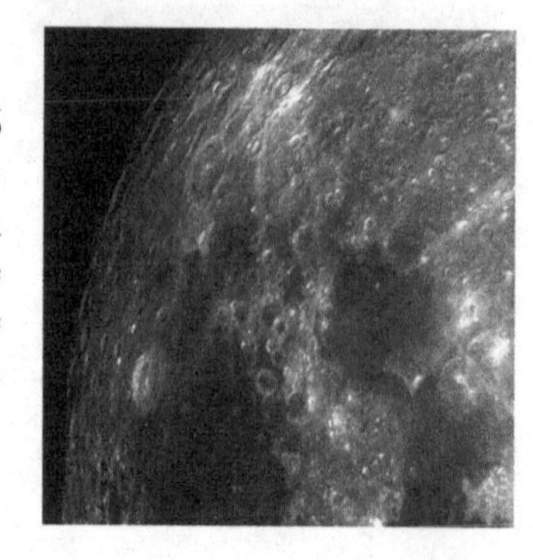

Plate 15e

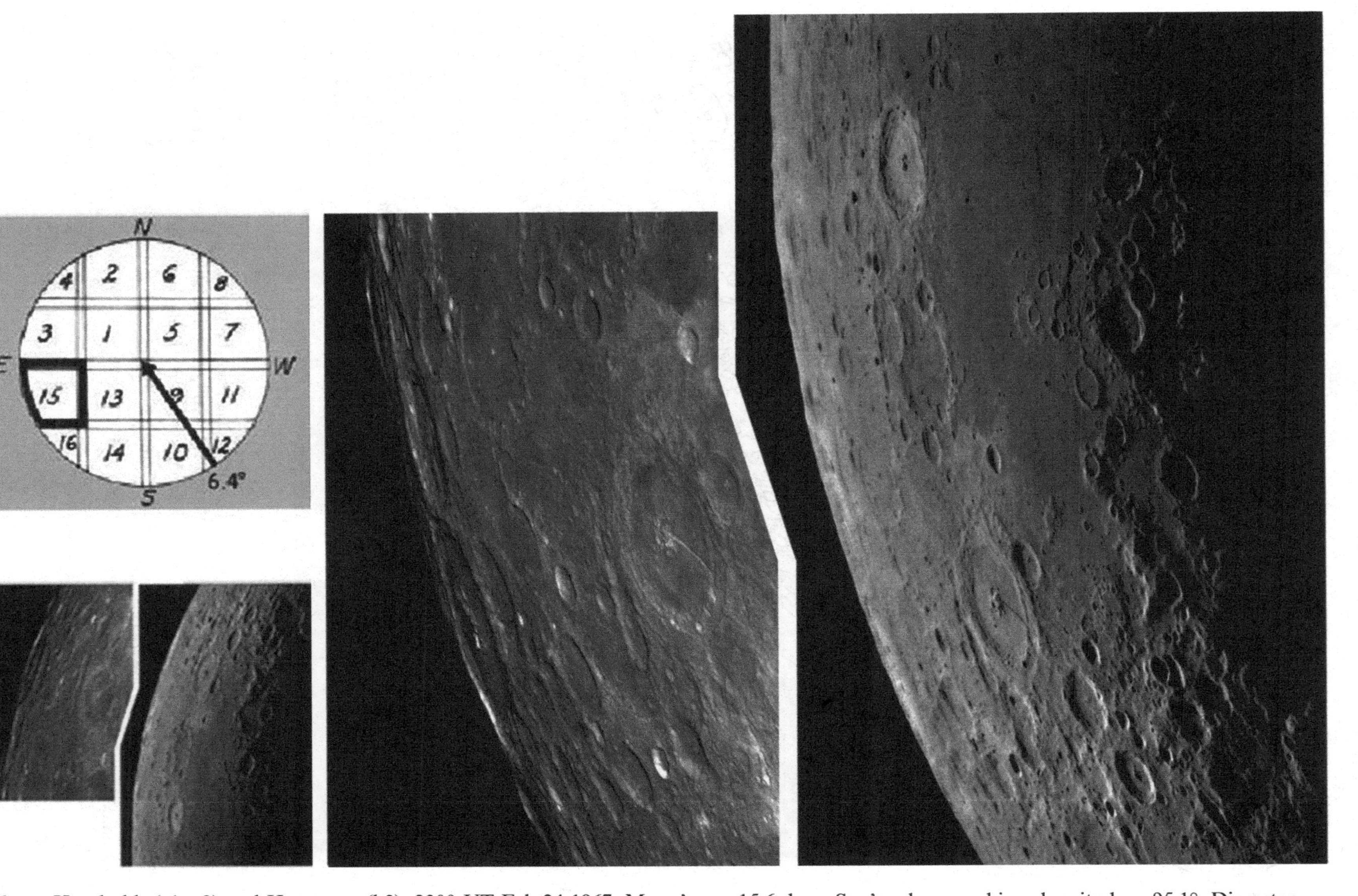

Above: Humboldt (c1, c2) and Hecataeus (b3). 2300 UT Feb 24 1967. Moon's age 15.6 days. Sun's selenographic colongitude = 95.1°. Diameter 99 cm approx. This is an enlargement of part of Plate 15d.

Right: 2033 UT May 23 1966. Moon's age 3.4 days. Sun's selenographic colongitude = 317.5°. Diameter 64 cm. Compare this with Plate 15d. The librations are very similar.

Map 16

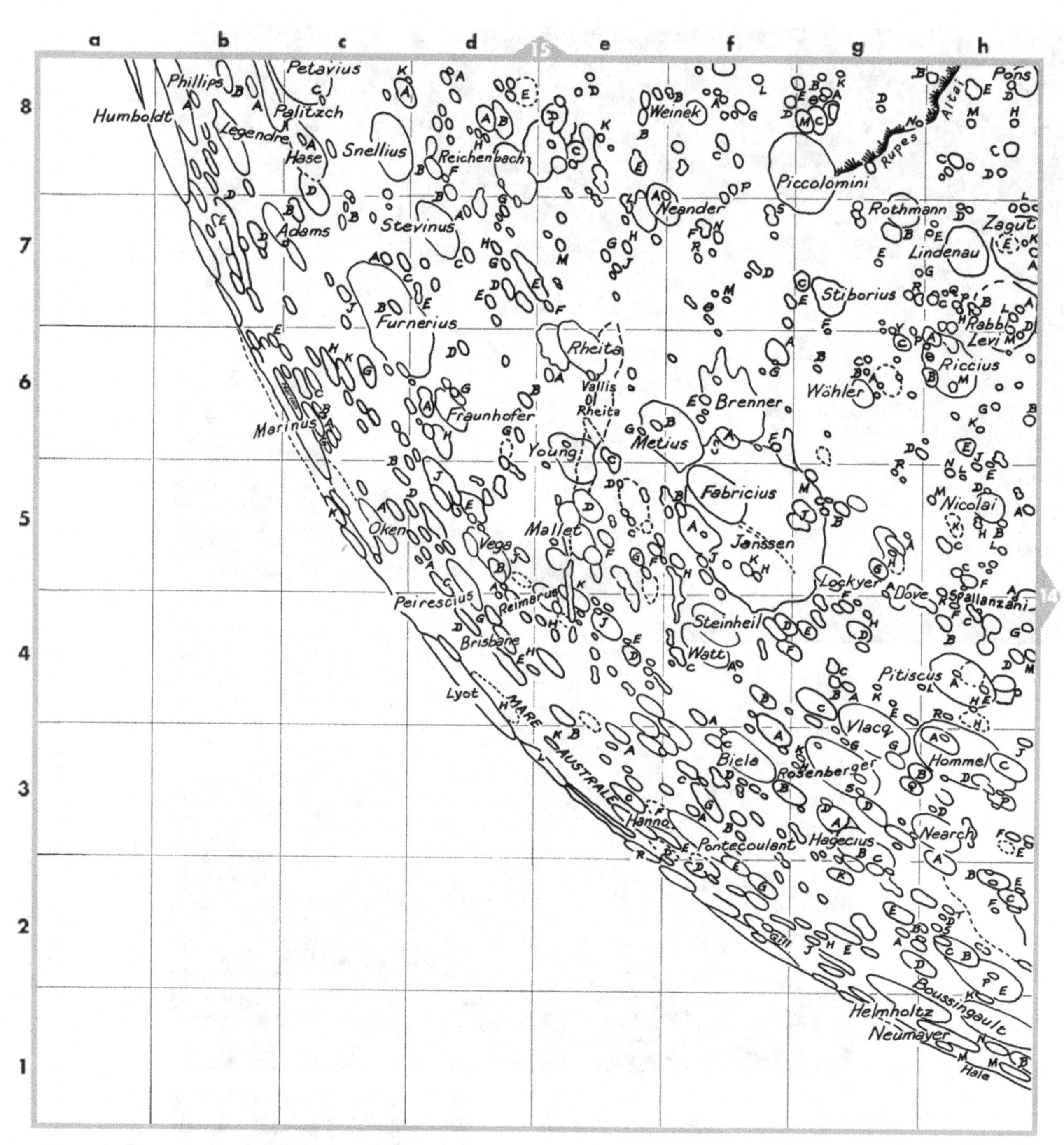

Crater Diameters

Crater	Diameter
Weinek	32 km (f8)
Snellius	86 km (c8)
Steinheil	65 km (f4)
Boussingault	128 km (h1)
Brisbane	44 km (d4)

This map has been prepared from Plates 16a and 16c. Plates 16b, 16d, and 16e show the same area under different lighting. Plate 16f shows a larger scale photograph of the Mare Australe area (d4).

Plate 16a

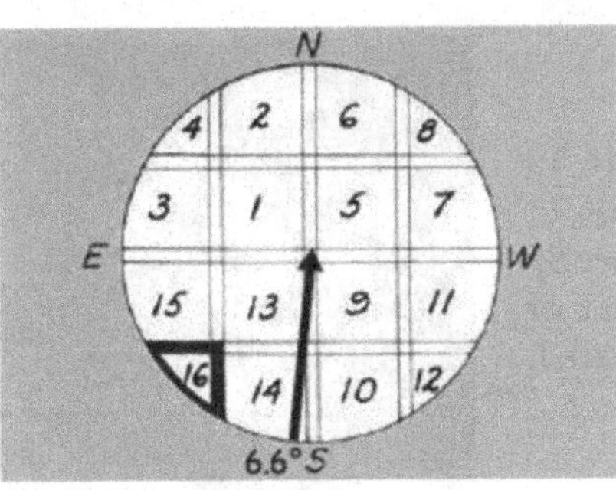

2109 UT Jun 23 1966. Moon's age 5.0 days. Sun's selenographic colongitude = 336.6°. Diameter 64 cm. Compare this with Plate 16c, which shows almost exactly the same area.

Plate 16b

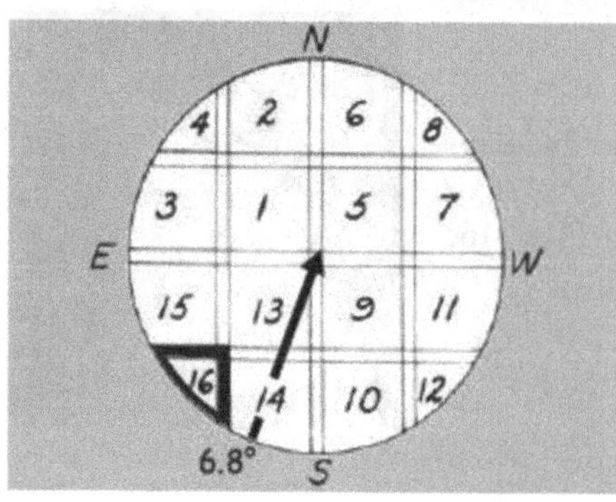

2326 UT Feb 06 1966. Moon's age 16.3 days. selenographic colongitude = 106.7°. Diameter 64 cm. This was taken nearly 1½ days after Full Moon, and so some of the limb detail has already disappeared, despite the favorable libration.

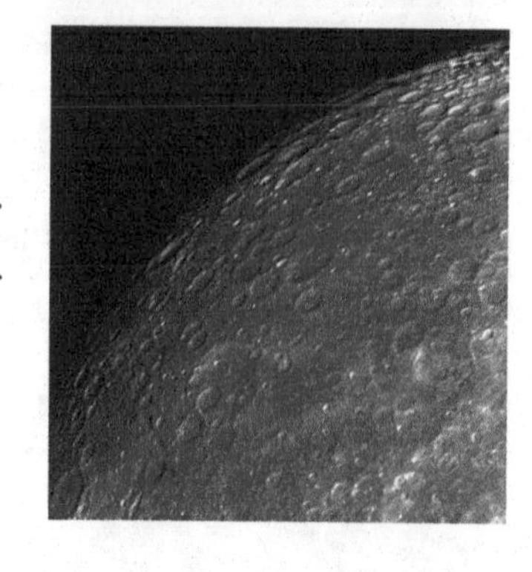

Plate 16c

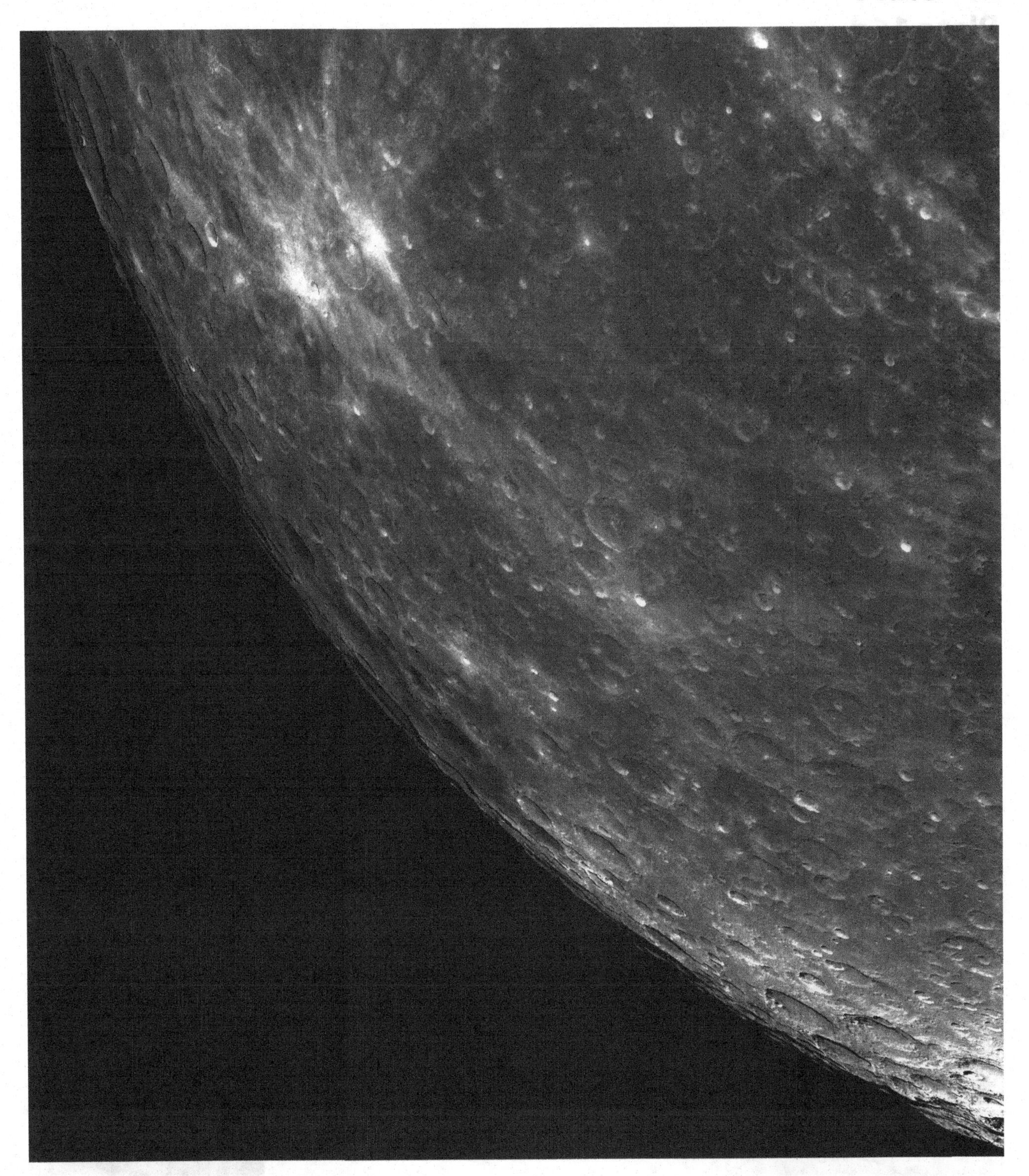

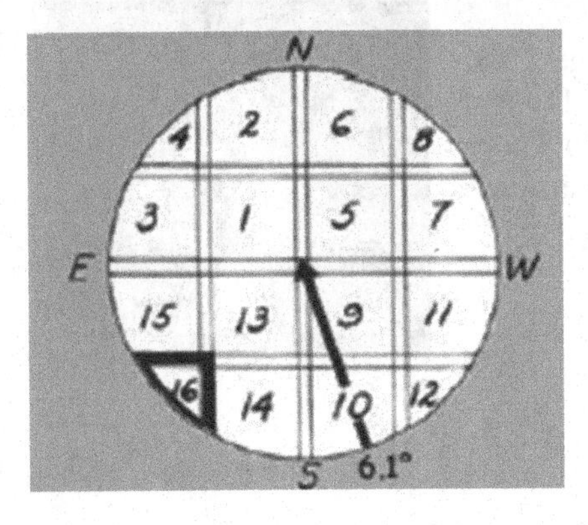

2259 UT Feb 24 1967. Moon's age 15.6 days. selenographic colongitude = 95.1°. Diameter 64 cm. This was taken about 5 hours after Full Moon. The libration was not favorable, but even so the craters on the limb near Brisbane (d4) will not often be seen like this.

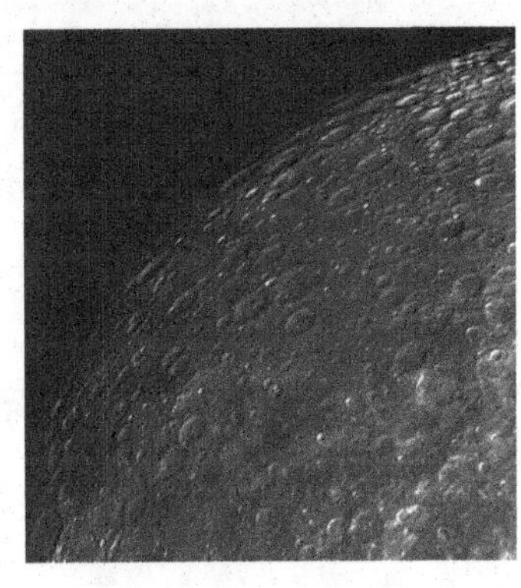

Plate 16d

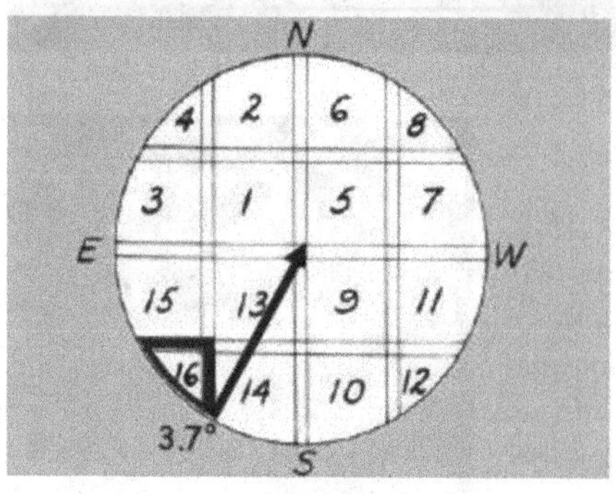

2334 UT Feb 26 1967. Moon's age 17.6 days. selenographic colongitude = 119.6°. Diameter 64 cm. Compare this with Plate 16b, which shows almost exactly the same area.

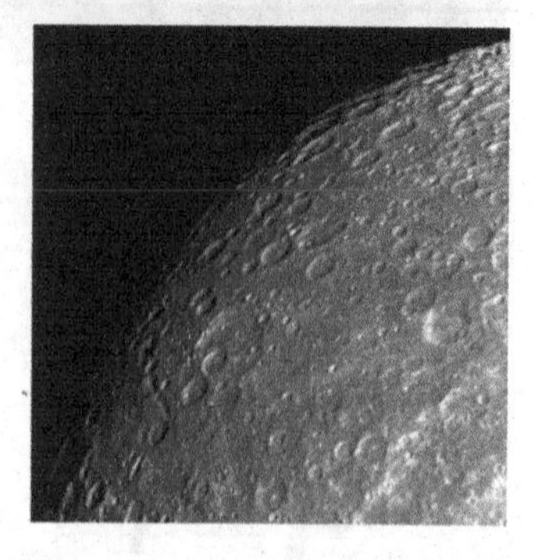

Plate 16e

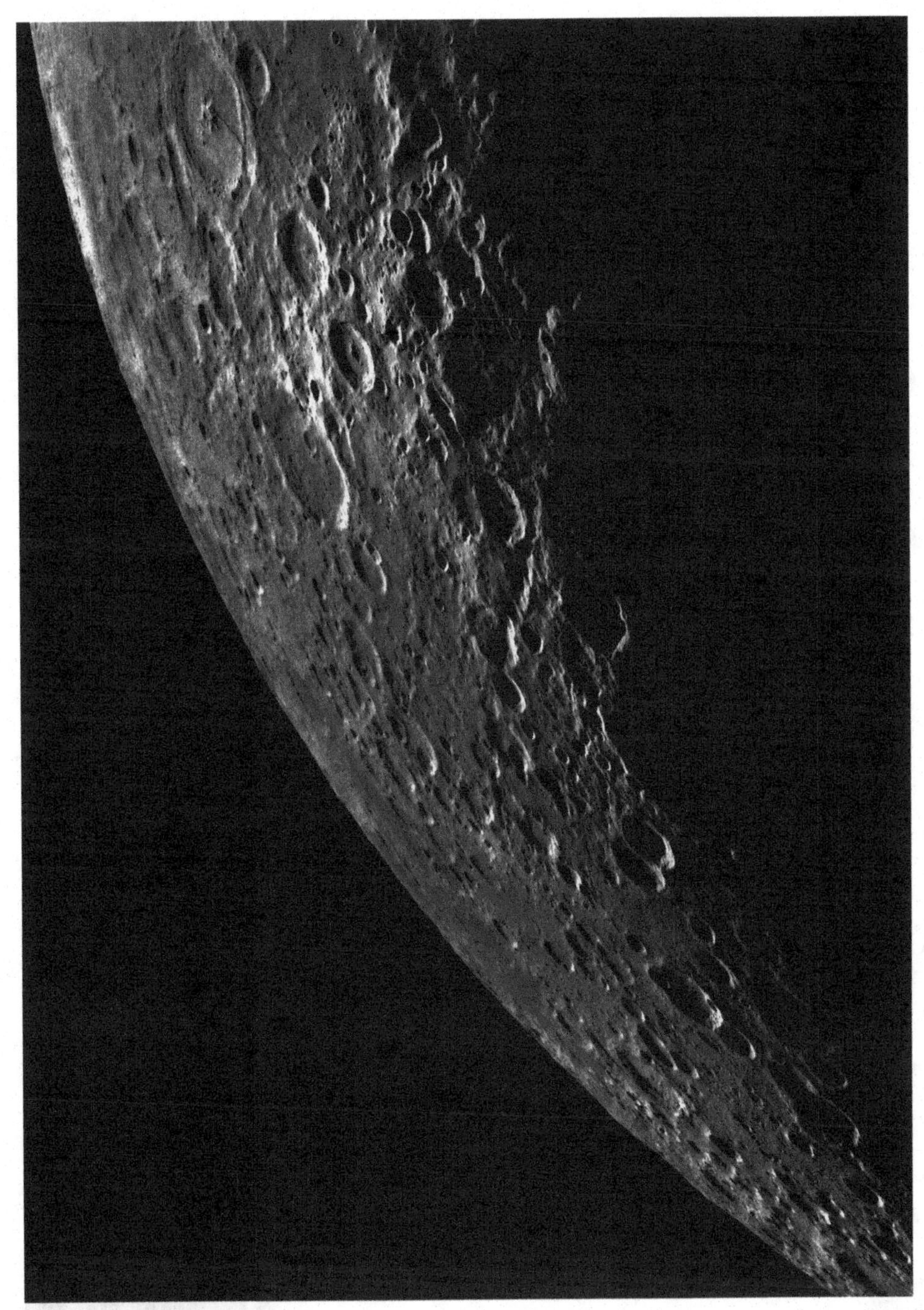

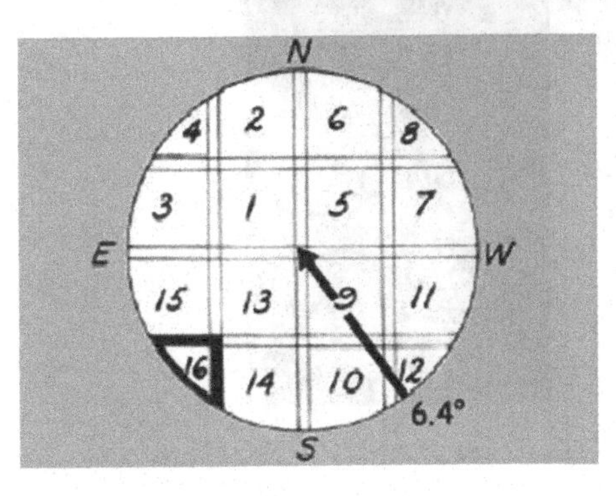

2036 UT May 23 1966. Moon's age 3.4 days. Selenographic Colongitude = 317.5°. Diameter 64 cm. Compare this with Plate 16c. The librations are very similar.

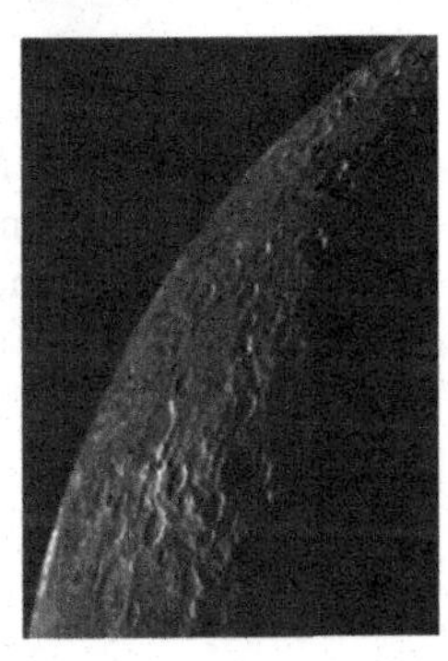

Plate 16f

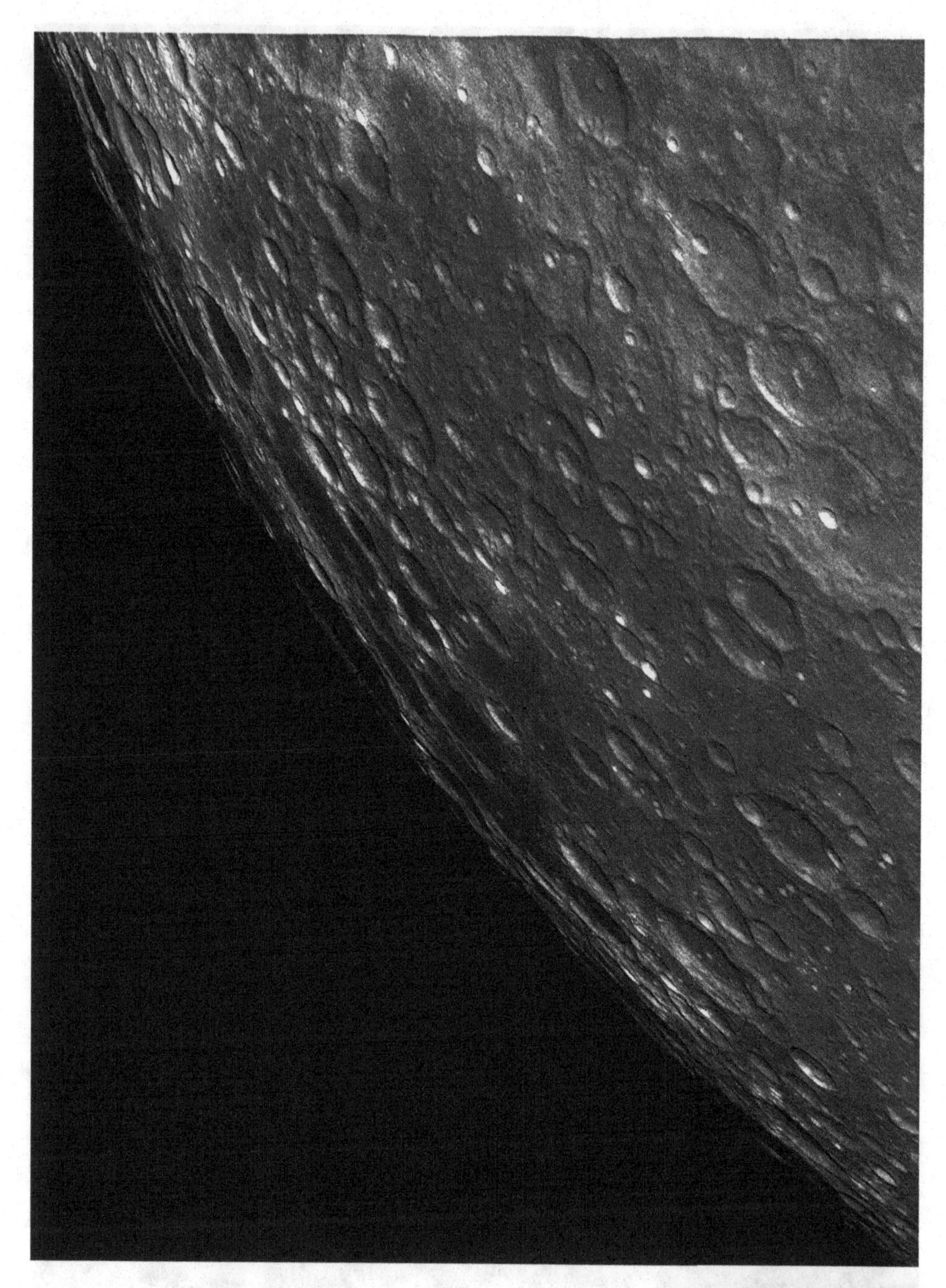

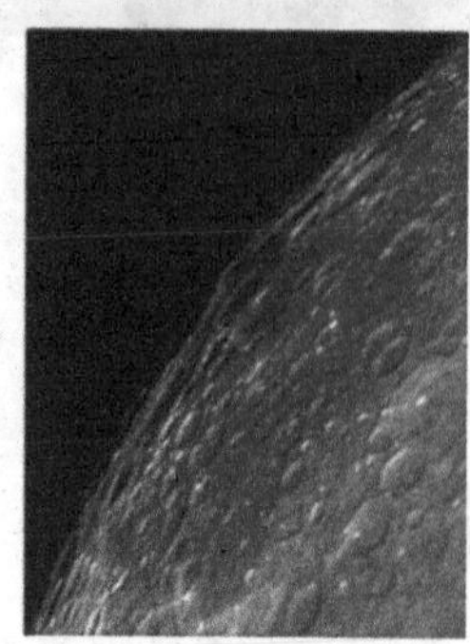

The Mare Australe area (e3). 2259 UT Feb 24 1967. Moon's age is 15.6 days. selenographic colongitude = 95.1°. Diameter 99 cm approx. This is an enlargement of part of Plate 16c.

Plate 17

Earthshine is the Moon's surface illuminated by light reflected off Earth. 2032 UT May 13 1967. Moon's age 4.3 days. Selenographic colongitude = 325.0°. Compare this with the Full Moon Key photograph, immediately preceding Map 1. The Earthshine needed an exposure of 10 seconds, and the part of the Moon illuminated by the Sun was therefore grossly over-exposed.

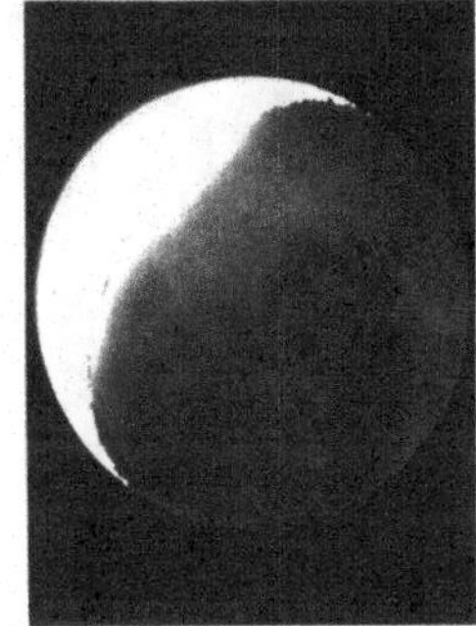

4 Simulations of Sunrise and Sunset over Selected Areas

The figures shown in this chapter are 100 % computer visualizations and consequently have been kept separate from the digitally augmented plates in the previous chapter. The simulations are illustrated in a double-page spread format, with the left page being sunrise and the right side sunset. In all these figures, north will be towards the top, east to the left, and west to the right – in other words an SCT type of image. Image scales may vary, but you can compare these with the craters in the plates in the previous chapter. The title numbering system corresponds to the map sheet numbers from last chapter. In all cases the simulated images are each separated 3° apart in terms of their selenographic colongitude, or approximately 6 h difference in Earth time. The sub-solar point has been kept fixed at 0° latitude, and the libration has been set to be on the lunar equator on a meridian line of 0° longitude (Fig. 4.1).

A.C. Cook, *The Hatfield SCT Lunar Atlas: Photographic Atlas for Meade, Celestron, and Other SCT Telescopes: A Digitally Re-Mastered Edition*, DOI 10.1007/978-1-4614-8639-8_4, © Springer Science+Business Media New York 2014

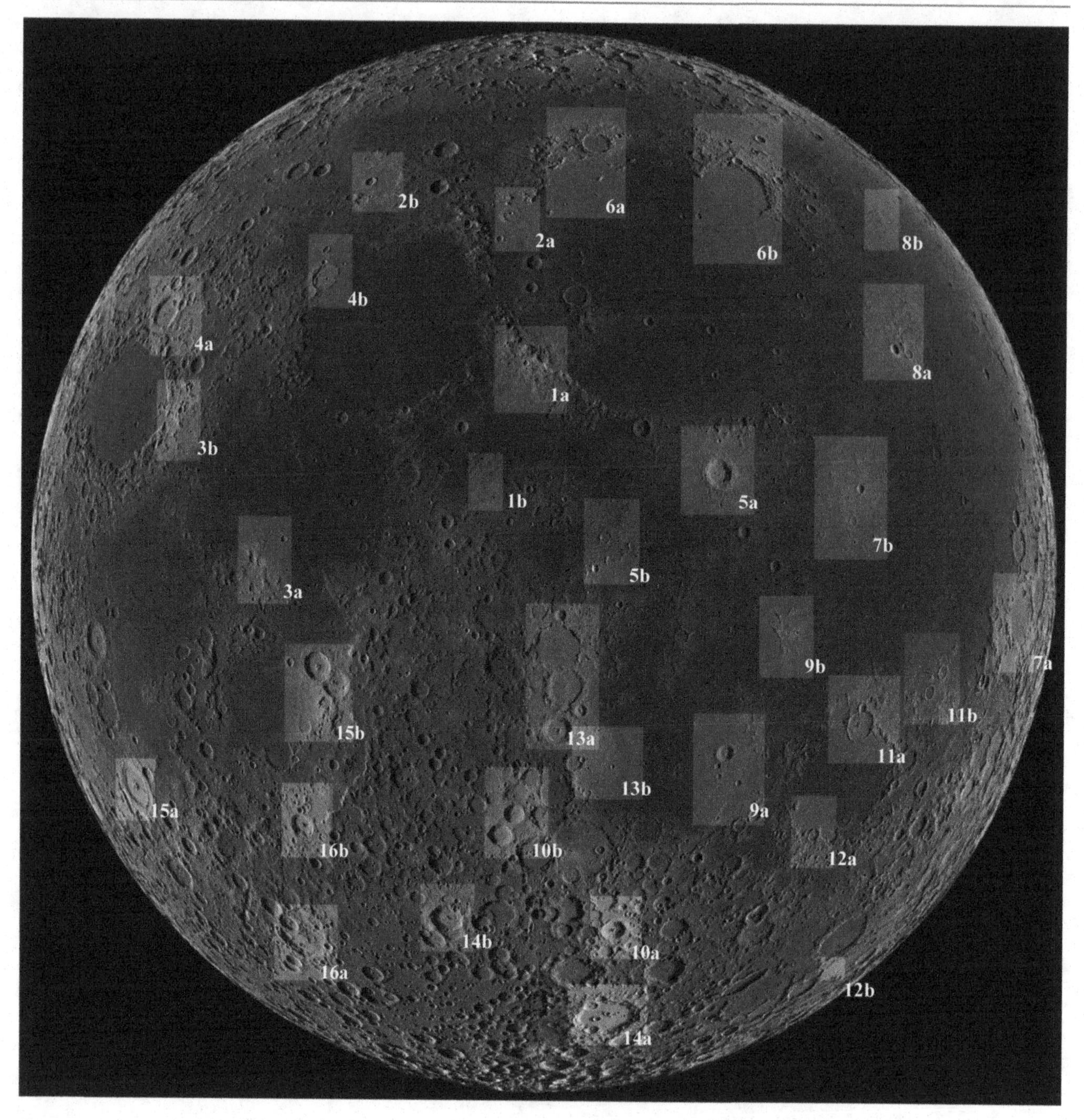

Fig. 4.1 A finder chart for the sunrise and sunset visualizations of the selected areas 1a to 16b in the remainder of this chapter

Selected Area 1a – Montes Apenninus (Sunrise)

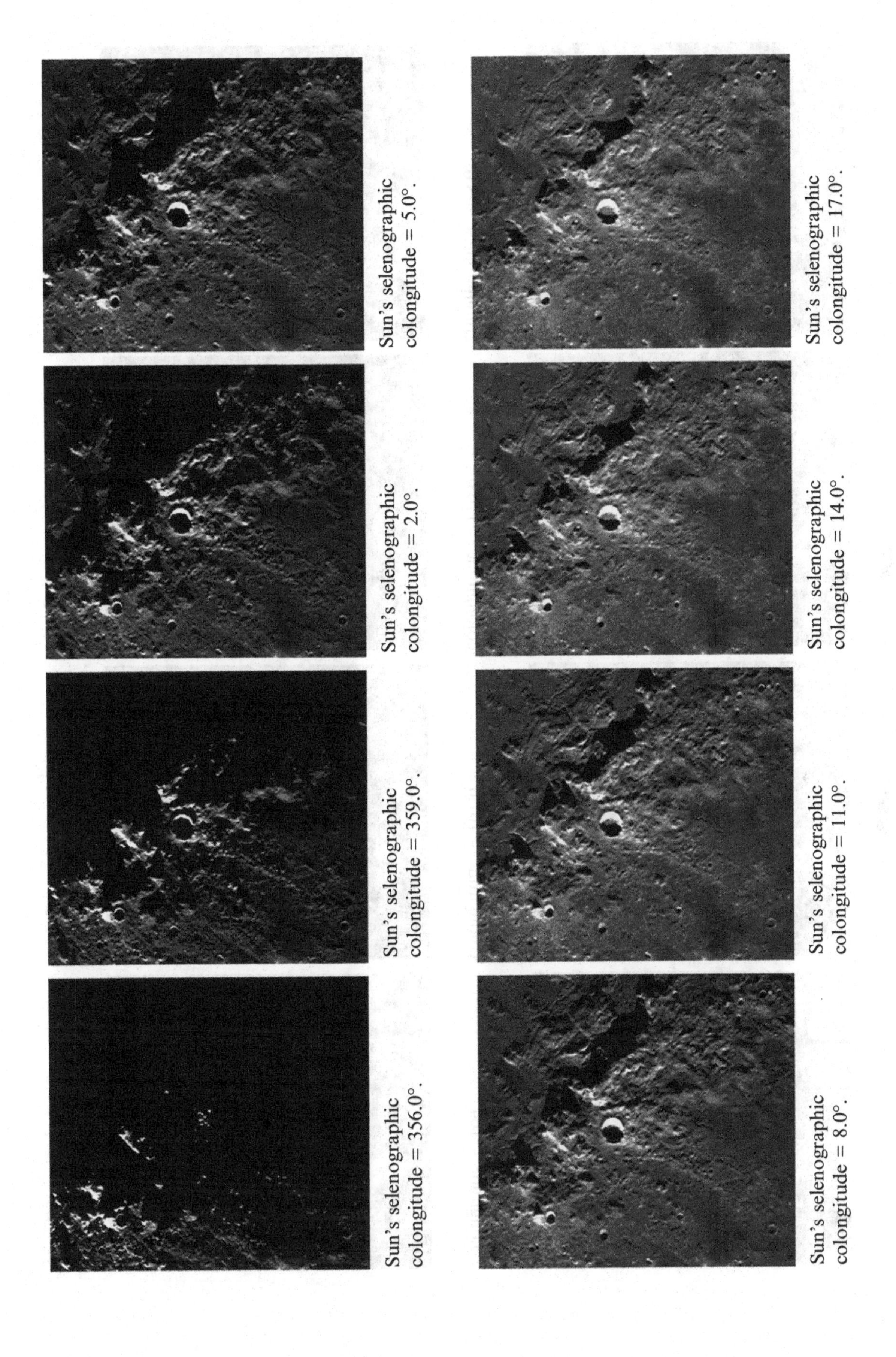

Sun's selenographic colongitude = 356.0°.

Sun's selenographic colongitude = 359.0°.

Sun's selenographic colongitude = 2.0°.

Sun's selenographic colongitude = 5.0°.

Sun's selenographic colongitude = 8.0°.

Sun's selenographic colongitude = 11.0°.

Sun's selenographic colongitude = 14.0°.

Sun's selenographic colongitude = 17.0°.

Selected Area 1a – Montes Apenninus (Sunset)

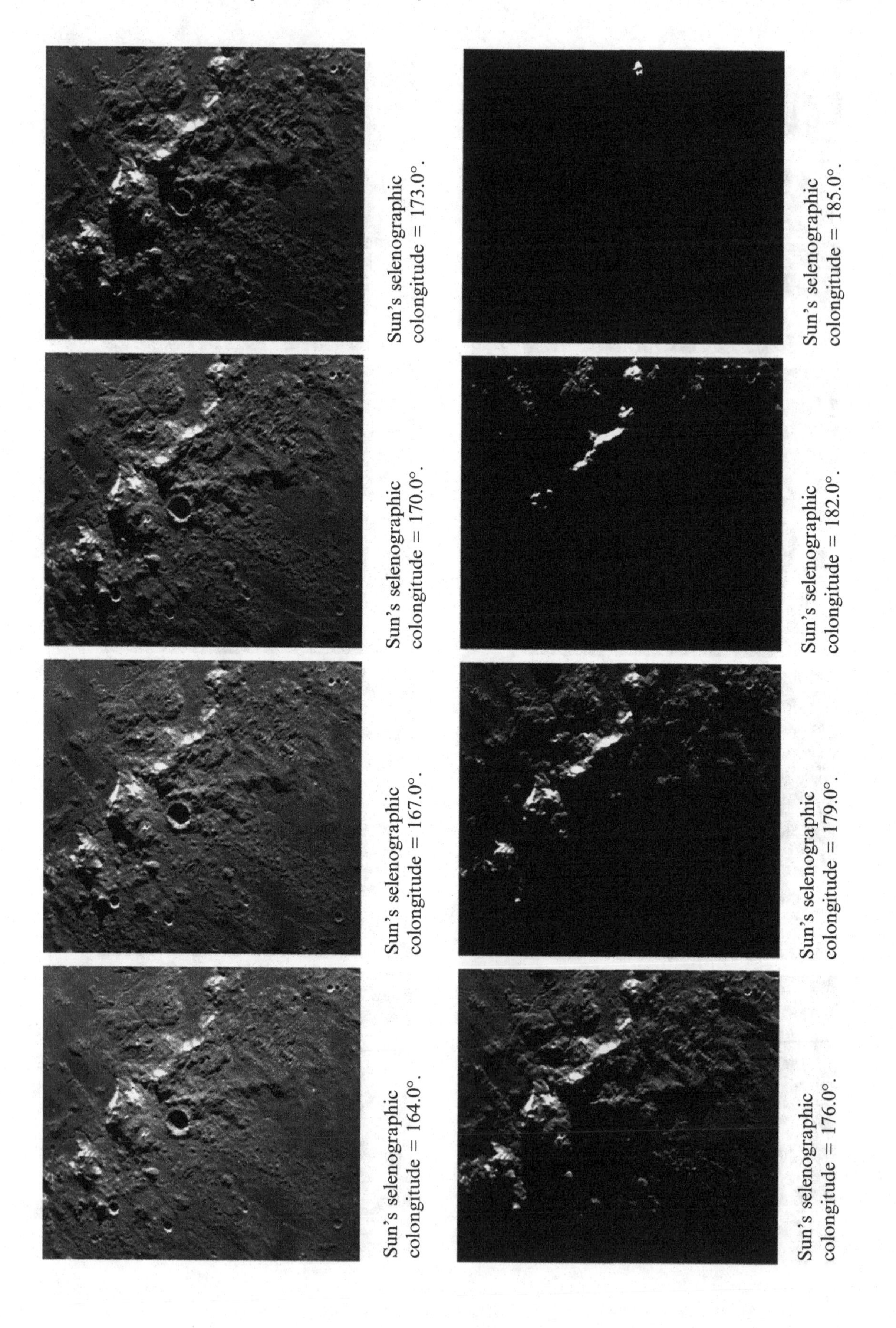

Sun's selenographic colongitude = 173.0°.

Sun's selenographic colongitude = 170.0°.

Sun's selenographic colongitude = 167.0°.

Sun's selenographic colongitude = 164.0°.

Sun's selenographic colongitude = 185.0°.

Sun's selenographic colongitude = 182.0°.

Sun's selenographic colongitude = 179.0°.

Sun's selenographic colongitude = 176.0°.

Selected Area 1b – Rima Hyginus (Sunrise)

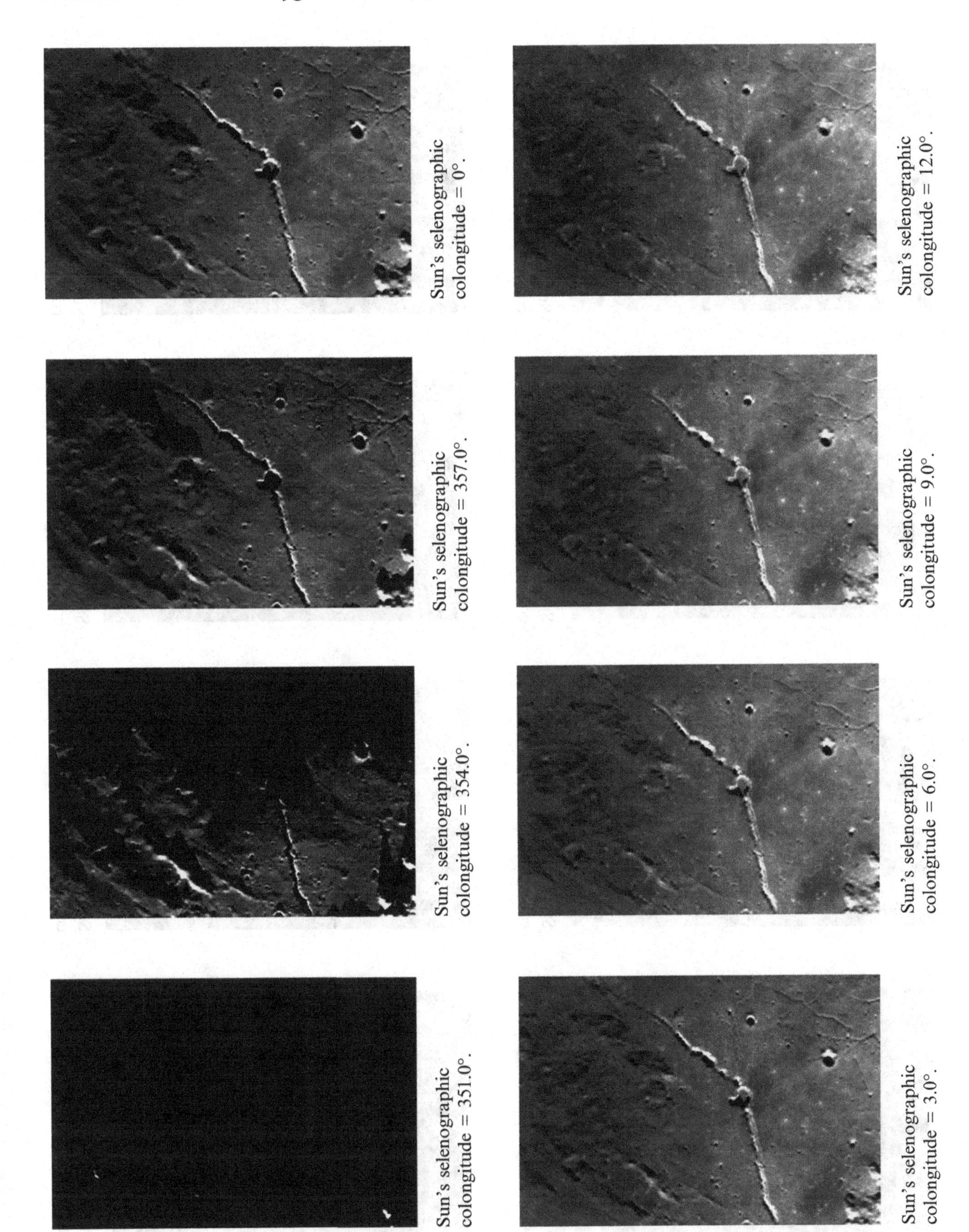

Sun's selenographic colongitude = 0°.

Sun's selenographic colongitude = 357.0°.

Sun's selenographic colongitude = 354.0°.

Sun's selenographic colongitude = 351.0°.

Sun's selenographic colongitude = 12.0°.

Sun's selenographic colongitude = 9.0°.

Sun's selenographic colongitude = 6.0°.

Sun's selenographic colongitude = 3.0°.

Selected Area 1b – Rima Hyginus (Sunset)

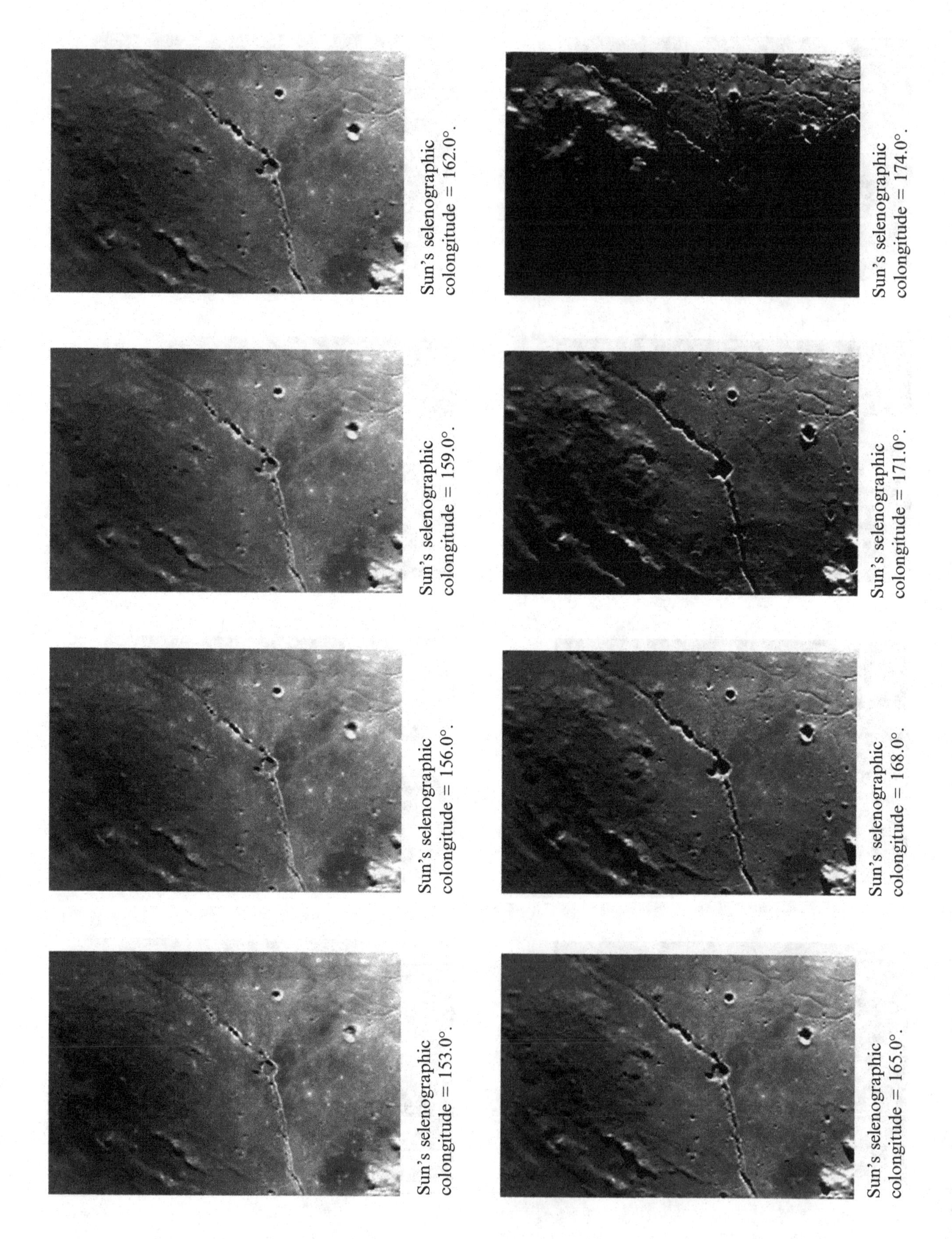

Sun's selenographic colongitude = 153.0°.

Sun's selenographic colongitude = 156.0°.

Sun's selenographic colongitude = 159.0°.

Sun's selenographic colongitude = 162.0°.

Sun's selenographic colongitude = 165.0°.

Sun's selenographic colongitude = 168.0°.

Sun's selenographic colongitude = 171.0°.

Sun's selenographic colongitude = 174.0°.

Selected Area 2a – Cassini (Sunrise)

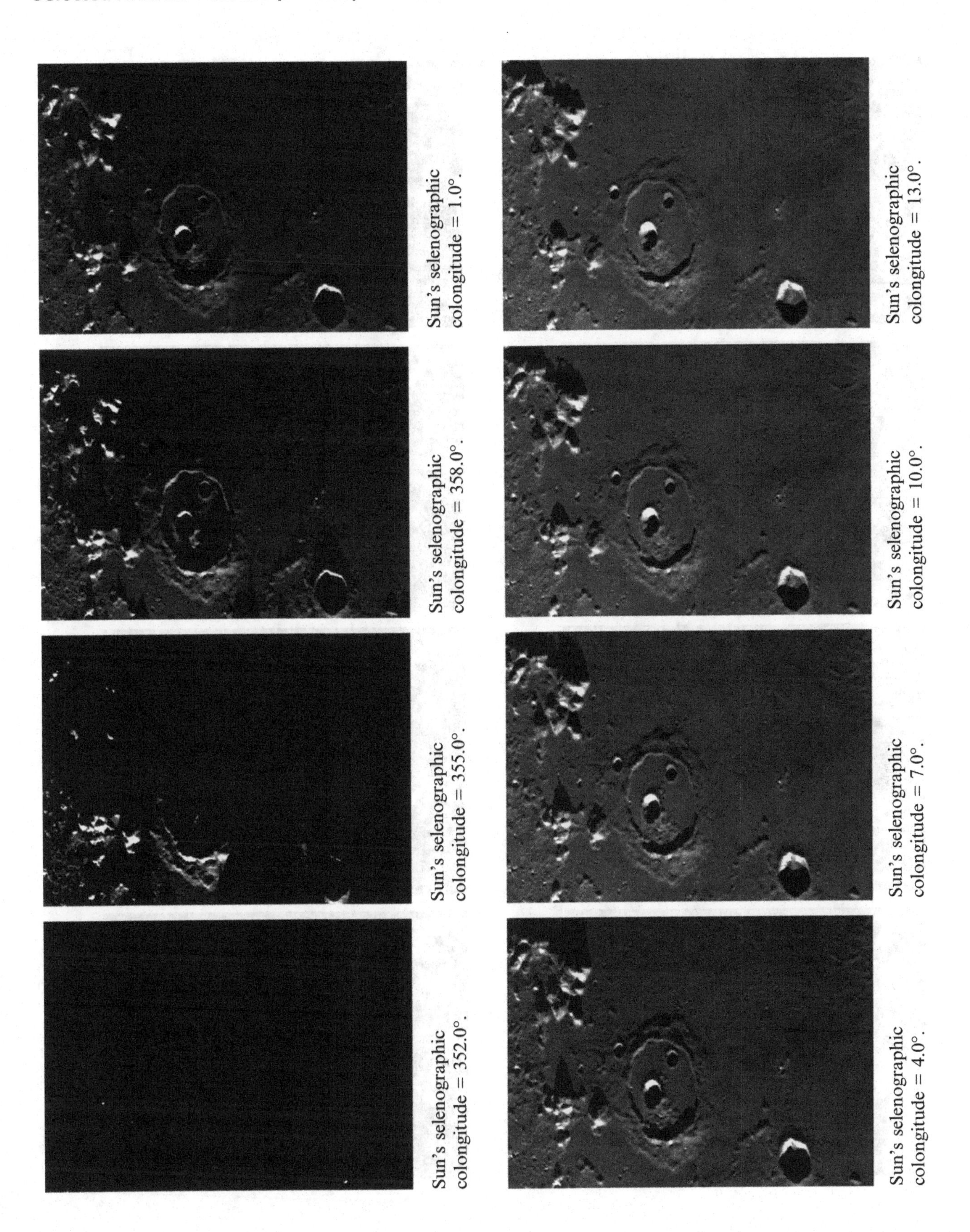

Sun's selenographic colongitude = 1.0°.

Sun's selenographic colongitude = 358.0°.

Sun's selenographic colongitude = 355.0°.

Sun's selenographic colongitude = 352.0°.

Sun's selenographic colongitude = 13.0°.

Sun's selenographic colongitude = 10.0°.

Sun's selenographic colongitude = 7.0°.

Sun's selenographic colongitude = 4.0°.

Selected Area 2a – Cassini (Sunset)

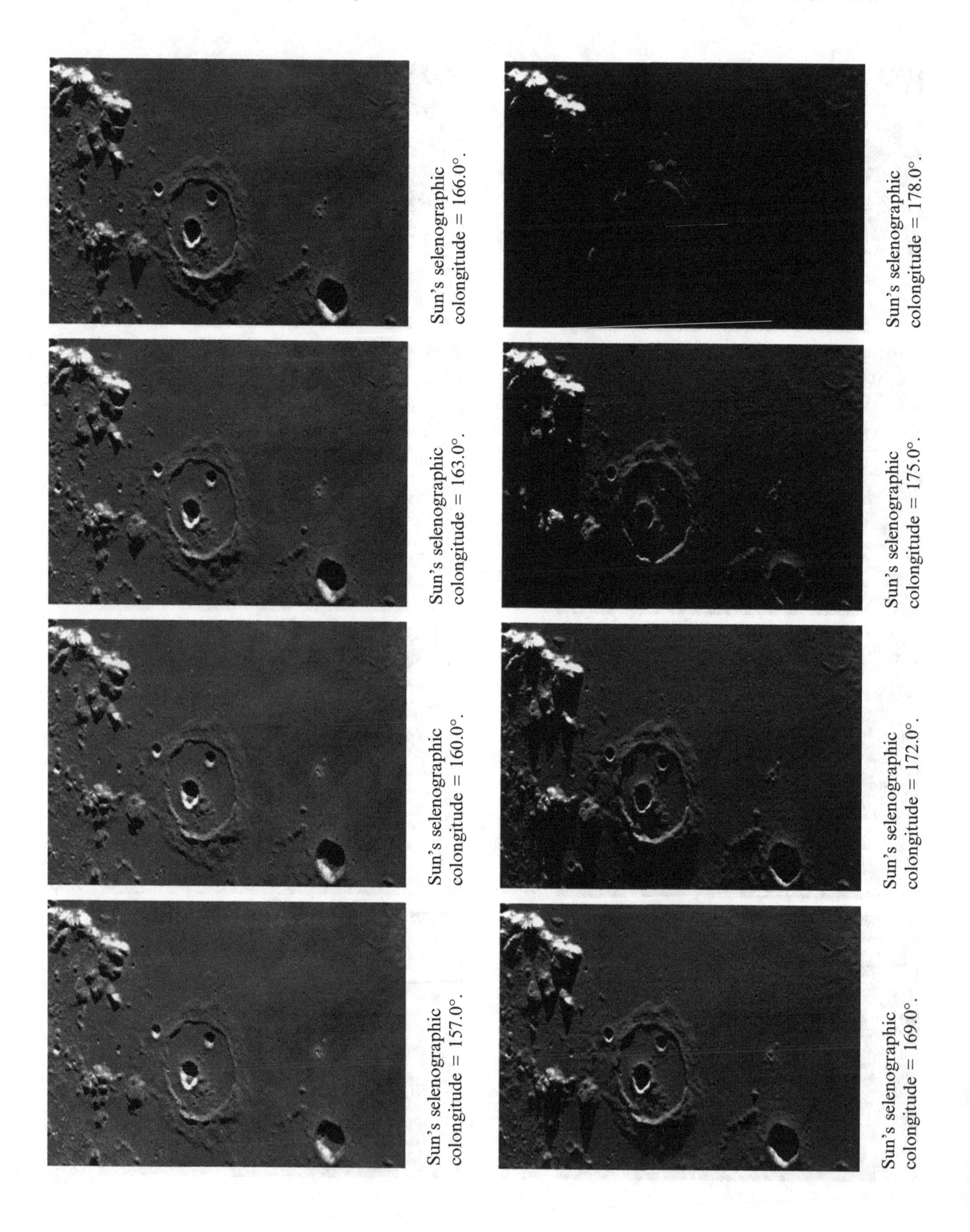

Sun's selenographic colongitude = 157.0°.

Sun's selenographic colongitude = 160.0°.

Sun's selenographic colongitude = 163.0°.

Sun's selenographic colongitude = 166.0°.

Sun's selenographic colongitude = 169.0°.

Sun's selenographic colongitude = 172.0°.

Sun's selenographic colongitude = 175.0°.

Sun's selenographic colongitude = 178.0°.

Selected Area 2b – Mortis Lacus (Sunrise)

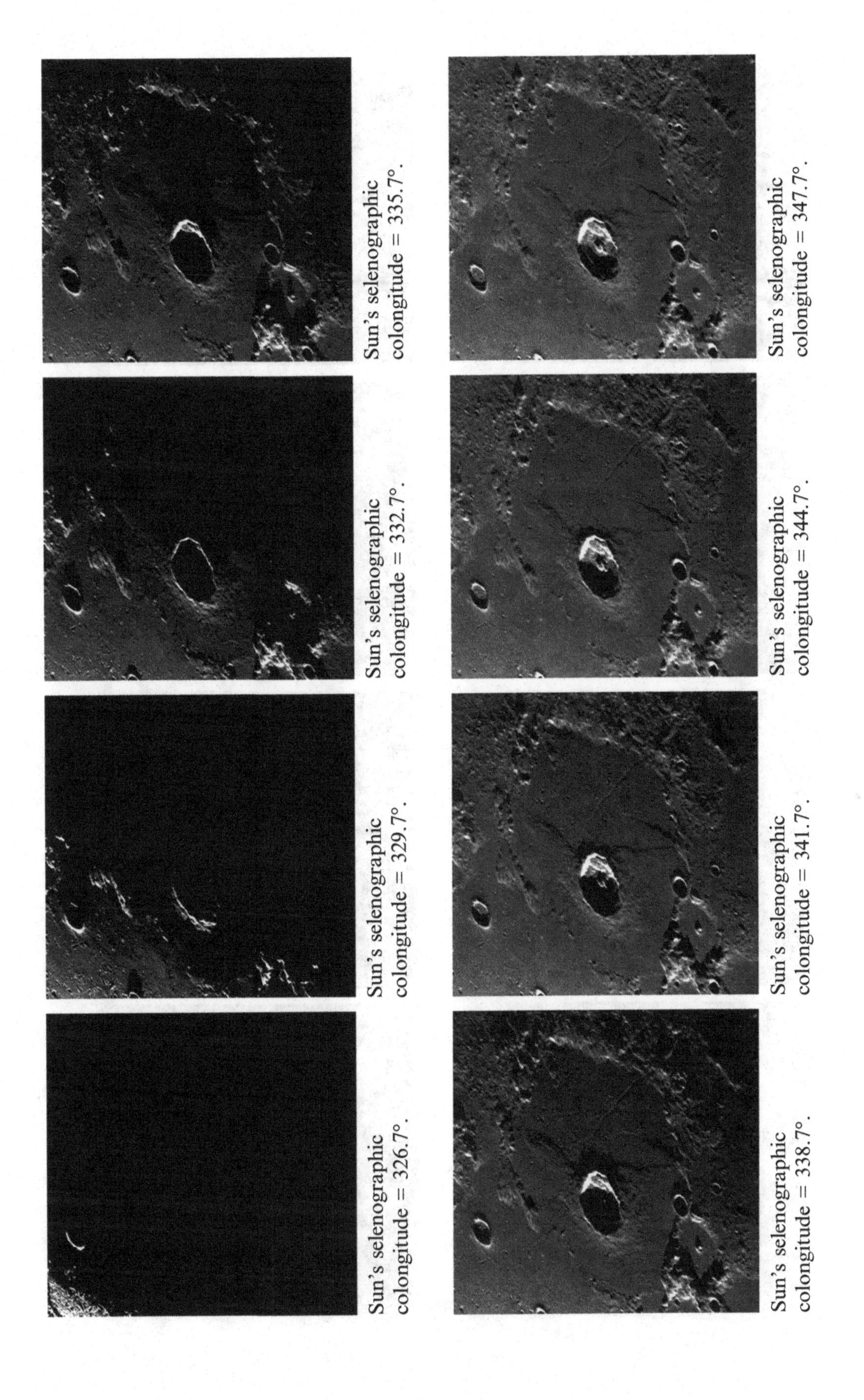

Sun's selenographic colongitude = 326.7°.

Sun's selenographic colongitude = 329.7°.

Sun's selenographic colongitude = 332.7°.

Sun's selenographic colongitude = 335.7°.

Sun's selenographic colongitude = 338.7°.

Sun's selenographic colongitude = 341.7°.

Sun's selenographic colongitude = 344.7°.

Sun's selenographic colongitude = 347.7°.

Selected Area 2b – Mortis Lacus (Sunset)

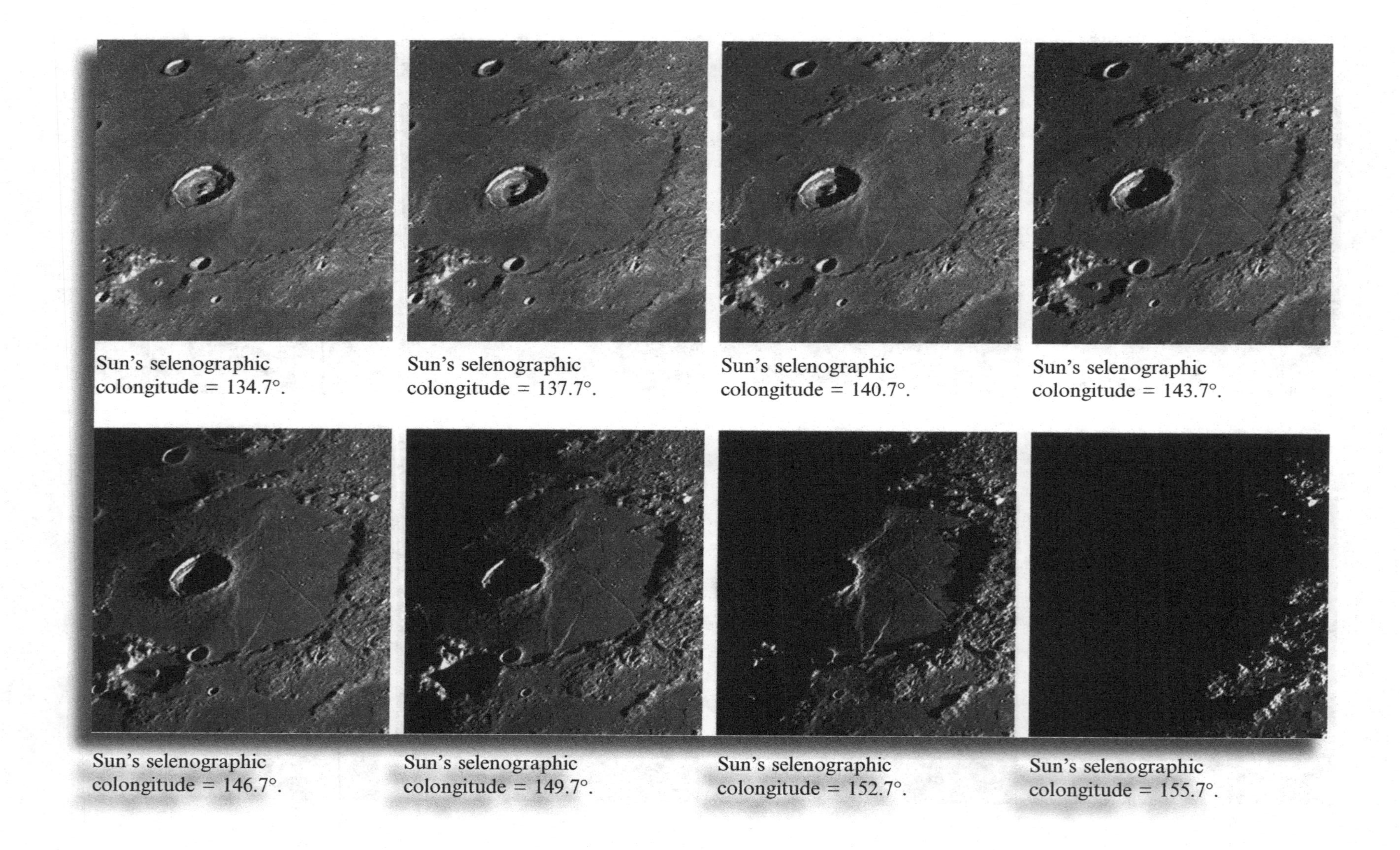

Sun's selenographic colongitude = 134.7°.

Sun's selenographic colongitude = 137.7°.

Sun's selenographic colongitude = 140.7°.

Sun's selenographic colongitude = 143.7°.

Sun's selenographic colongitude = 146.7°.

Sun's selenographic colongitude = 149.7°.

Sun's selenographic colongitude = 152.7°.

Sun's selenographic colongitude = 155.7°.

Selected Area 3a – Censorinus (Sunrise)

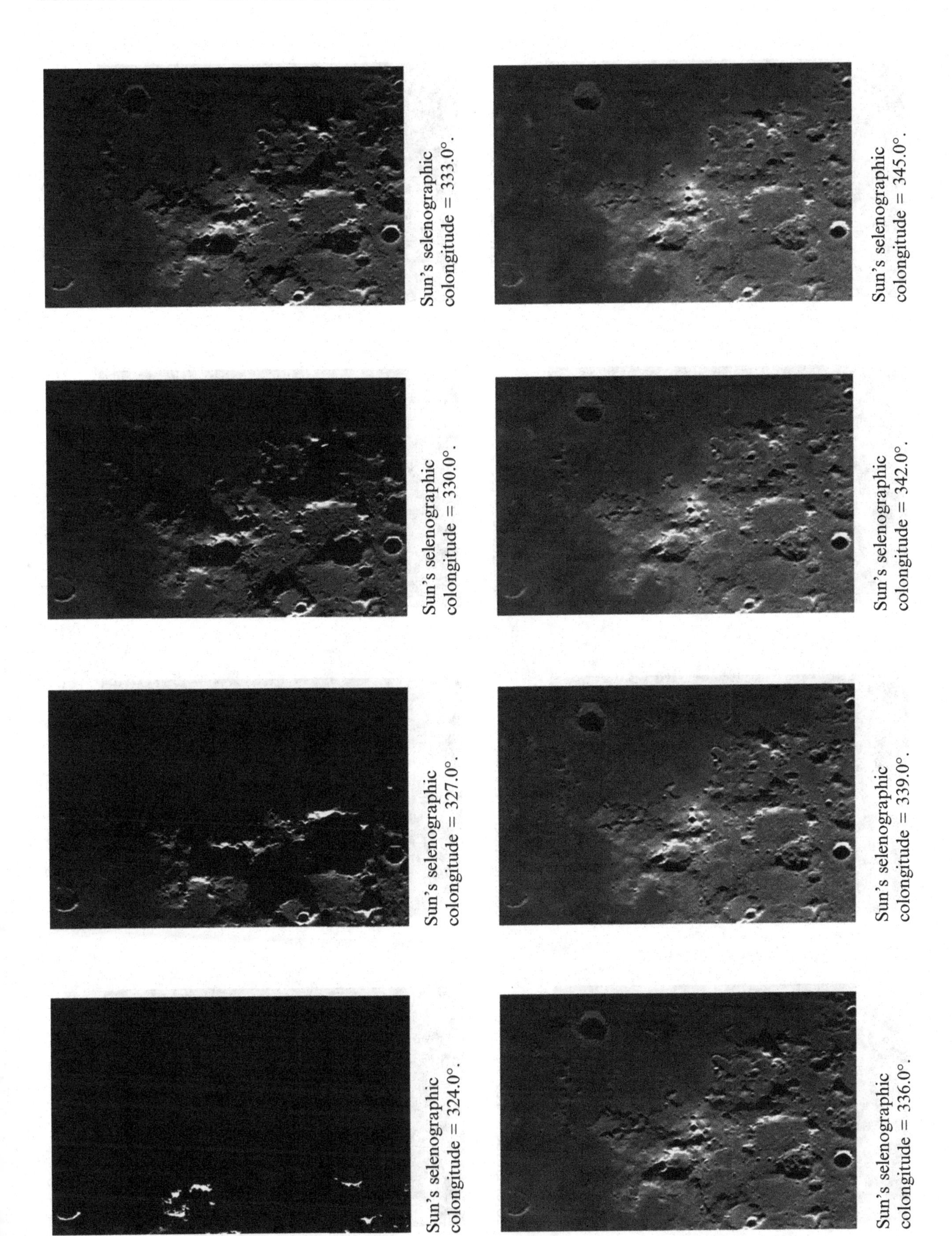

Sun's selenographic colongitude = 324.0°.

Sun's selenographic colongitude = 327.0°.

Sun's selenographic colongitude = 330.0°.

Sun's selenographic colongitude = 333.0°.

Sun's selenographic colongitude = 336.0°.

Sun's selenographic colongitude = 339.0°.

Sun's selenographic colongitude = 342.0°.

Sun's selenographic colongitude = 345.0°.

Selected Area 3a – Censorinus (Sunset)

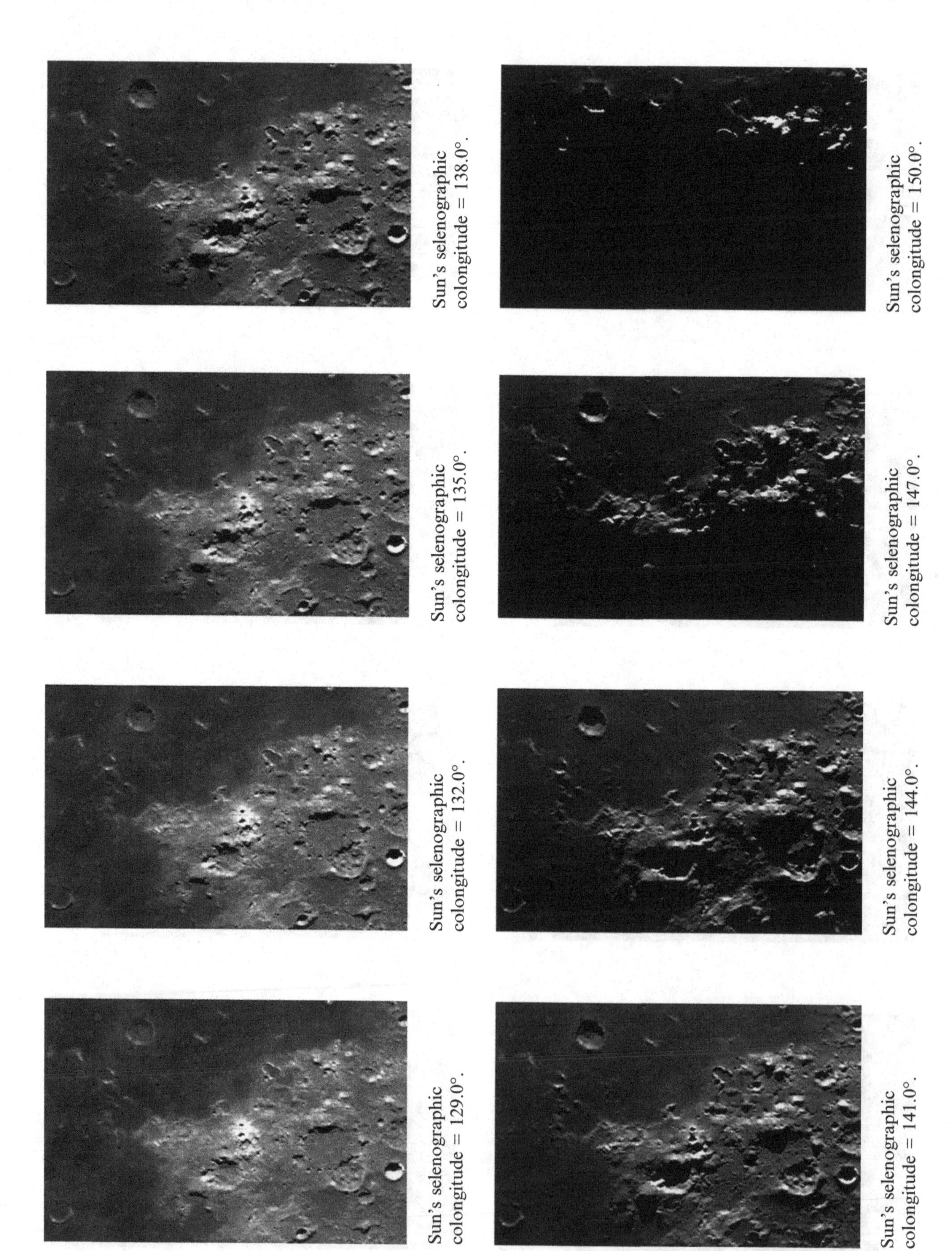

Sun's selenographic colongitude = 138.0°.

Sun's selenographic colongitude = 135.0°.

Sun's selenographic colongitude = 132.0°.

Sun's selenographic colongitude = 129.0°.

Sun's selenographic colongitude = 150.0°.

Sun's selenographic colongitude = 147.0°.

Sun's selenographic colongitude = 144.0°.

Sun's selenographic colongitude = 141.0°.

Selected Area 3b – Proclus (Sunrise)

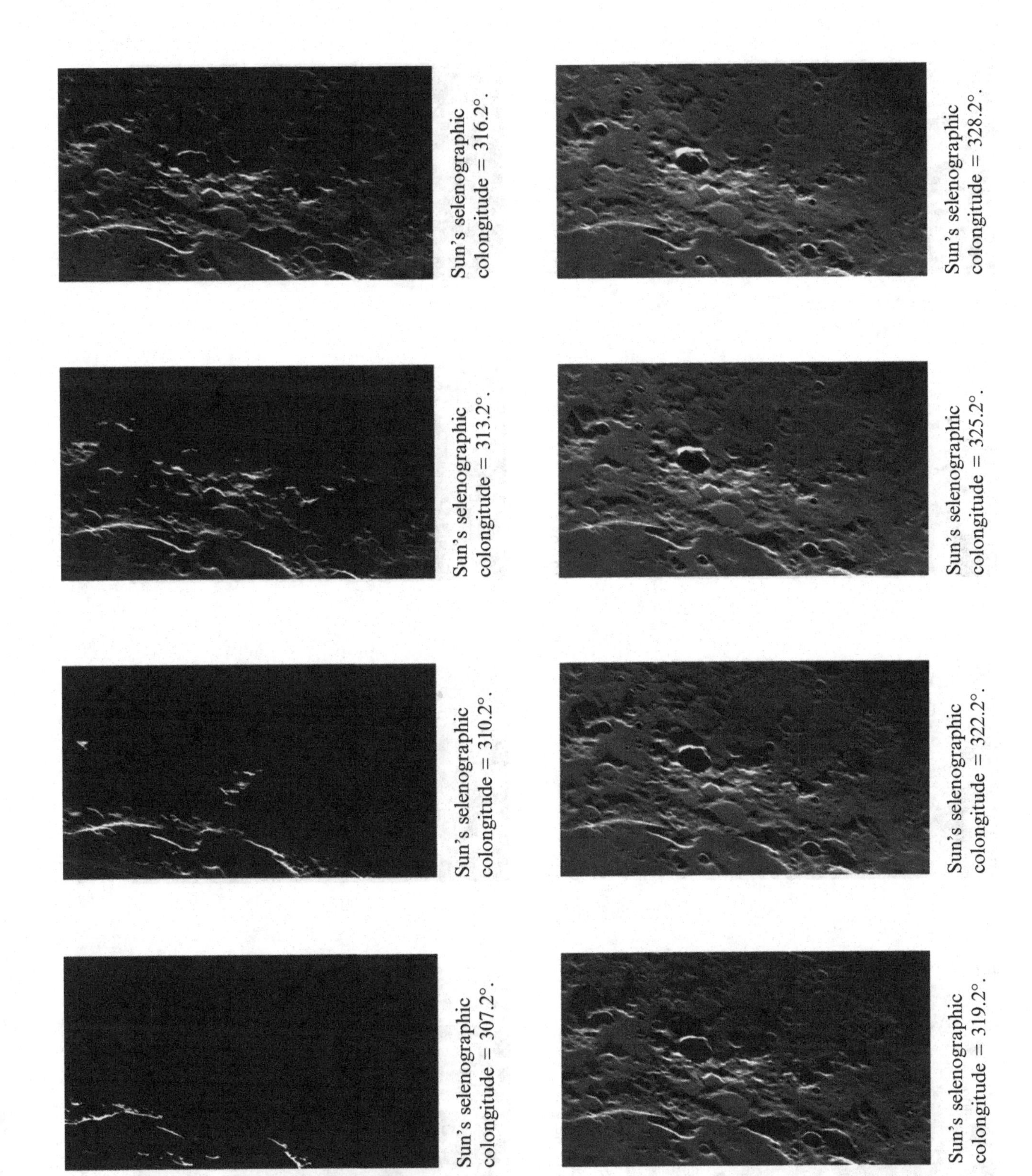

Sun's selenographic colongitude = 316.2°.

Sun's selenographic colongitude = 328.2°.

Sun's selenographic colongitude = 313.2°.

Sun's selenographic colongitude = 325.2°.

Sun's selenographic colongitude = 310.2°.

Sun's selenographic colongitude = 322.2°.

Sun's selenographic colongitude = 307.2°.

Sun's selenographic colongitude = 319.2°.

Selected Area 3b – Proclus (Sunset)

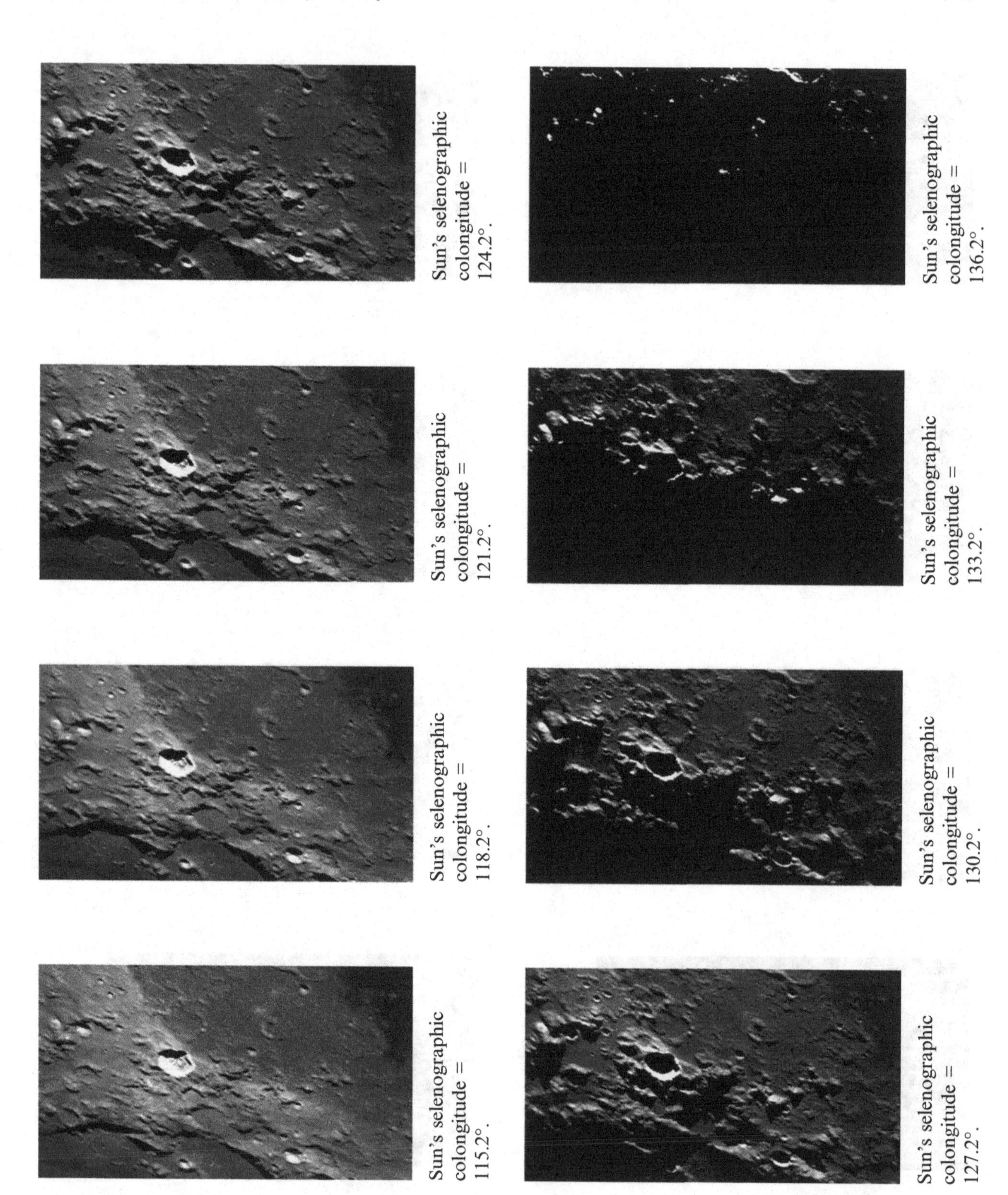

Selected Area 4a – Cleomedes (Sunrise)

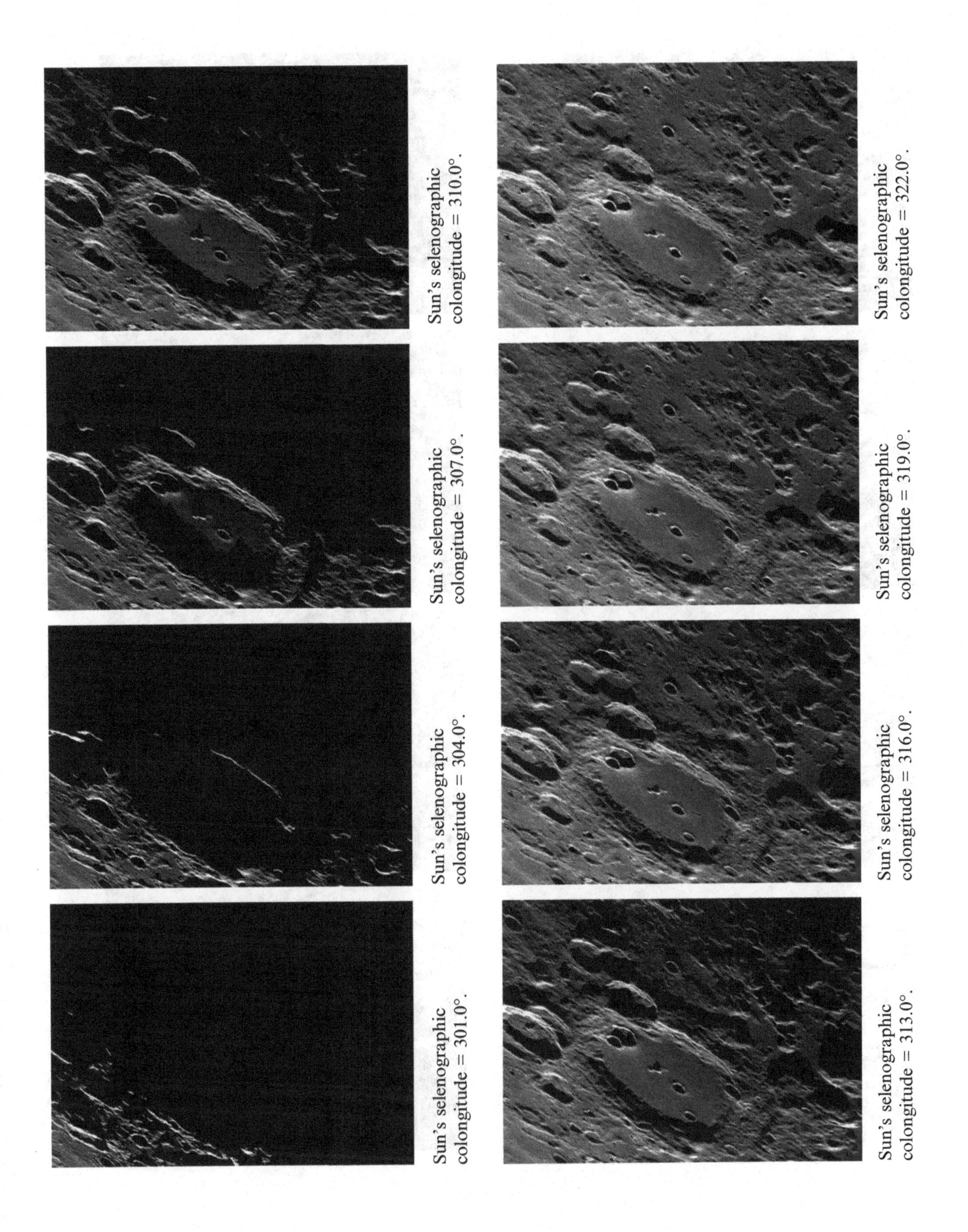

Sun's selenographic colongitude = 301.0°.

Sun's selenographic colongitude = 304.0°.

Sun's selenographic colongitude = 307.0°.

Sun's selenographic colongitude = 310.0°.

Sun's selenographic colongitude = 313.0°.

Sun's selenographic colongitude = 316.0°.

Sun's selenographic colongitude = 319.0°.

Sun's selenographic colongitude = 322.0°.

Selected Area 4a – Cleomedes (Sunset)

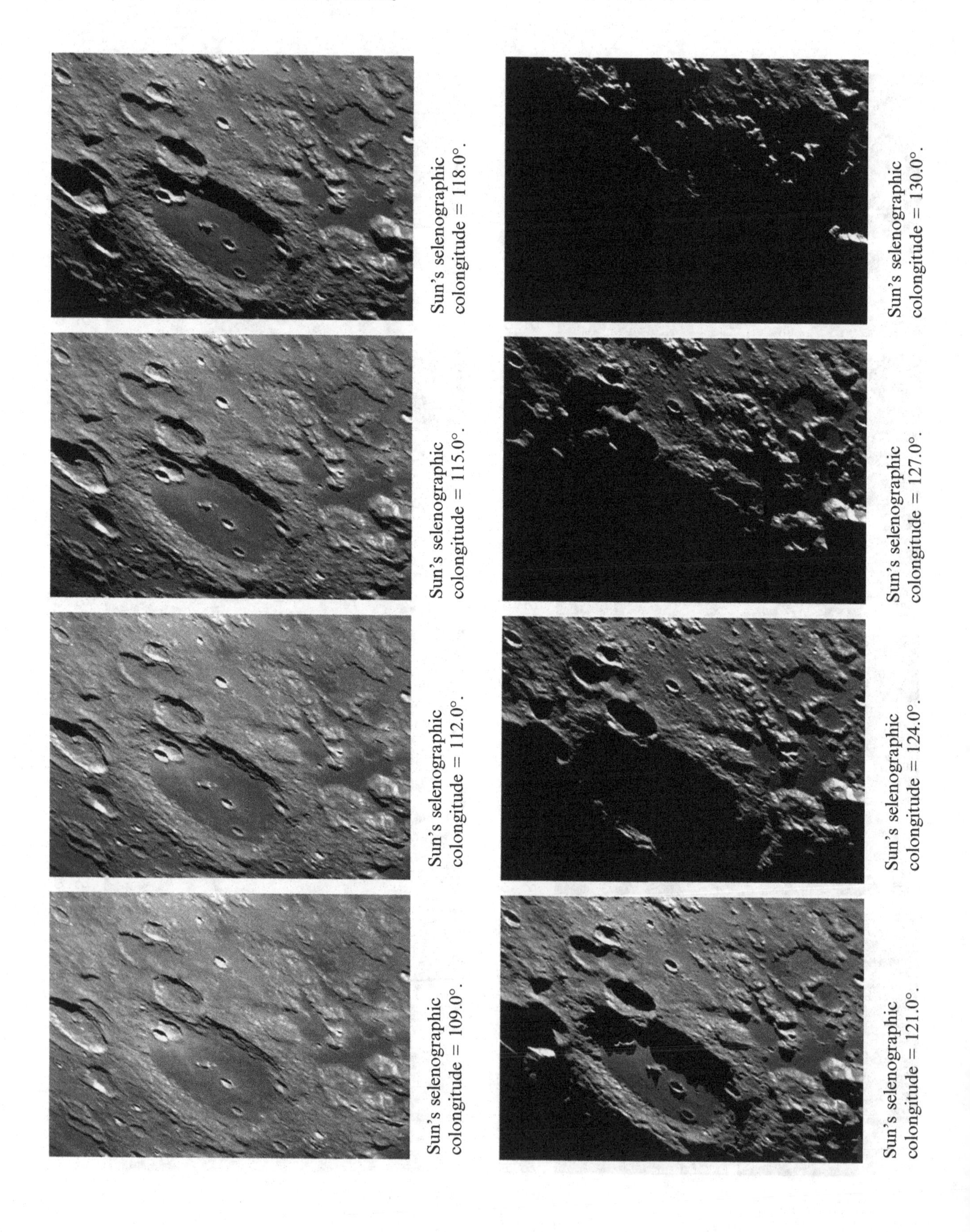

Sun's selenographic colongitude = 109.0°.

Sun's selenographic colongitude = 112.0°.

Sun's selenographic colongitude = 115.0°.

Sun's selenographic colongitude = 118.0°.

Sun's selenographic colongitude = 121.0°.

Sun's selenographic colongitude = 124.0°.

Sun's selenographic colongitude = 127.0°.

Sun's selenographic colongitude = 130.0°.

Selected Area 4b – Posidonius (Sunrise)

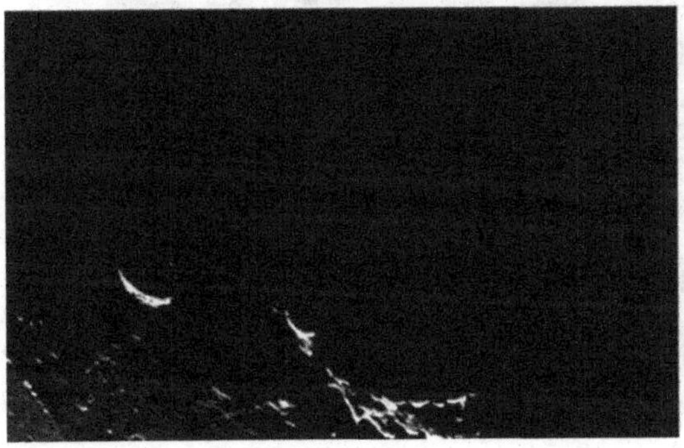

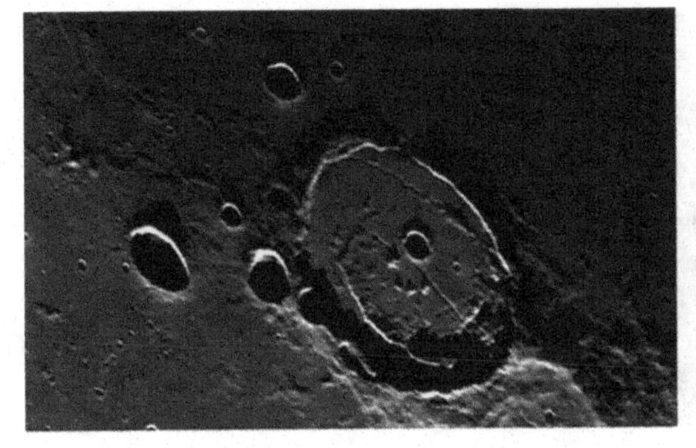

Sun's selenographic colongitude = 336.0°.

Sun's selenographic colongitude = 333.0°.

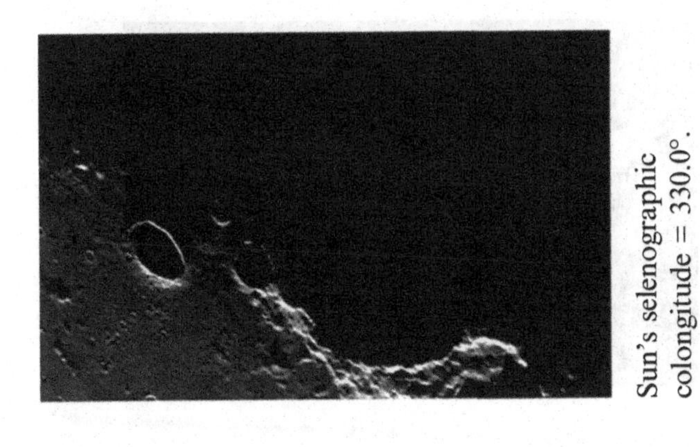

Sun's selenographic colongitude = 330.0°.

Sun's selenographic colongitude = 327.0°.

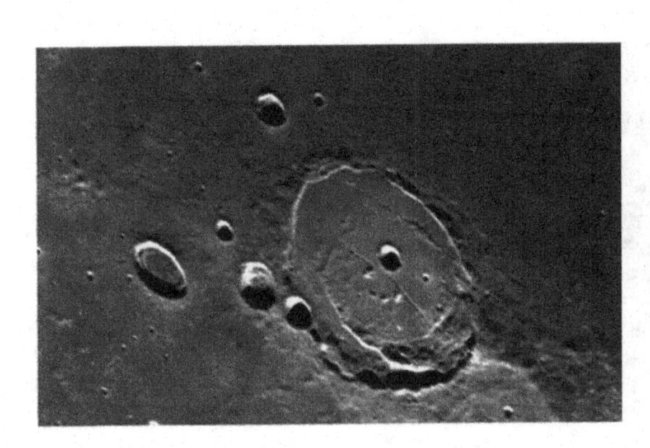

Sun's selenographic colongitude = 348.0°.

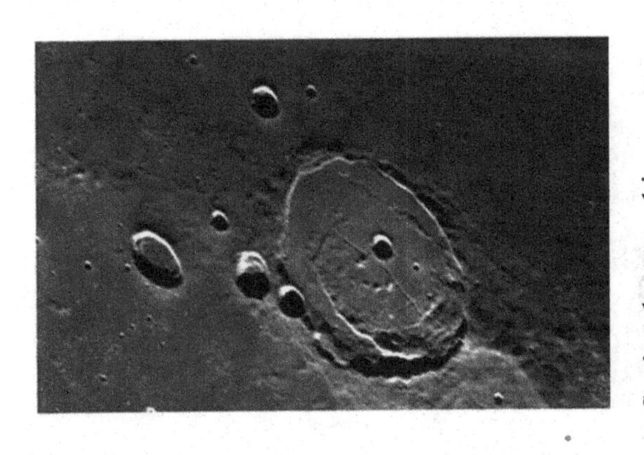

Sun's selenographic colongitude = 345.0°.

Sun's selenographic colongitude = 342.0°.

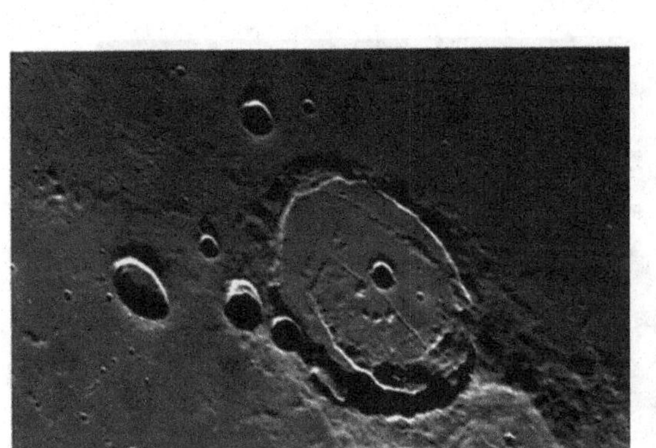

Sun's selenographic colongitude = 339.0°.

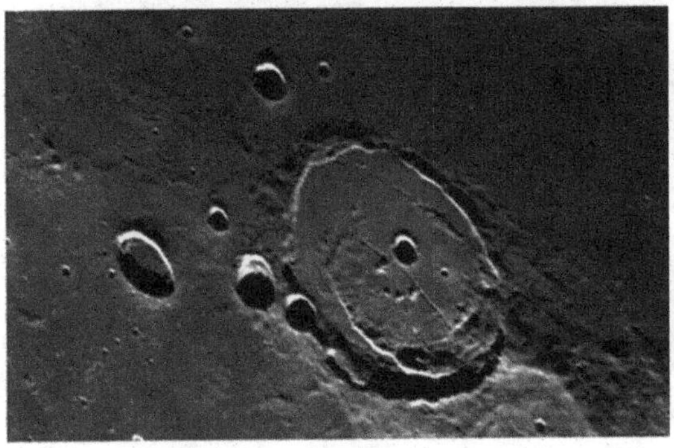

Selected Area 4b – Posidonius (Sunset)

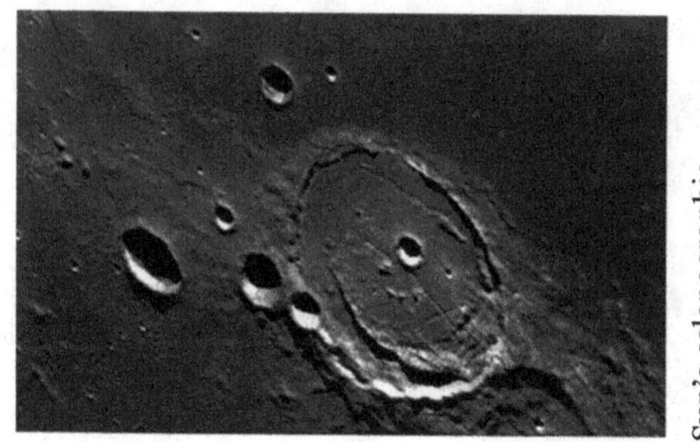

Sun's selenographic colongitude = 141.0°.

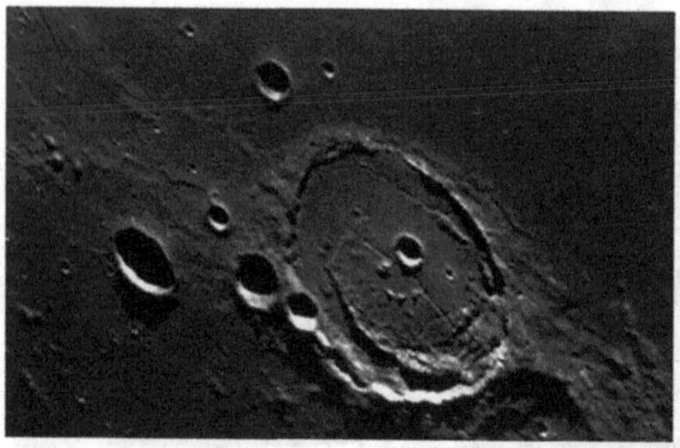

Sun's selenographic colongitude = 153.0°.

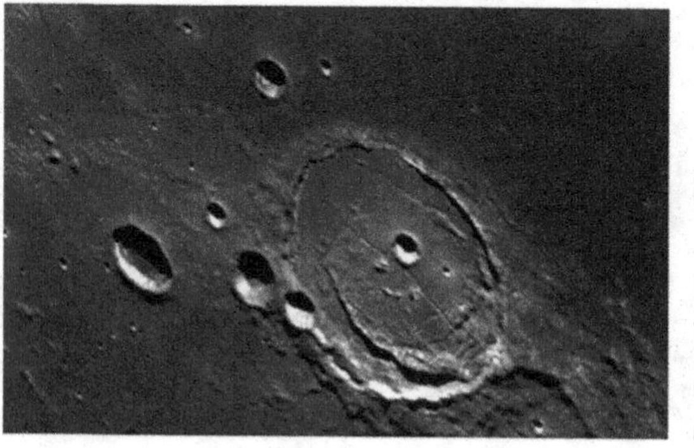

Sun's selenographic colongitude = 138.0°.

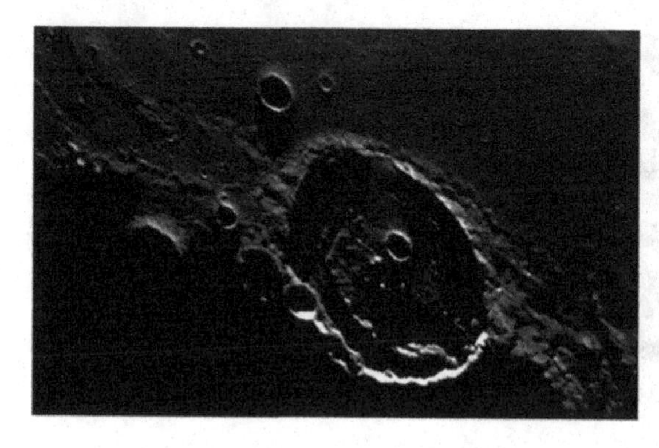

Sun's selenographic colongitude = 150.0°.

Sun's selenographic colongitude = 135.0°.

Sun's selenographic colongitude = 147.0°.

Sun's selenographic colongitude = 132.0°.

Sun's selenographic colongitude = 144.0°.

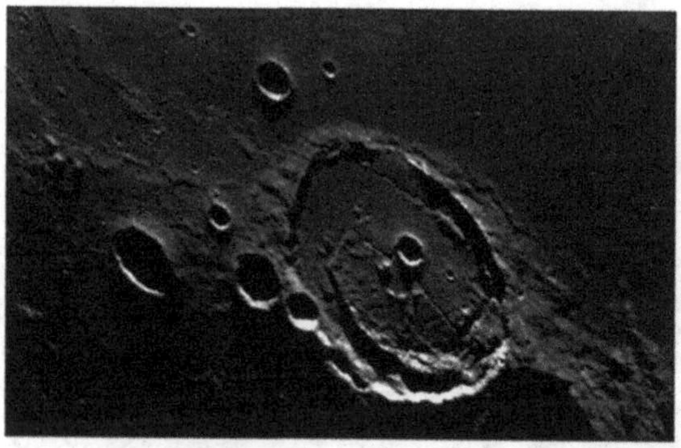

Selected Area 5a – Copernicus (Sunrise)

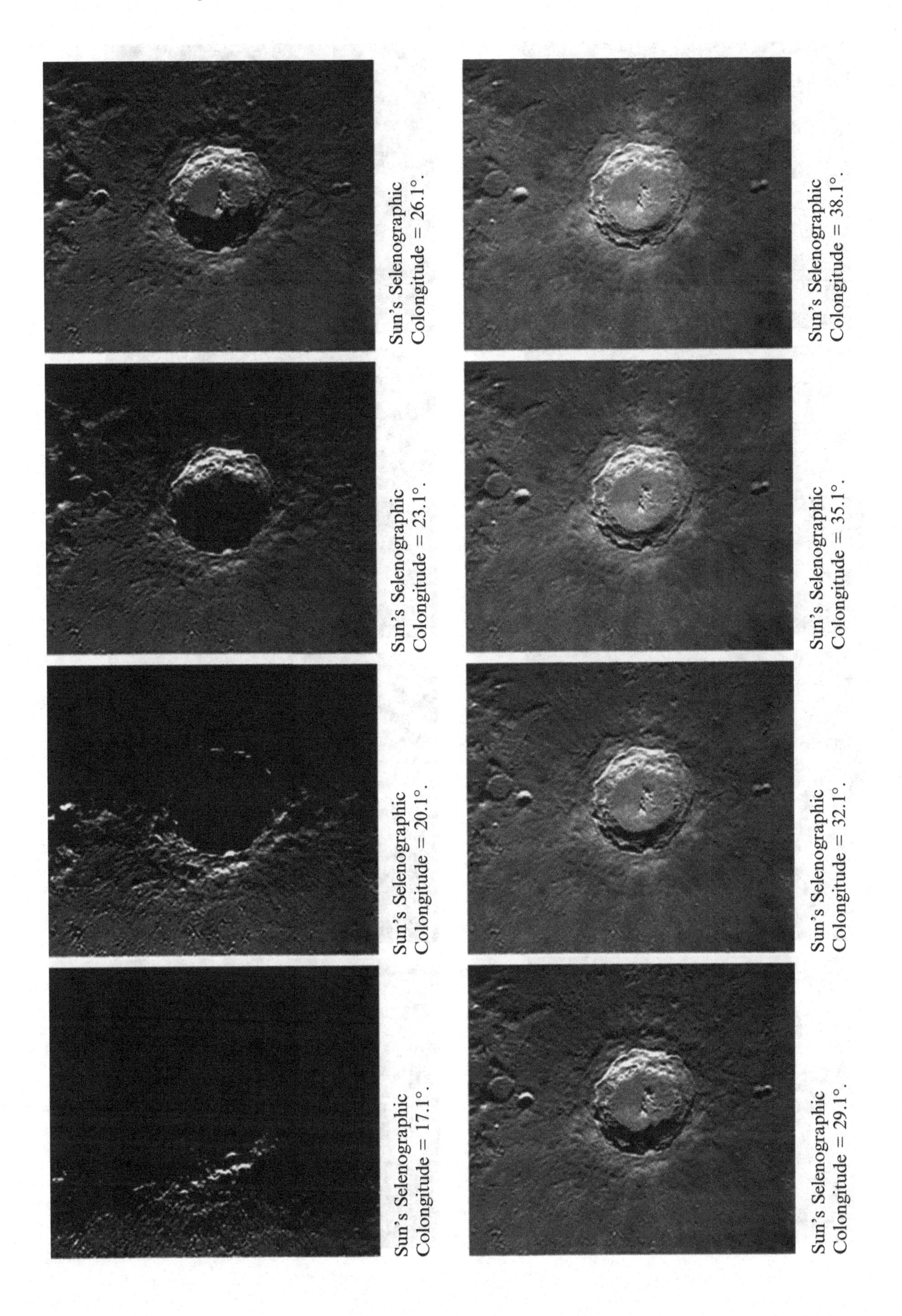

Sun's Selenographic Colongitude = 17.1°.

Sun's Selenographic Colongitude = 20.1°.

Sun's Selenographic Colongitude = 23.1°.

Sun's Selenographic Colongitude = 26.1°.

Sun's Selenographic Colongitude = 29.1°.

Sun's Selenographic Colongitude = 32.1°.

Sun's Selenographic Colongitude = 35.1°.

Sun's Selenographic Colongitude = 38.1°.

Selected Area 5a – Copernicus (Sunset)

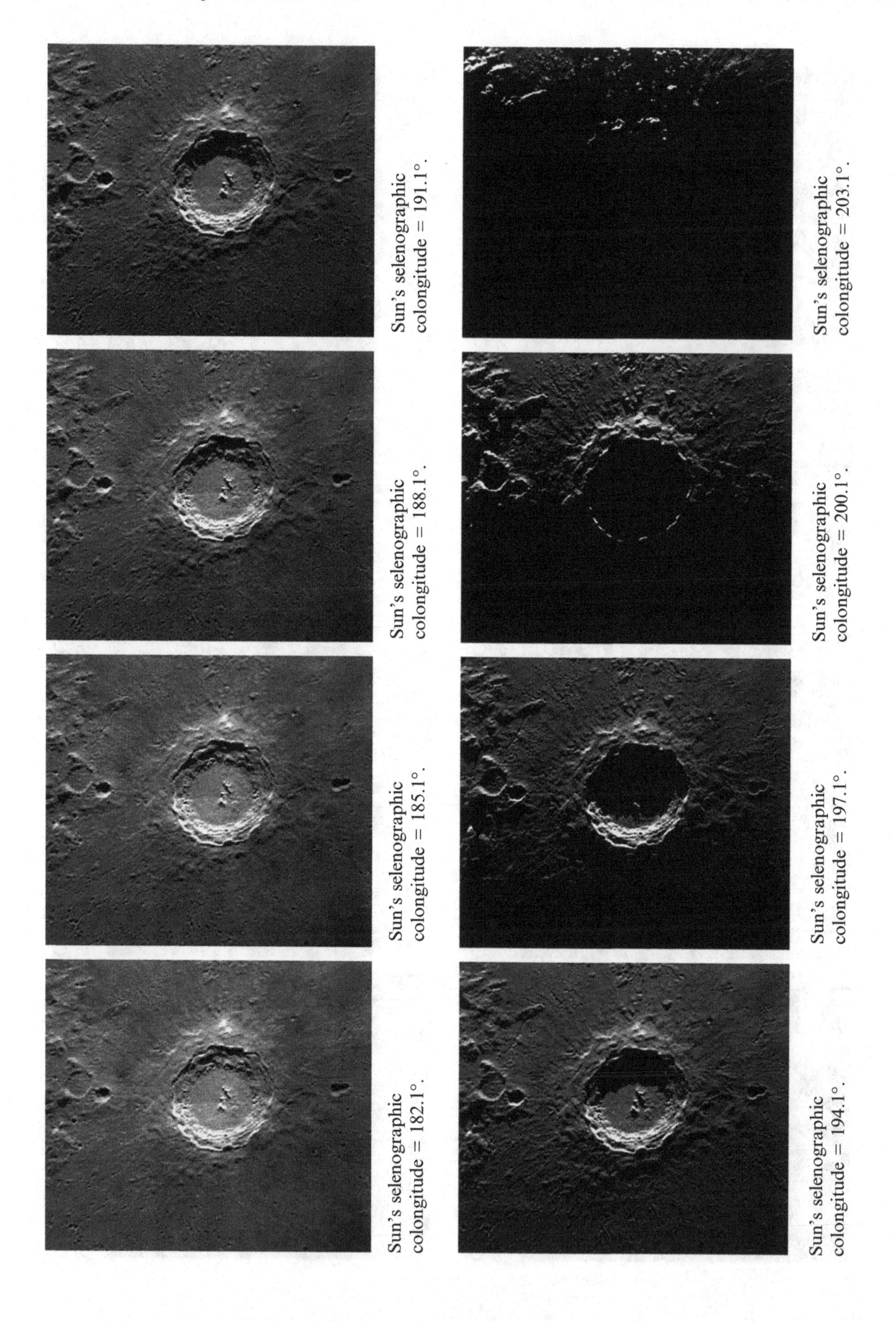

Sun's selenographic colongitude = 182.1°.

Sun's selenographic colongitude = 185.1°.

Sun's selenographic colongitude = 188.1°.

Sun's selenographic colongitude = 191.1°.

Sun's selenographic colongitude = 194.1°.

Sun's selenographic colongitude = 197.1°.

Sun's selenographic colongitude = 200.1°.

Sun's selenographic colongitude = 203.1°.

Selected Area 5b – Schröter (Sunrise)

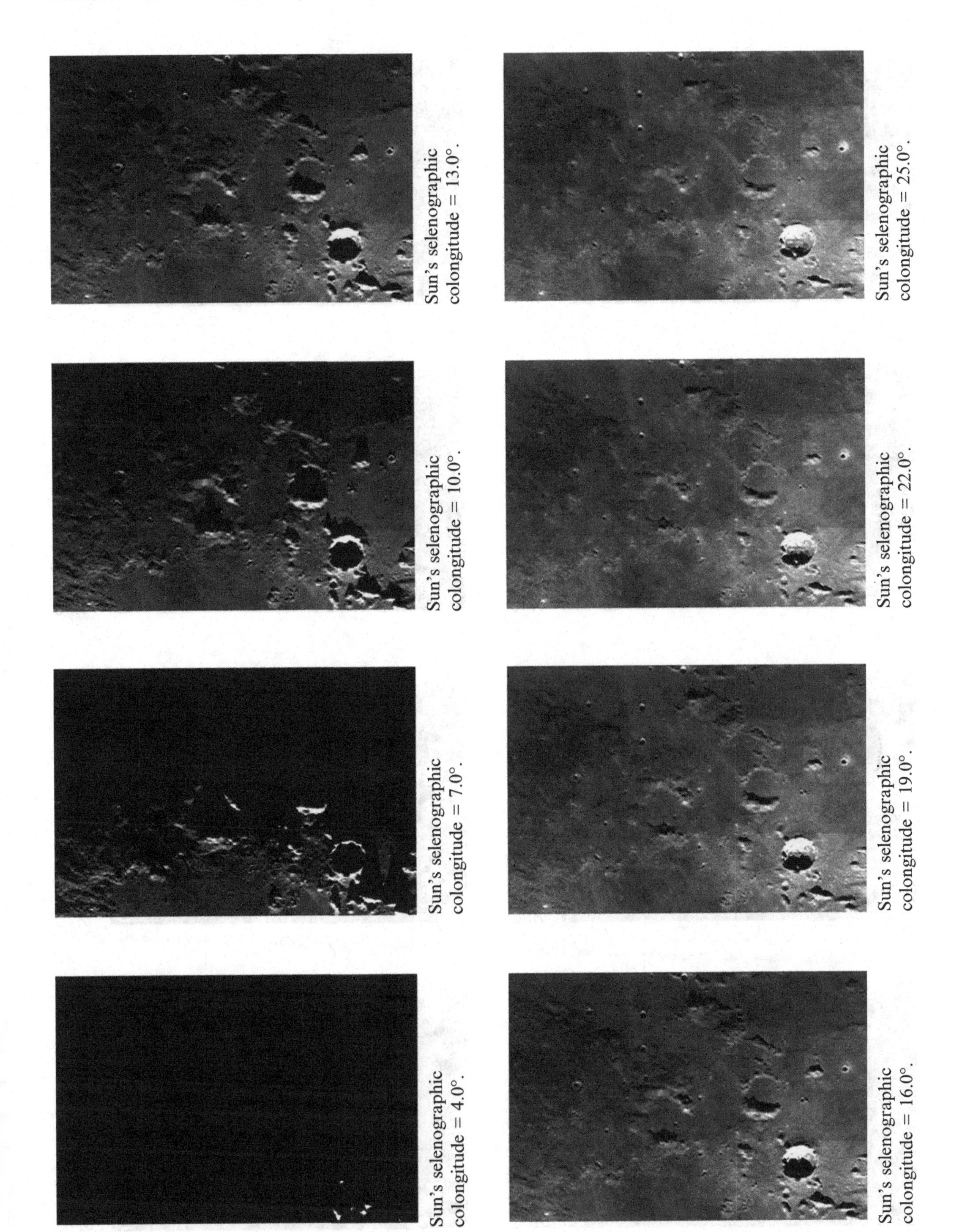

Sun's selenographic colongitude = 4.0°.

Sun's selenographic colongitude = 7.0°.

Sun's selenographic colongitude = 10.0°.

Sun's selenographic colongitude = 13.0°.

Sun's selenographic colongitude = 16.0°.

Sun's selenographic colongitude = 19.0°.

Sun's selenographic colongitude = 22.0°.

Sun's selenographic colongitude = 25.0°.

Selected Area 5b – Schröter (Sunset)

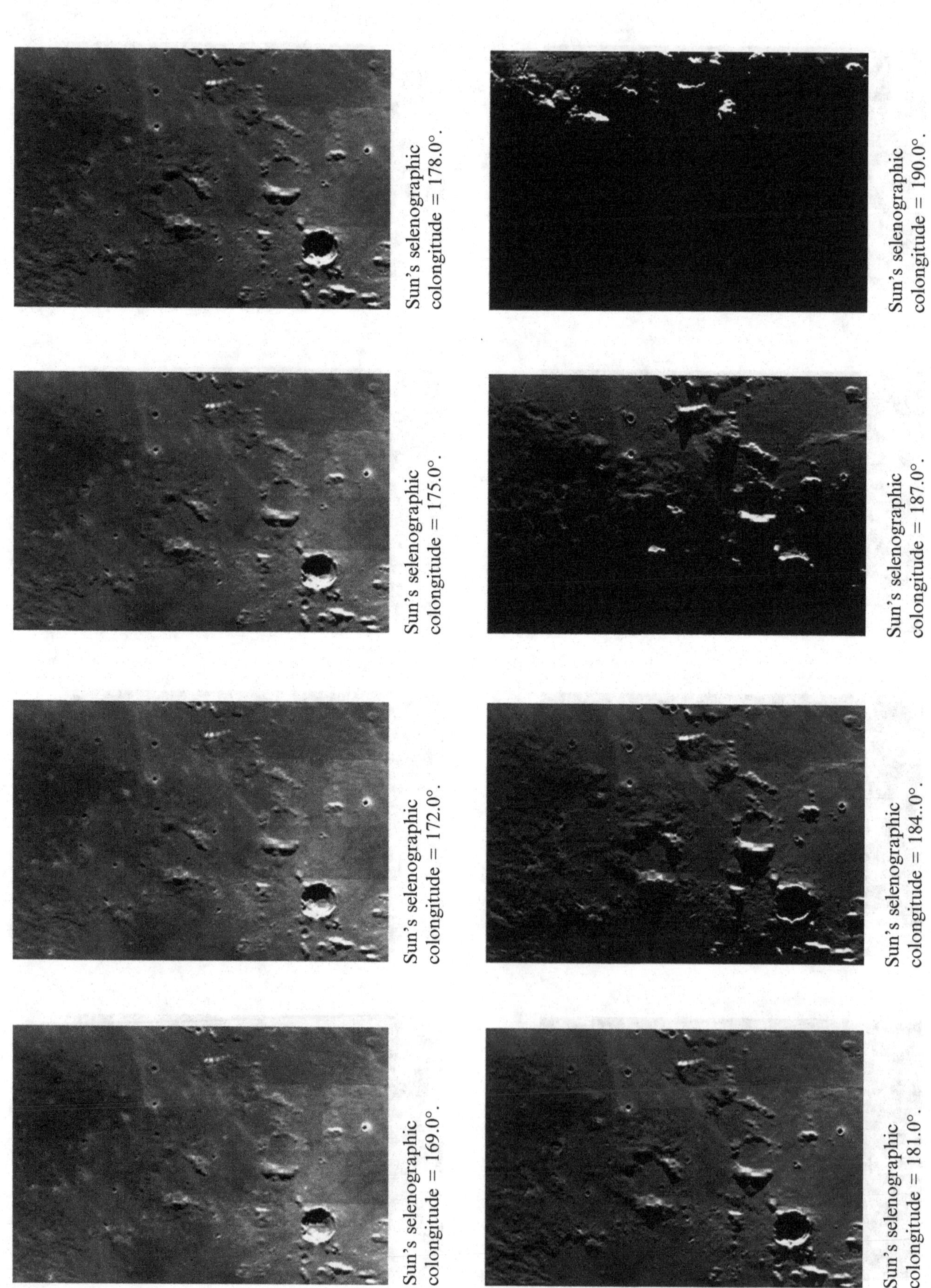

Sun's selenographic colongitude = 178.0°.

Sun's selenographic colongitude = 190.0°.

Sun's selenographic colongitude = 175.0°.

Sun's selenographic colongitude = 187.0°.

Sun's selenographic colongitude = 172.0°.

Sun's selenographic colongitude = 184..0°.

Sun's selenographic colongitude = 169.0°.

Sun's selenographic colongitude = 181.0°.

Selected Area 6a – Plato (Sunrise)

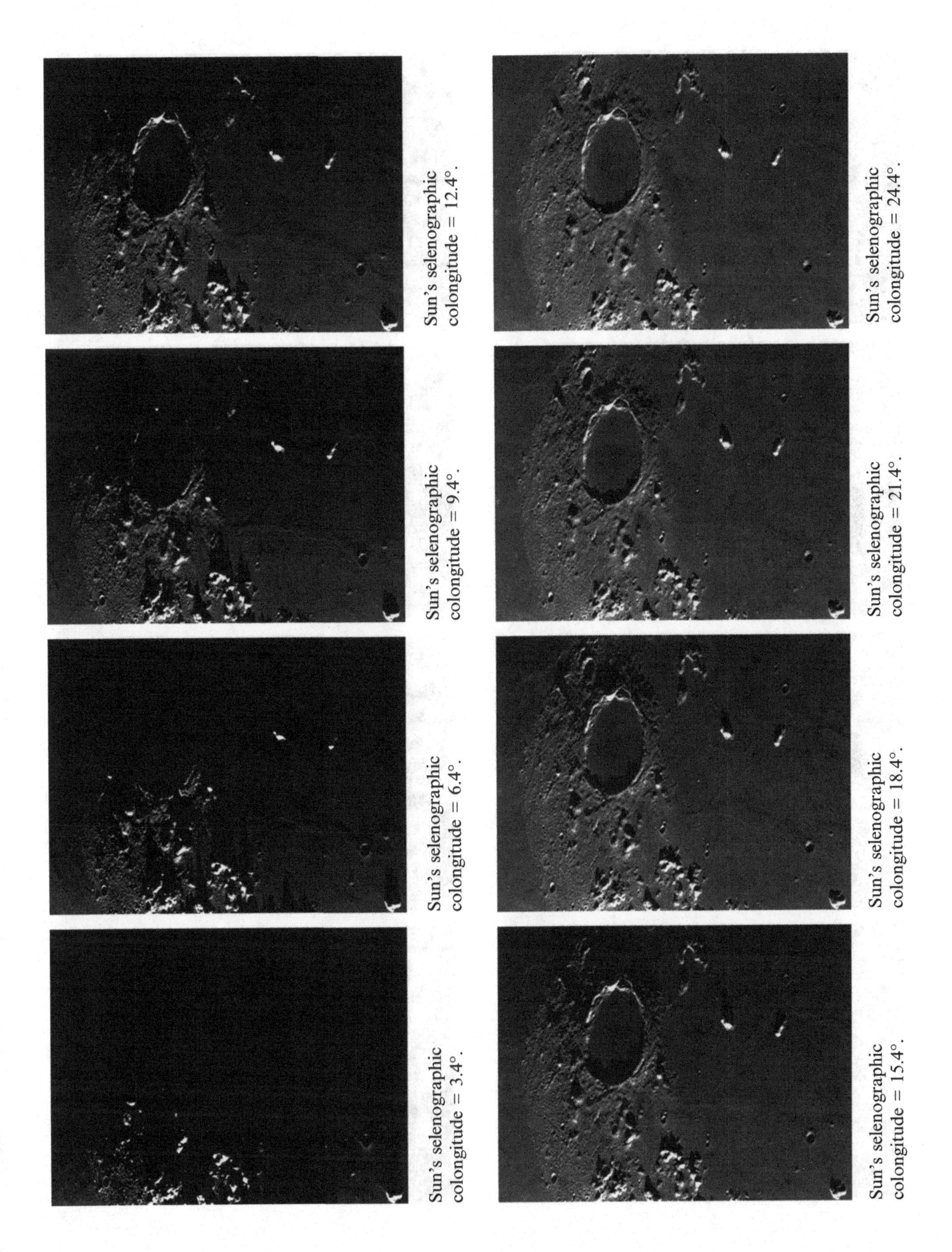

Sun's selenographic colongitude = 3.4°.

Sun's selenographic colongitude = 6.4°.

Sun's selenographic colongitude = 9.4°.

Sun's selenographic colongitude = 12.4°.

Sun's selenographic colongitude = 15.4°.

Sun's selenographic colongitude = 18.4°.

Sun's selenographic colongitude = 21.4°.

Sun's selenographic colongitude = 24.4°.

Selected Area 6a – Plato (Sunset)

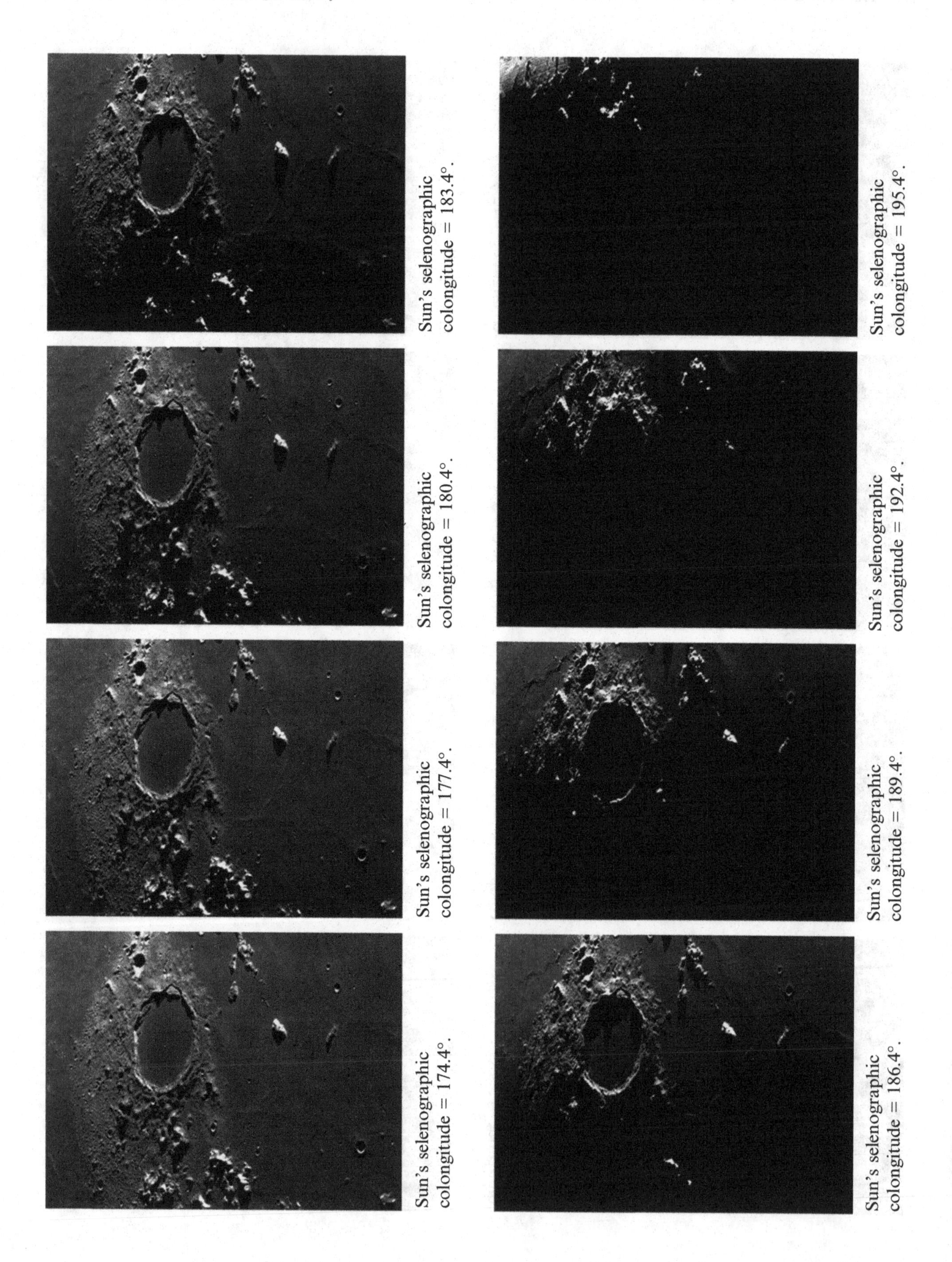

Sun's selenographic colongitude = 174.4°.

Sun's selenographic colongitude = 177.4°.

Sun's selenographic colongitude = 180.4°.

Sun's selenographic colongitude = 183.4°.

Sun's selenographic colongitude = 186.4°.

Sun's selenographic colongitude = 189.4°.

Sun's selenographic colongitude = 192.4°.

Sun's selenographic colongitude = 195.4°.

Selected Area 6b – Sinus Iridum (Sunrise)

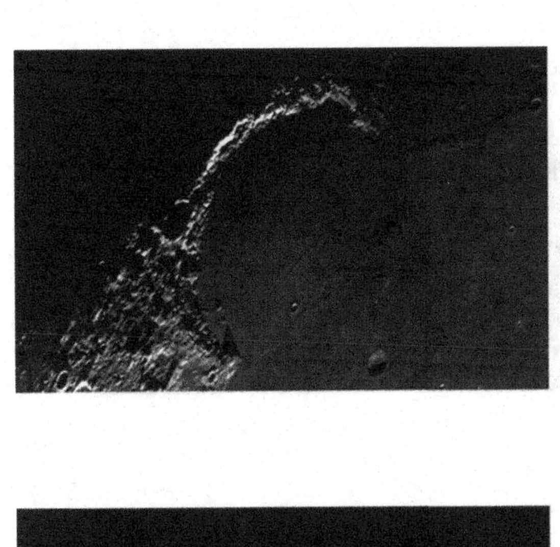

Sun's selenographic colongitude = 34.7°.

Sun's selenographic colongitude = 46.7°.

Sun's selenographic colongitude = 31.8°.

Sun's selenographic colongitude = 43.7°.

Sun's selenographic colongitude = 28.7°.

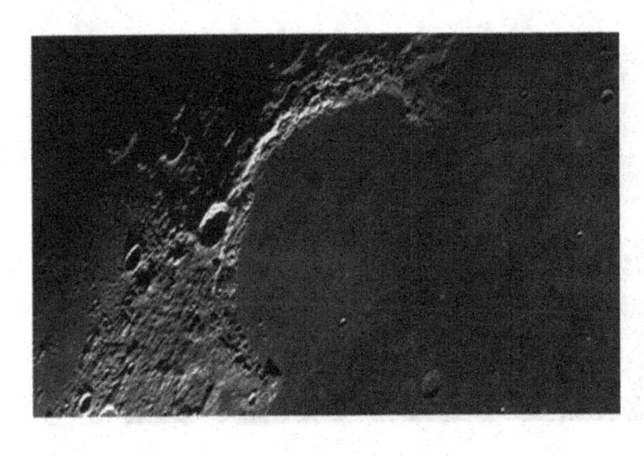

Sun's selenographic colongitude = 40.7°.

Sun's selenographic colongitude = 25.7°.

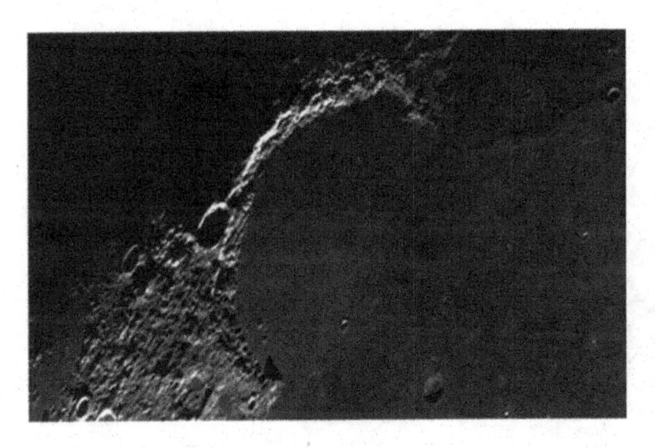

Sun's selenographic colongitude = 37.7°.

Selected Area 6b – Sinus Iridum (Sunset)

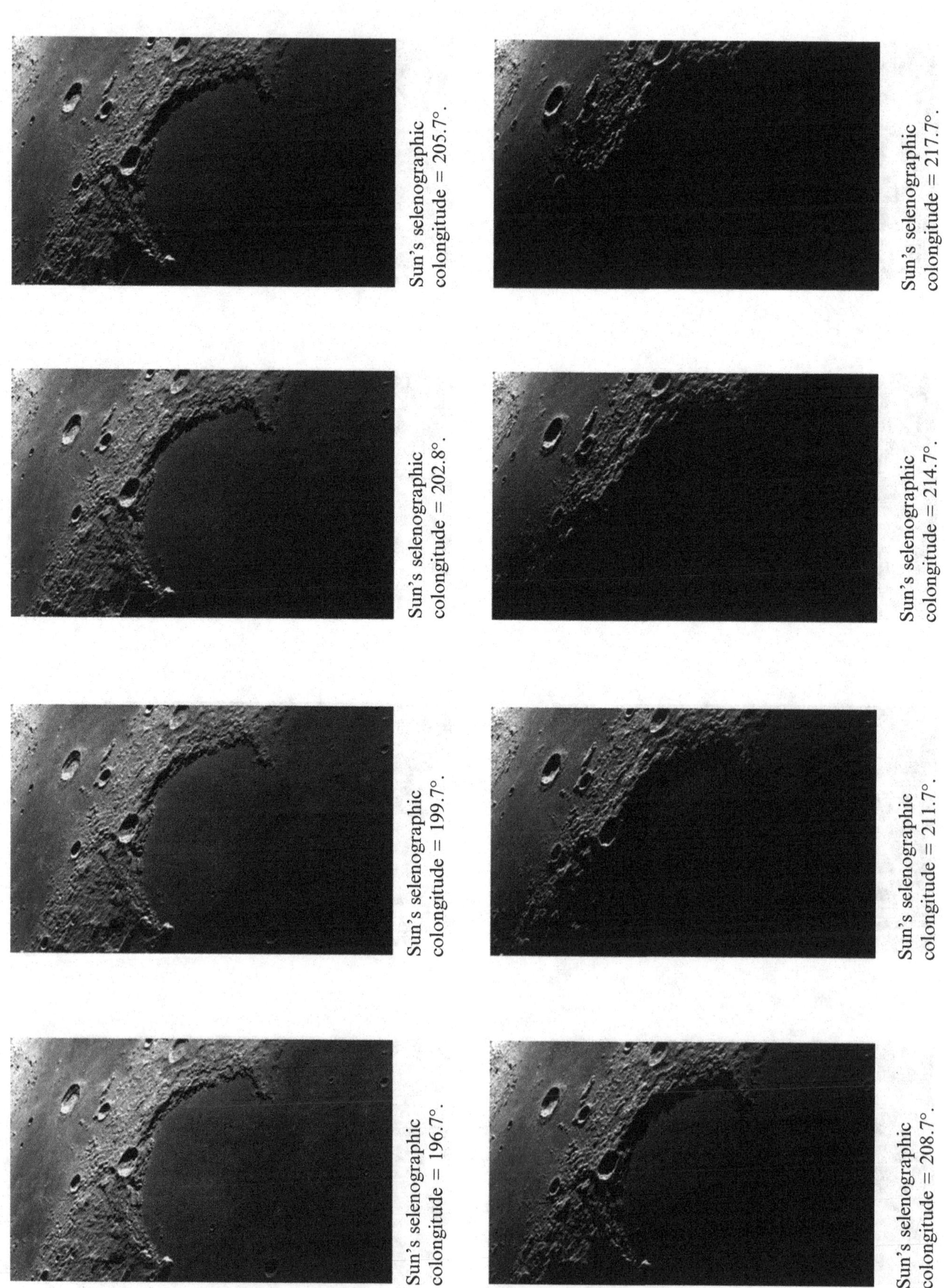

Sun's selenographic colongitude = 205.7°.

Sun's selenographic colongitude = 202.8°.

Sun's selenographic colongitude = 199.7°.

Sun's selenographic colongitude = 196.7°.

Sun's selenographic colongitude = 217.7°.

Sun's selenographic colongitude = 214.7°.

Sun's selenographic colongitude = 211.7°.

Sun's selenographic colongitude = 208.7°.

Selected Area 7a – Grimaldi (Sunrise)

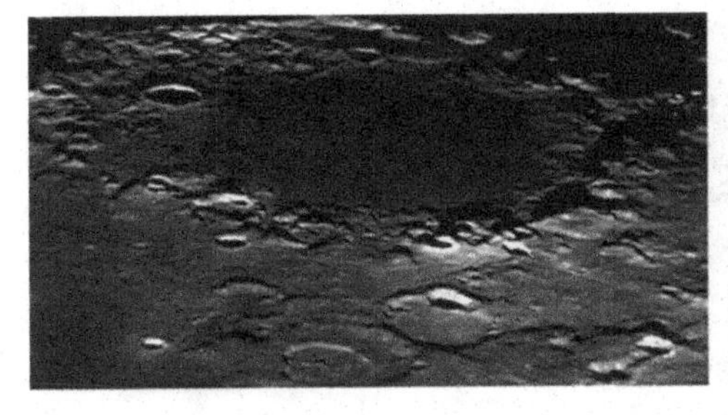

Sun's selenographic colongitude = 71.4°.

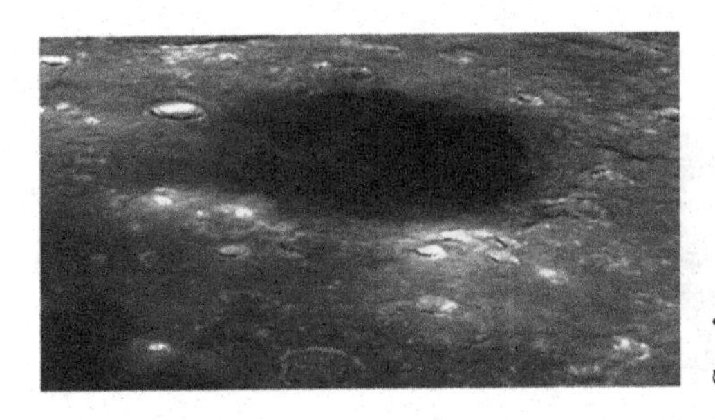

Sun's selenographic colongitude = 83.4°.

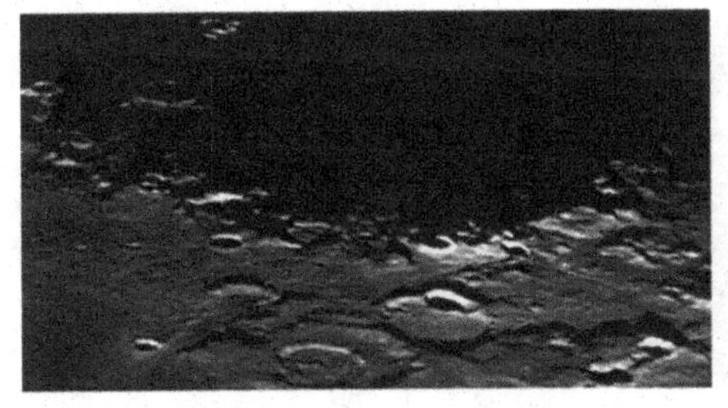

Sun's selenographic colongitude = 68.4°.

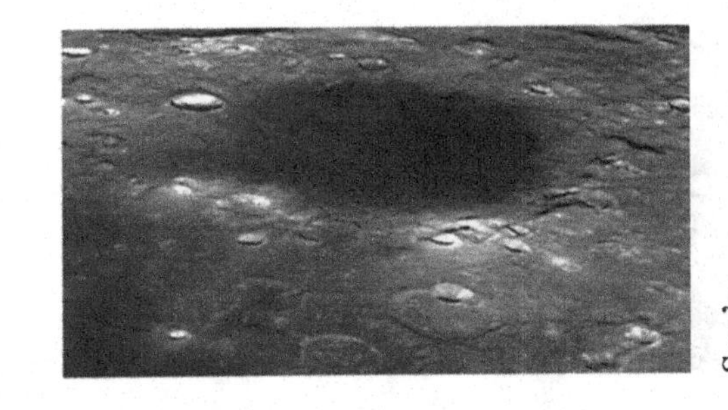

Sun's selenographic colongitude = 80.4°.

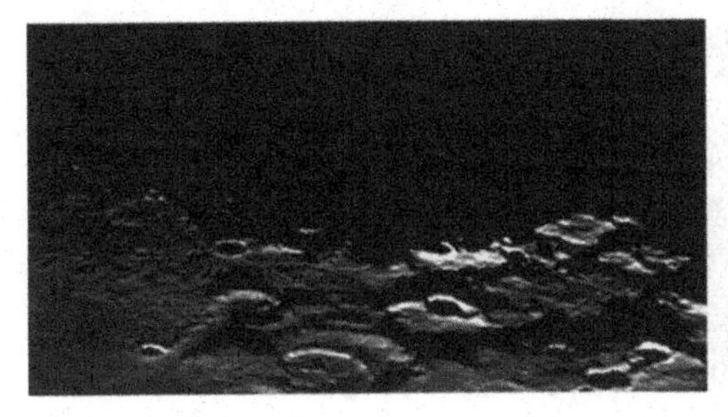

Sun's selenographic colongitude = 65.4°.

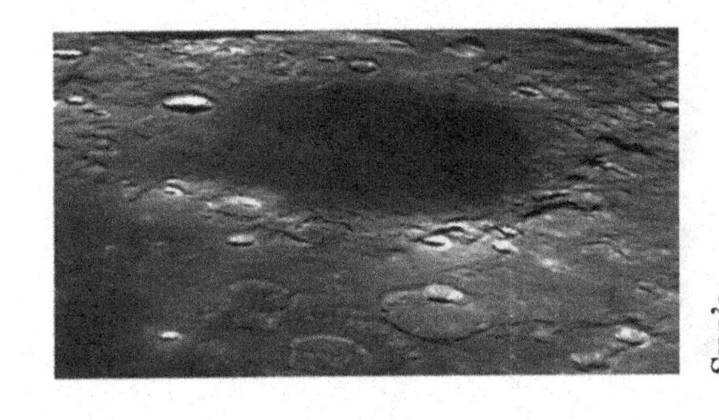

Sun's selenographic colongitude = 77.4°.

Sun's selenographic colongitude = 62.4°.

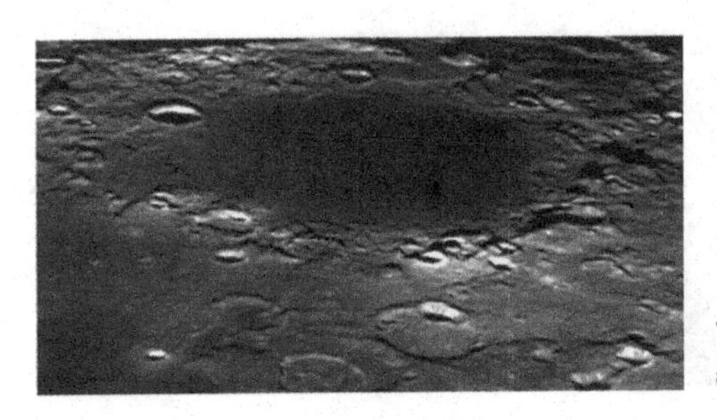

Sun's selenographic colongitude = 74.4°.

Selected Area 7a – Grimaldi (Sunset)

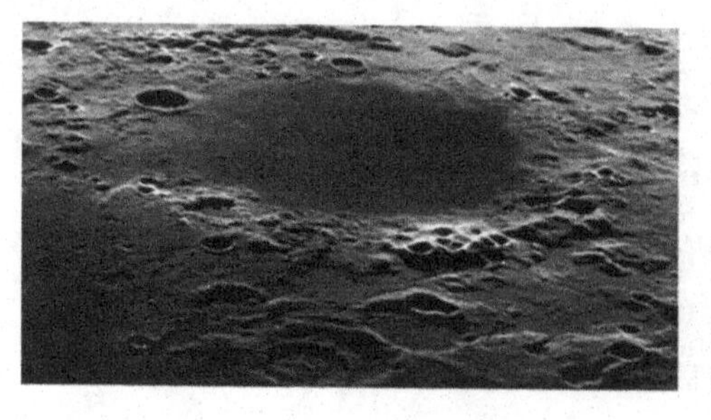

Sun's selenographic colongitude = 236.4°.

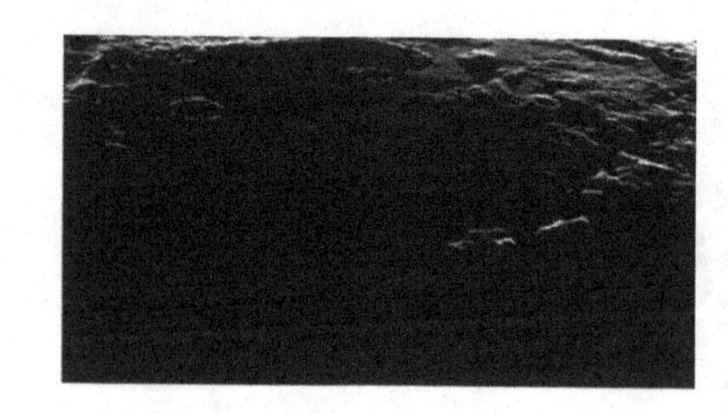

Sun's selenographic colongitude = 248.4°.

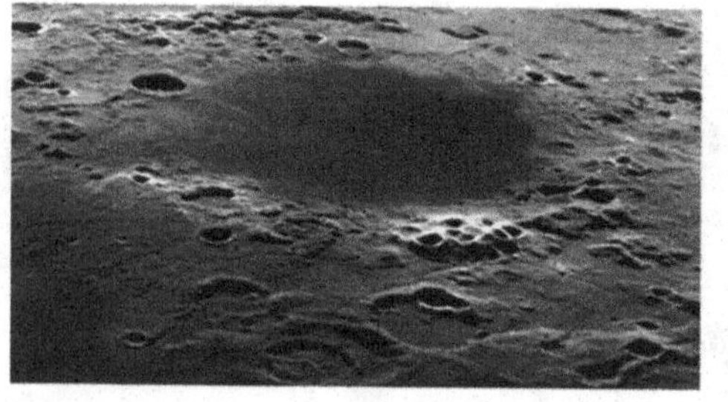

Sun's selenographic colongitude = 233.4°.

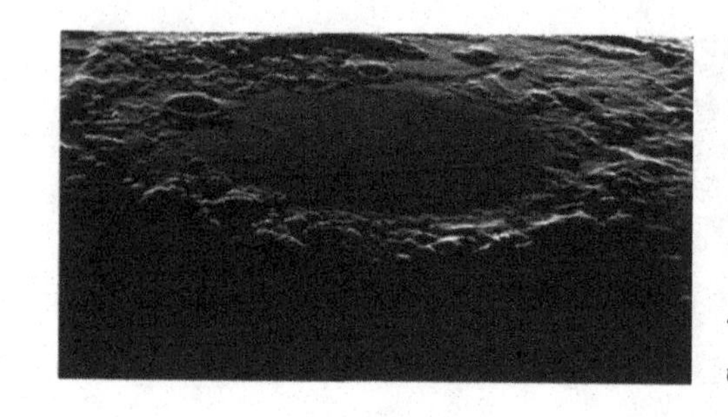

Sun's selenographic colongitude = 245.4°.

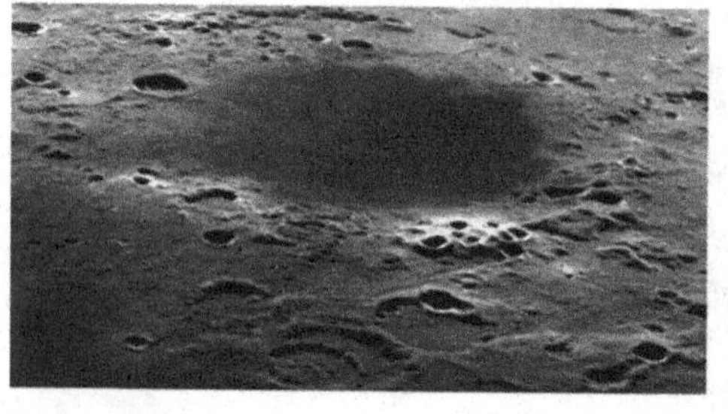

Sun's selenographic colongitude = 230.4°.

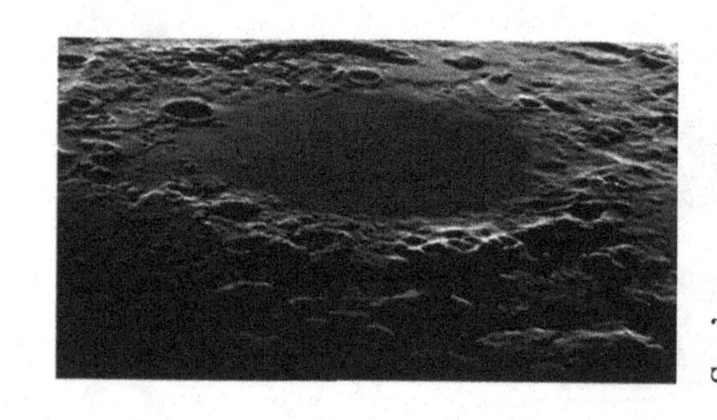

Sun's selenographic colongitude = 242.4°.

Sun's selenographic colongitude = 227.4°.

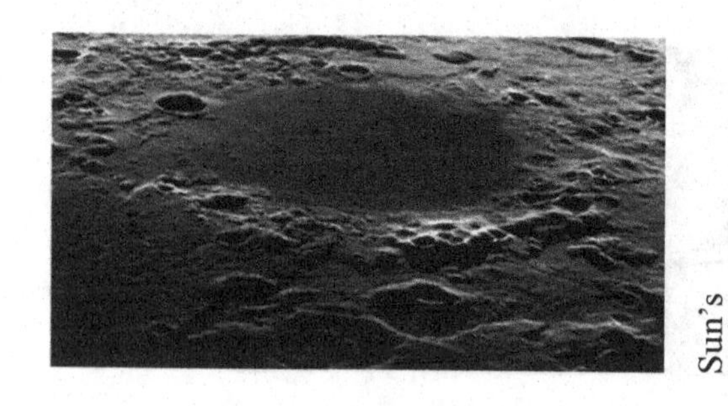

Sun's selenographic colongitude = 239.4°.

Selected Area 7b – Kepler (Sunrise)

Sun's selenographic colongitude = 44.0°.

Sun's selenographic colongitude = 56.0°.

Sun's selenographic colongitude = 41.0°.

Sun's selenographic colongitude = 53.0°.

Sun's selenographic colongitude = 38.0°.

Sun's selenographic colongitude = 50.0°.

Sun's selenographic colongitude = 35.0°.

Sun's selenographic colongitude = 47.0°.

Selected Area 7b – Kepler (Sunset)

Sun's selenographic colongitude = 209.0°.

Sun's selenographic colongitude = 206.0°.

Sun's selenographic colongitude = 203.0°.

Sun's selenographic colongitude = 200.0°.

Sun's selenographic colongitude = 221.0°.

Sun's selenographic colongitude = 218.0°.

Sun's selenographic colongitude = 215.0°.

Sun's selenographic colongitude = 212.0°.

Selected Area 8a – Aristarchus (Sunrise)

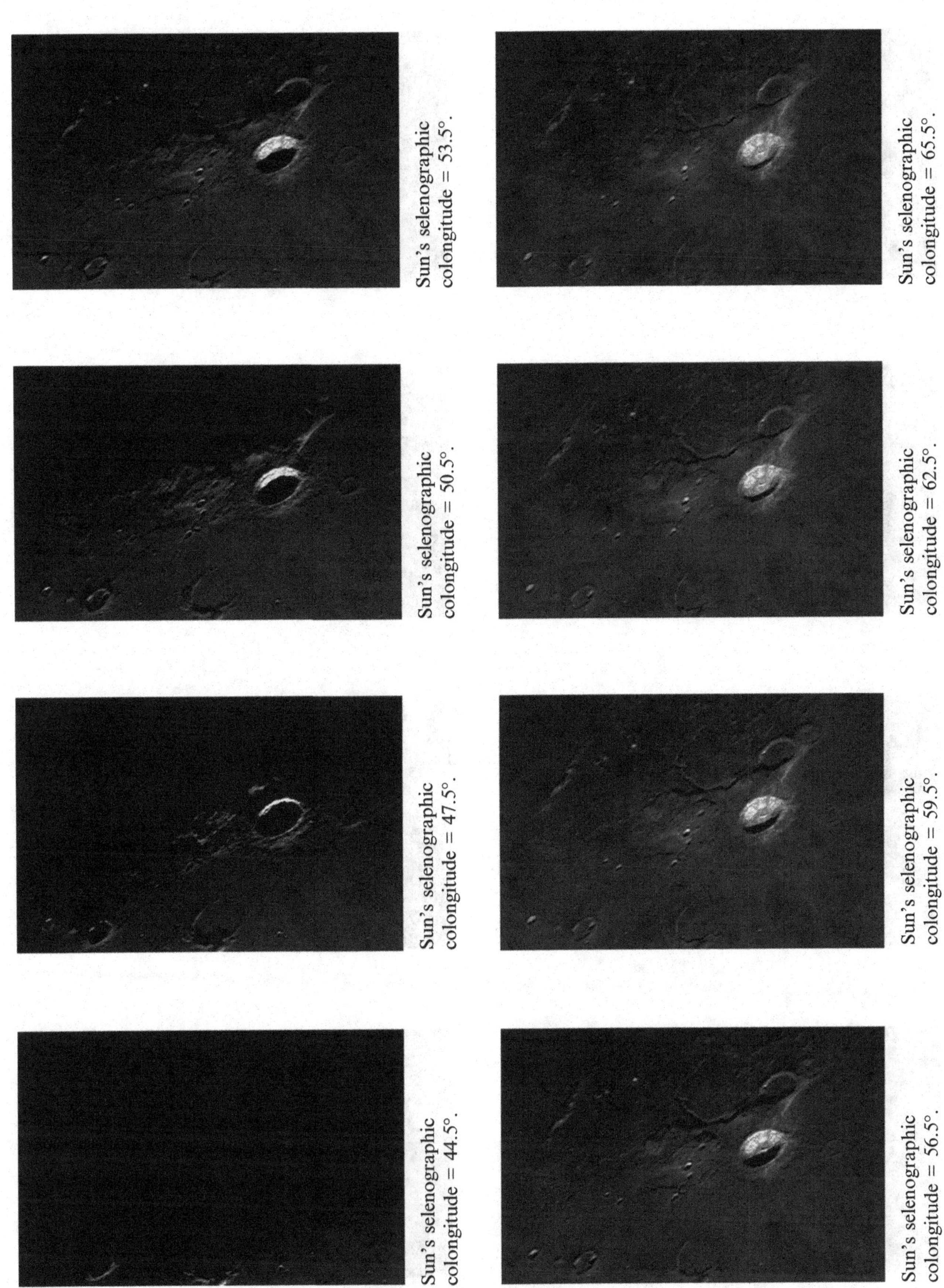

Sun's selenographic colongitude = 44.5°.

Sun's selenographic colongitude = 47.5°.

Sun's selenographic colongitude = 50.5°.

Sun's selenographic colongitude = 53.5°.

Sun's selenographic colongitude = 56.5°.

Sun's selenographic colongitude = 59.5°.

Sun's selenographic colongitude = 62.5°.

Sun's selenographic colongitude = 65.5°.

Selected Area 8a – Aristarchus (Sunset)

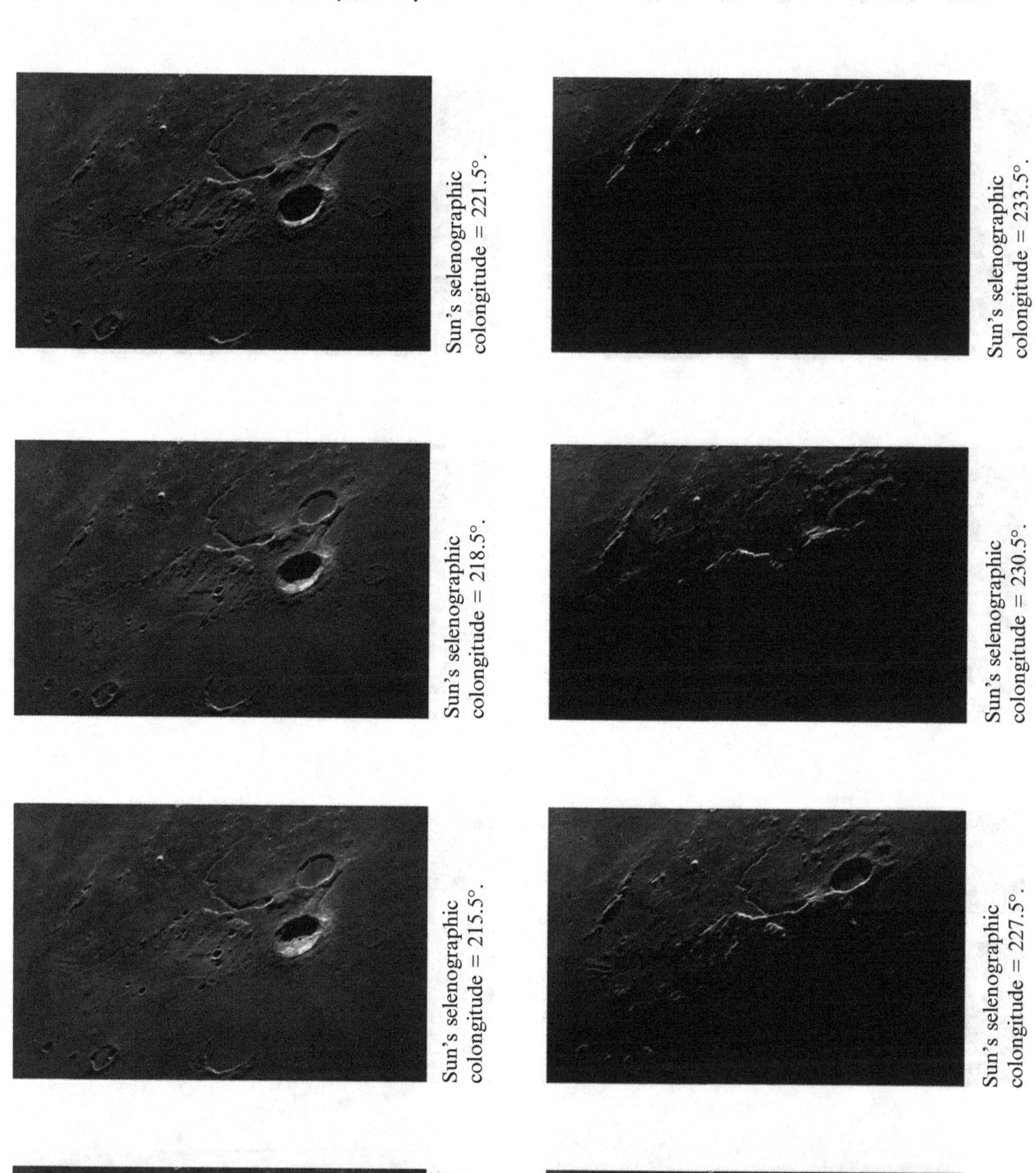

Sun's selenographic colongitude = 221.5°.

Sun's selenographic colongitude = 233.5°.

Sun's selenographic colongitude = 218.5°.

Sun's selenographic colongitude = 230.5°.

Sun's selenographic colongitude = 215.5°.

Sun's selenographic colongitude = 227.5°.

Sun's selenographic colongitude = 212.5°.

Sun's selenographic colongitude = 224.5°.

Selected Area 8b – Mons Rümker (Sunrise)

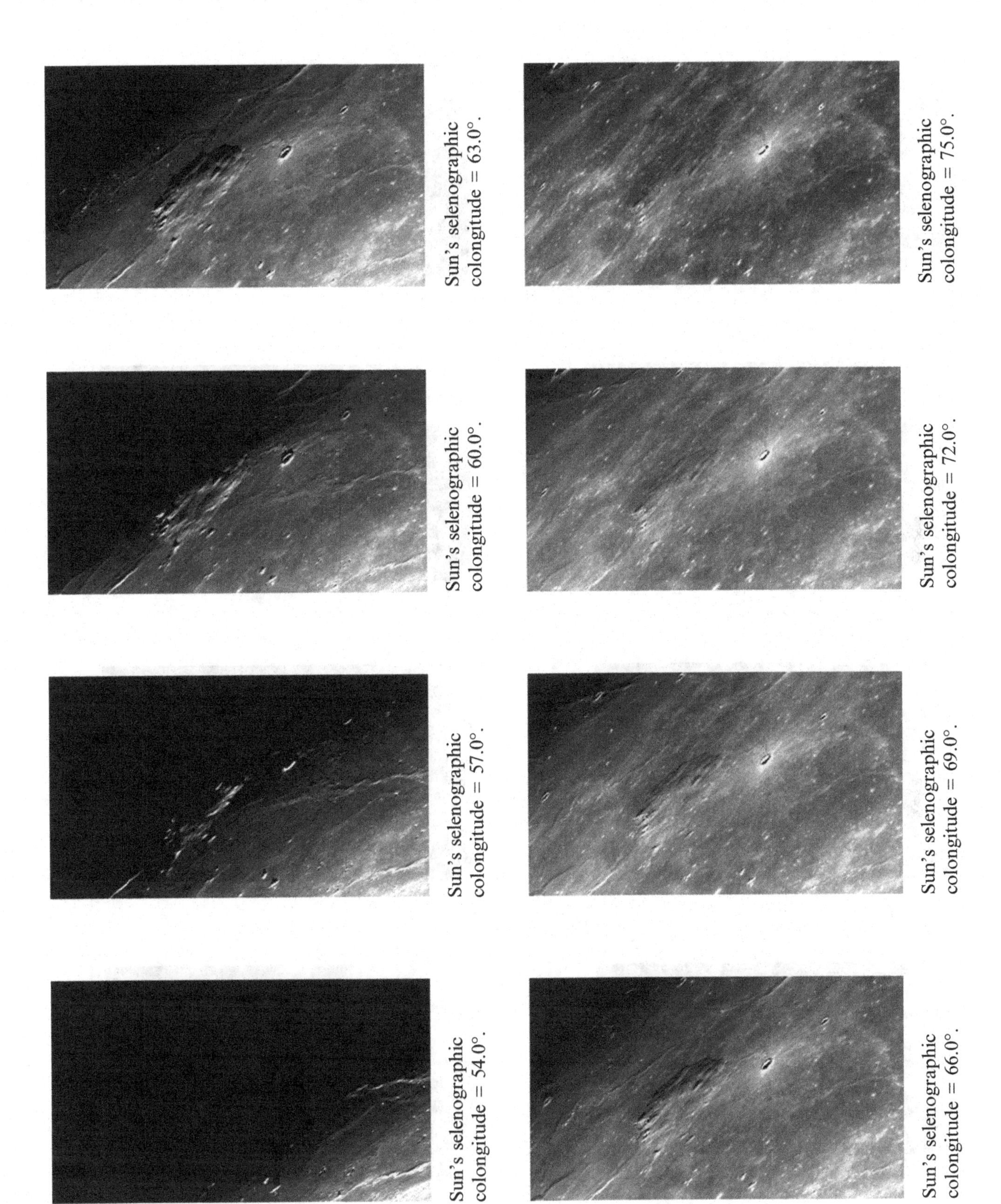

Sun's selenographic colongitude = 54.0°.

Sun's selenographic colongitude = 57.0°.

Sun's selenographic colongitude = 60.0°.

Sun's selenographic colongitude = 63.0°.

Sun's selenographic colongitude = 66.0°.

Sun's selenographic colongitude = 69.0°.

Sun's selenographic colongitude = 72.0°.

Sun's selenographic colongitude = 75.0°.

Selected Area 8b – Mons Rümker (Sunset)

Sun's selenographic colongitude = 222.0°.

Sun's selenographic colongitude = 225.0°.

Sun's selenographic colongitude = 228.0°.

Sun's selenographic colongitude = 231.0°.

Sun's selenographic colongitude = 234.0°.

Sun's selenographic colongitude = 237.0°.

Sun's selenographic colongitude = 240.0°.

Sun's selenographic colongitude = 243.0°.

Selected Area 9a – Bullialdus (Sunrise)

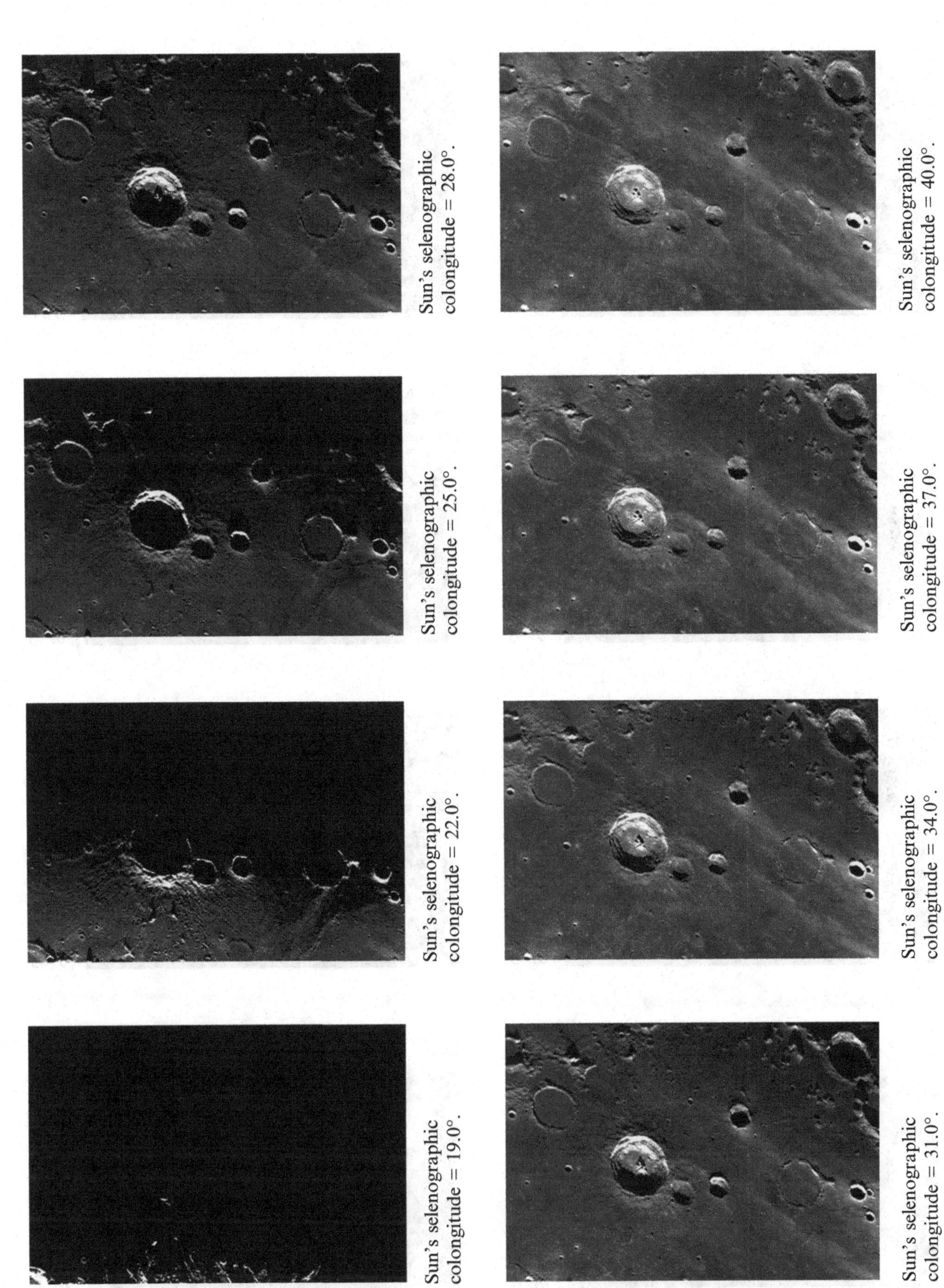

Sun's selenographic colongitude = 28.0°.

Sun's selenographic colongitude = 25.0°.

Sun's selenographic colongitude = 22.0°.

Sun's selenographic colongitude = 19.0°.

Sun's selenographic colongitude = 40.0°.

Sun's selenographic colongitude = 37.0°.

Sun's selenographic colongitude = 34.0°.

Sun's selenographic colongitude = 31.0°.

Selected Area 9a – Bullialdus (Sunset)

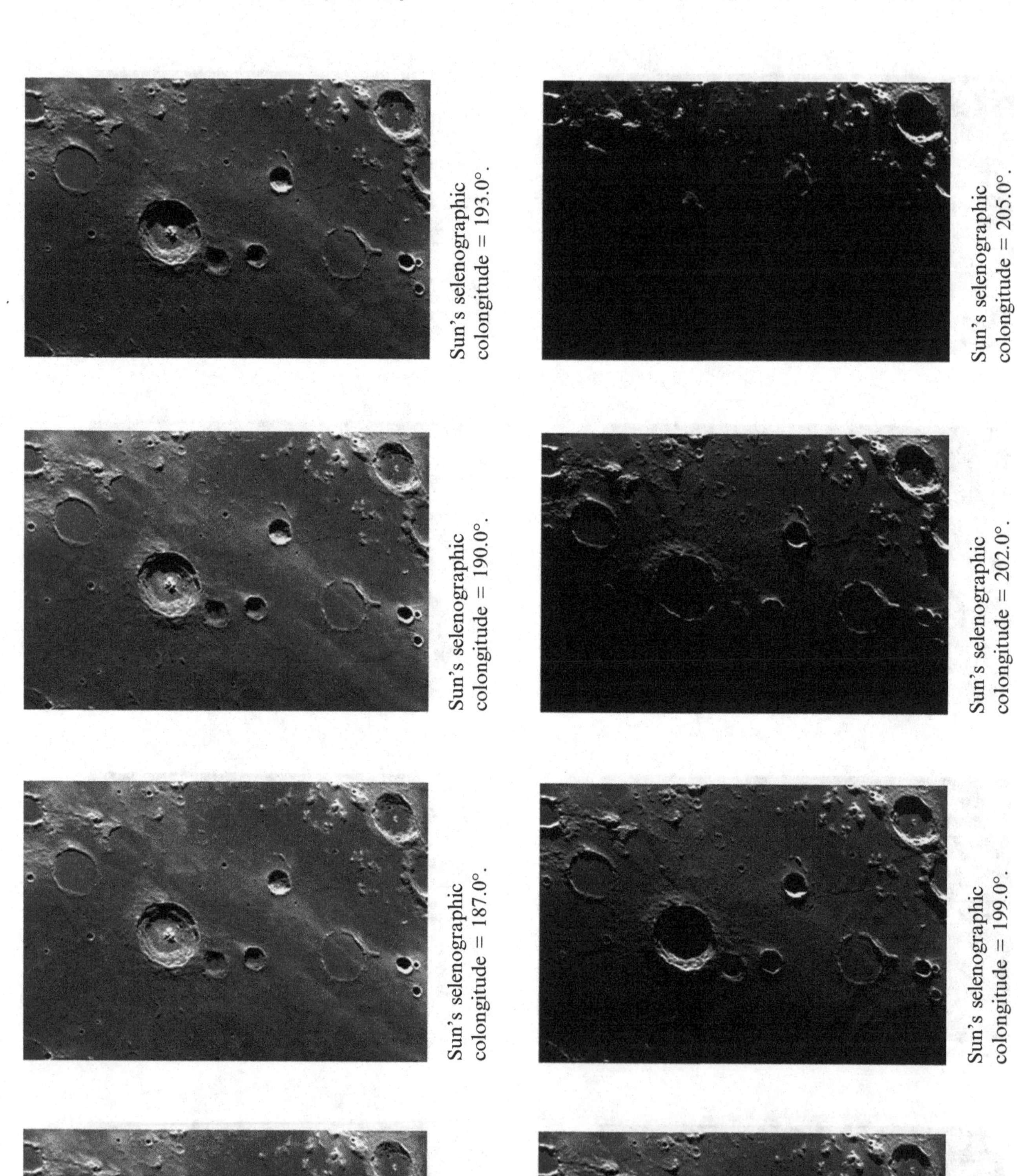

Sun's selenographic colongitude = 193.0°.

Sun's selenographic colongitude = 205.0°.

Sun's selenographic colongitude = 190.0°.

Sun's selenographic colongitude = 202.0°.

Sun's selenographic colongitude = 187.0°.

Sun's selenographic colongitude = 199.0°.

Sun's selenographic colongitude = 184.0°.

Sun's selenographic colongitude = 196.0°.

Selected Area 9b – Montes Riphaeus (Sunrise)

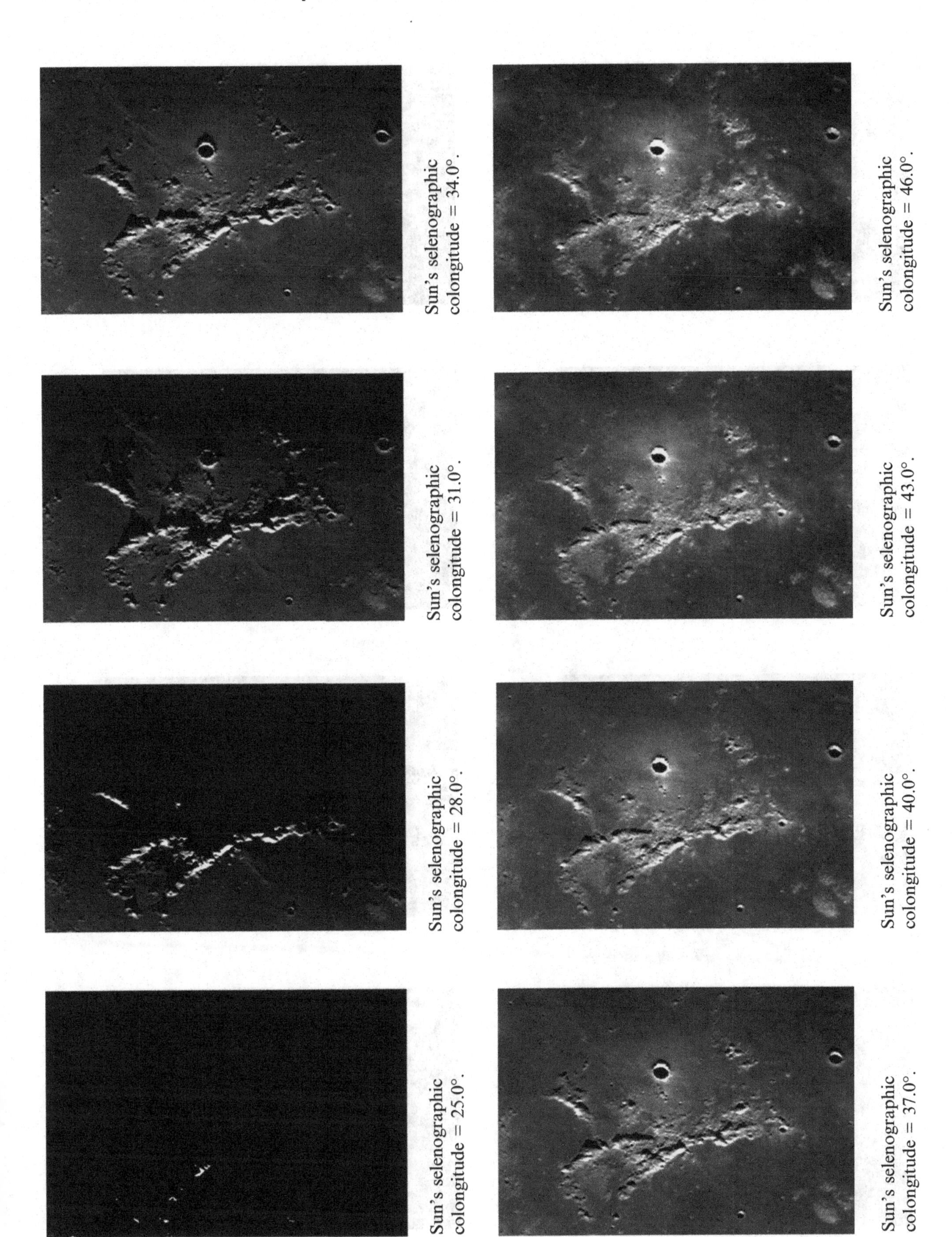

Sun's selenographic colongitude = 34.0°.

Sun's selenographic colongitude = 46.0°.

Sun's selenographic colongitude = 31.0°.

Sun's selenographic colongitude = 43.0°.

Sun's selenographic colongitude = 28.0°.

Sun's selenographic colongitude = 40.0°.

Sun's selenographic colongitude = 25.0°.

Sun's selenographic colongitude = 37.0°.

Selected Area 9b – Montes Riphaeus (Sunset)

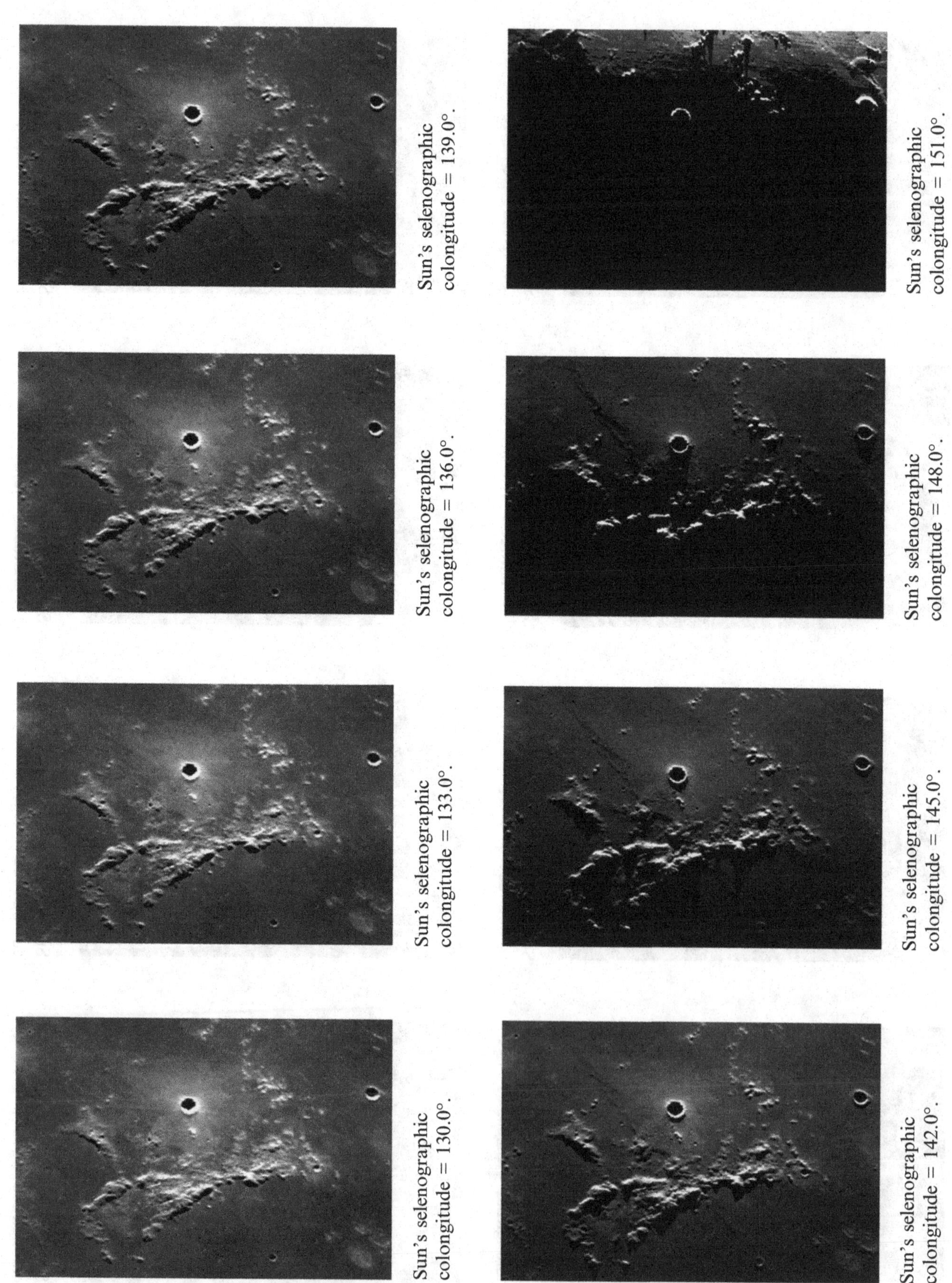

Sun's selenographic colongitude = 139.0°.

Sun's selenographic colongitude = 151.0°.

Sun's selenographic colongitude = 136.0°.

Sun's selenographic colongitude = 148.0°.

Sun's selenographic colongitude = 133.0°.

Sun's selenographic colongitude = 145.0°.

Sun's selenographic colongitude = 130.0°.

Sun's selenographic colongitude = 142.0°.

Selected Area 10a – Tycho (Sunrise)

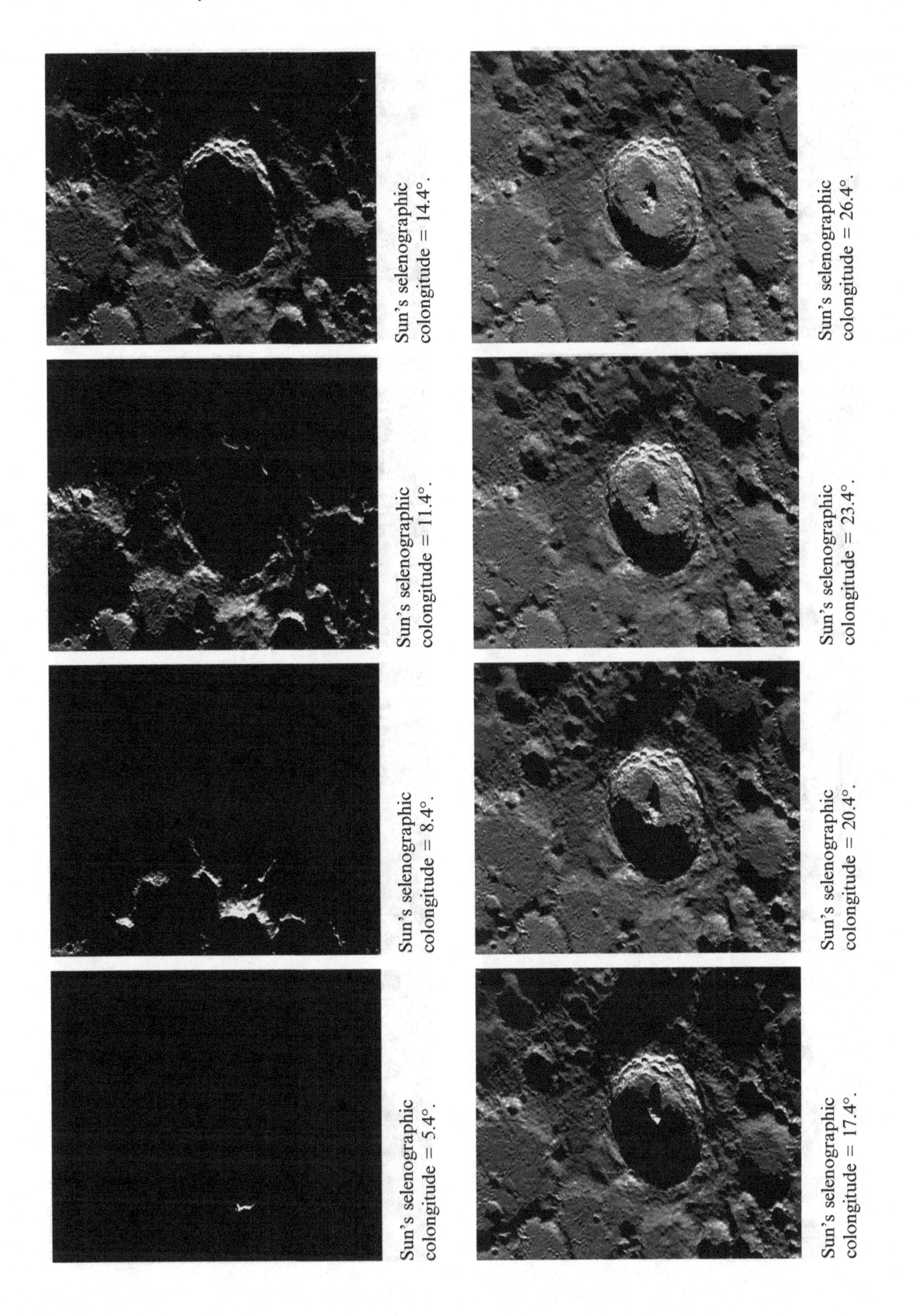

Sun's selenographic colongitude = 5.4°.

Sun's selenographic colongitude = 8.4°.

Sun's selenographic colongitude = 11.4°.

Sun's selenographic colongitude = 14.4°.

Sun's selenographic colongitude = 17.4°.

Sun's selenographic colongitude = 20.4°.

Sun's selenographic colongitude = 23.4°.

Sun's selenographic colongitude = 26.4°.

Selected Area 10a – Tycho (Sunset)

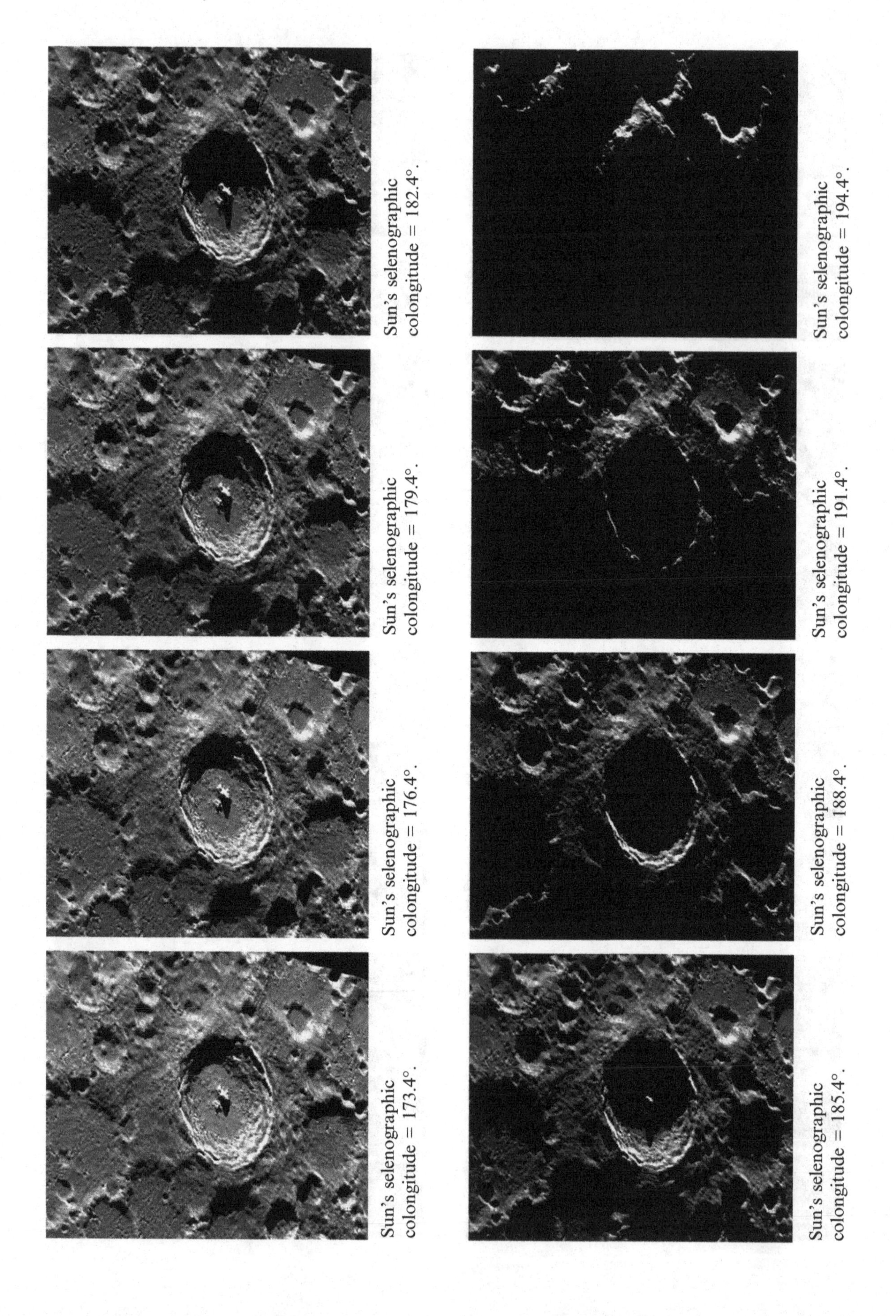

Sun's selenographic colongitude = 173.4°.

Sun's selenographic colongitude = 176.4°.

Sun's selenographic colongitude = 179.4°.

Sun's selenographic colongitude = 182.4°.

Sun's selenographic colongitude = 185.4°.

Sun's selenographic colongitude = 188.4°.

Sun's selenographic colongitude = 191.4°.

Sun's selenographic colongitude = 194.4°.

Selected Area 10b – Werner (Sunrise)

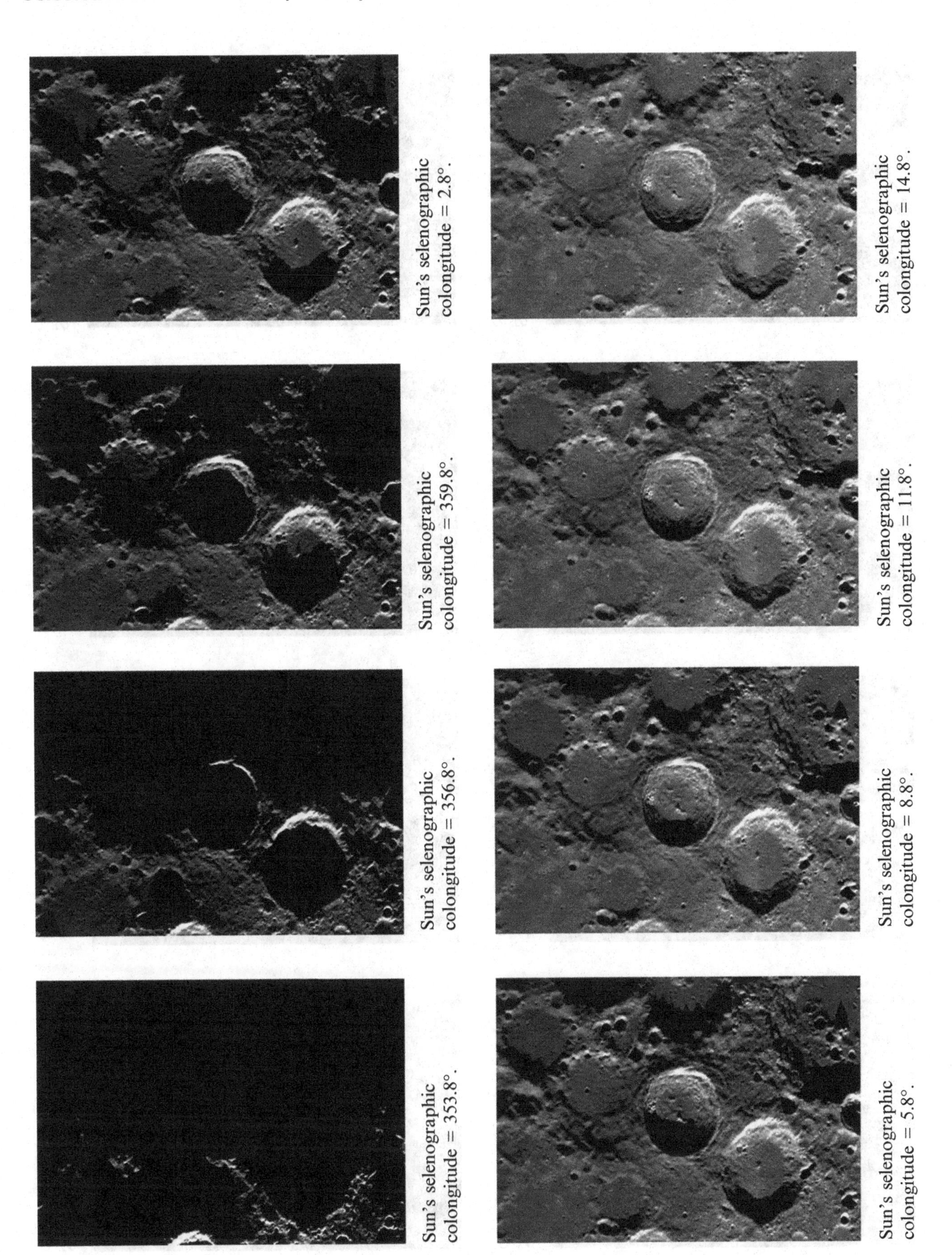

Sun's selenographic colongitude = 2.8°.

Sun's selenographic colongitude = 359.8°.

Sun's selenographic colongitude = 356.8°.

Sun's selenographic colongitude = 353.8°.

Sun's selenographic colongitude = 14.8°.

Sun's selenographic colongitude = 11.8°.

Sun's selenographic colongitude = 8.8°.

Sun's selenographic colongitude = 5.8°.

Selected Area 10b – Werner (Sunset)

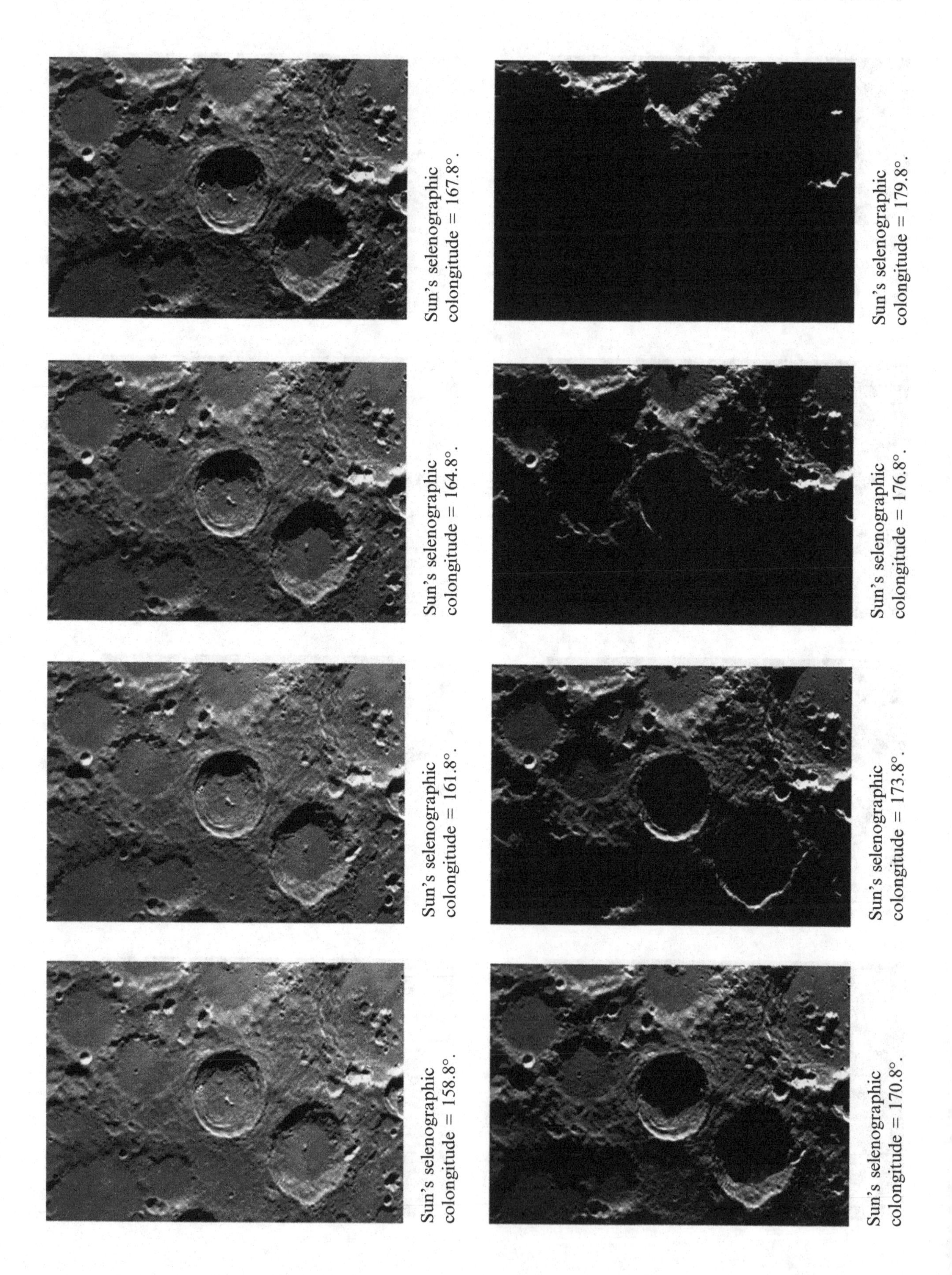

Sun's selenographic colongitude = 158.8°.

Sun's selenographic colongitude = 161.8°.

Sun's selenographic colongitude = 164.8°.

Sun's selenographic colongitude = 167.8°.

Sun's selenographic colongitude = 170.8°.

Sun's selenographic colongitude = 173.8°.

Sun's selenographic colongitude = 176.8°.

Sun's selenographic colongitude = 179.8°.

Selected Area 11a – Gassendi (Sunrise)

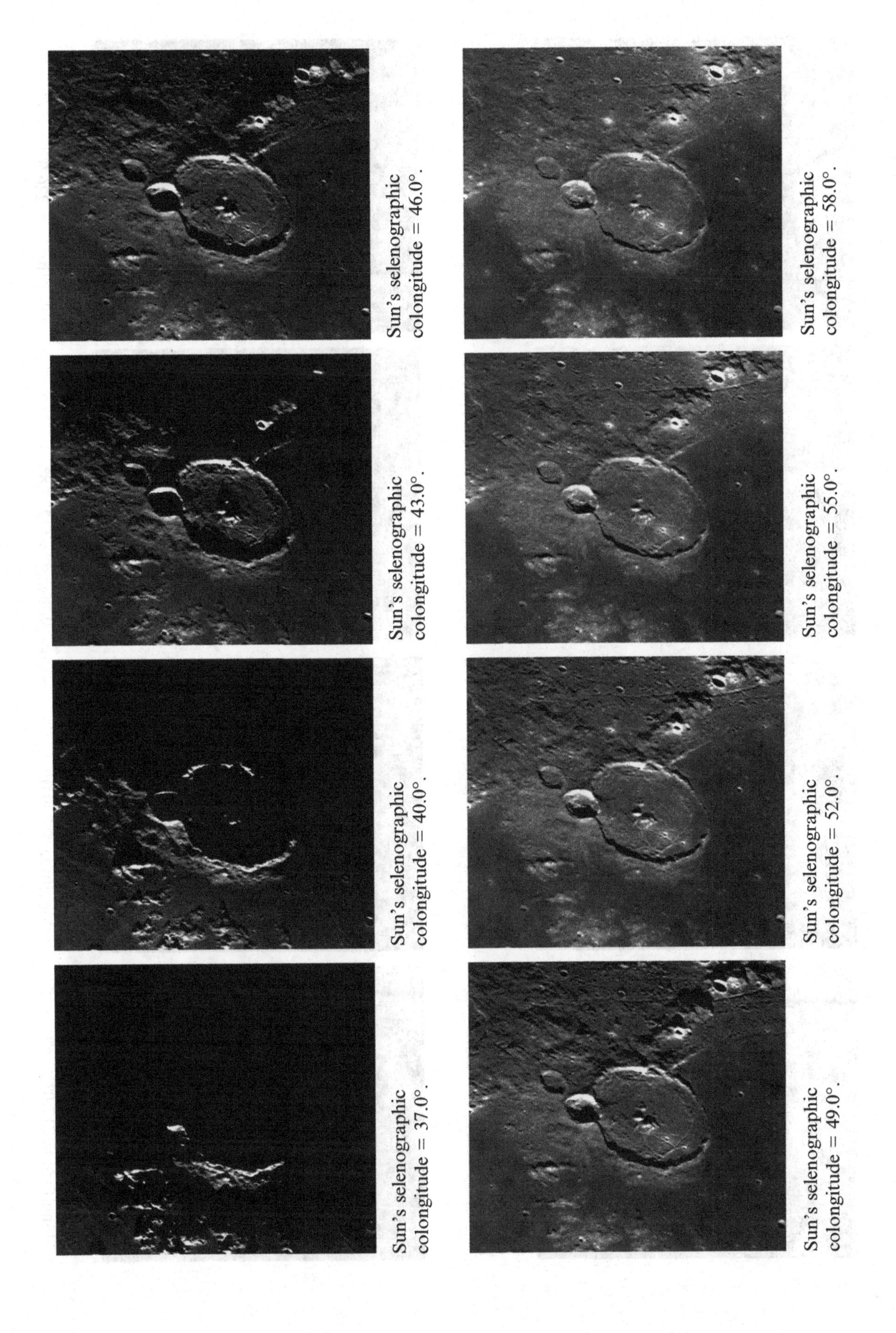

Sun's selenographic colongitude = 37.0°.

Sun's selenographic colongitude = 40.0°.

Sun's selenographic colongitude = 43.0°.

Sun's selenographic colongitude = 46.0°.

Sun's selenographic colongitude = 49.0°.

Sun's selenographic colongitude = 52.0°.

Sun's selenographic colongitude = 55.0°.

Sun's selenographic colongitude = 58.0°.

Selected Area 11a – Gassendi (Sunset)

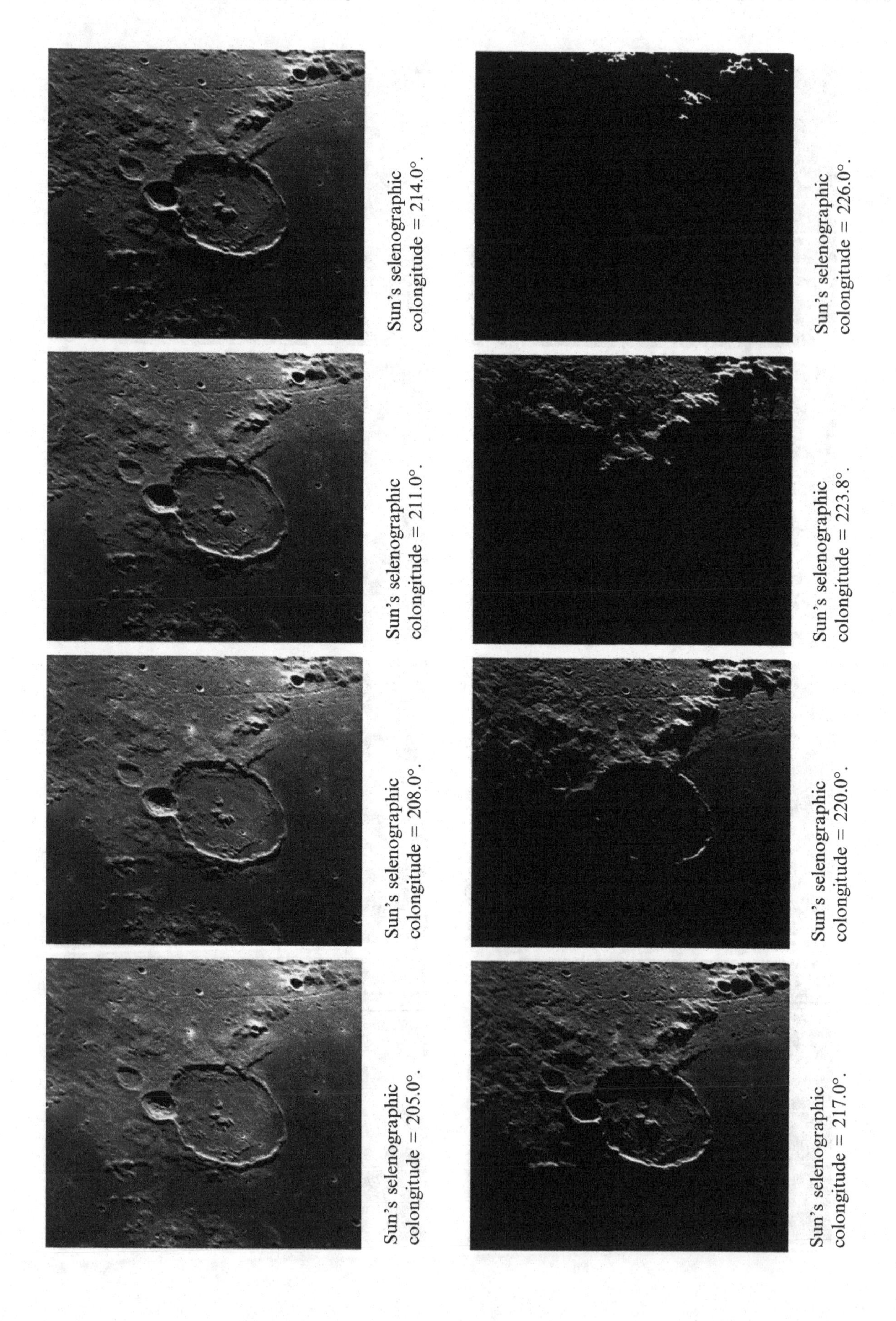
Sun's selenographic colongitude = 214.0°.

Sun's selenographic colongitude = 211.0°.

Sun's selenographic colongitude = 208.0°.

Sun's selenographic colongitude = 205.0°.

Sun's selenographic colongitude = 226.0°.

Sun's selenographic colongitude = 223.8°.

Sun's selenographic colongitude = 220.0°.

Sun's selenographic colongitude = 217.0°.

Selected Area 11b – Mons Hansteen (Sunrise)

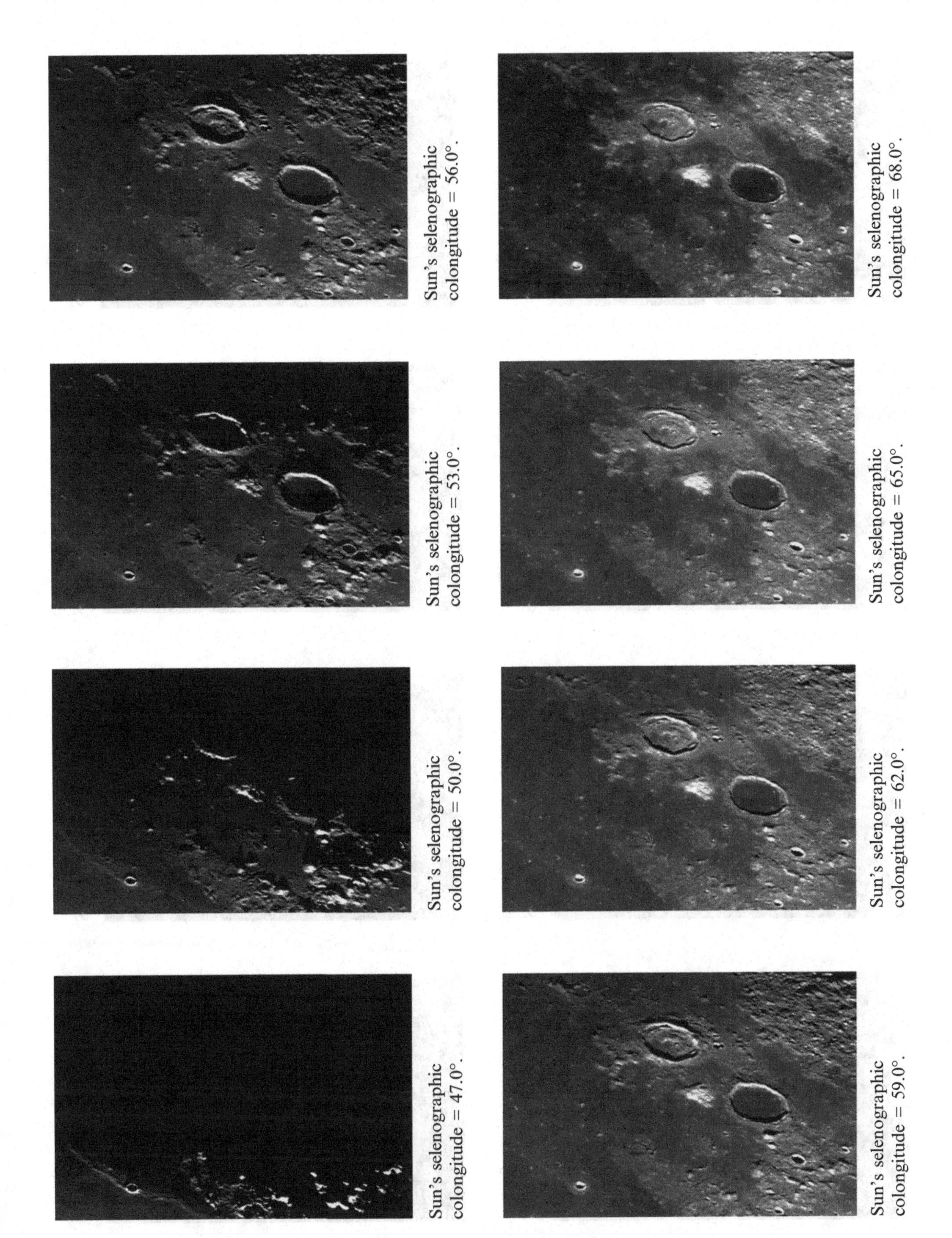

Sun's selenographic colongitude = 56.0°.

Sun's selenographic colongitude = 53.0°.

Sun's selenographic colongitude = 50.0°.

Sun's selenographic colongitude = 47.0°.

Sun's selenographic colongitude = 68.0°.

Sun's selenographic colongitude = 65.0°.

Sun's selenographic colongitude = 62.0°.

Sun's selenographic colongitude = 59.0°.

Selected Area 11b – Mons Hansteen (Sunset)

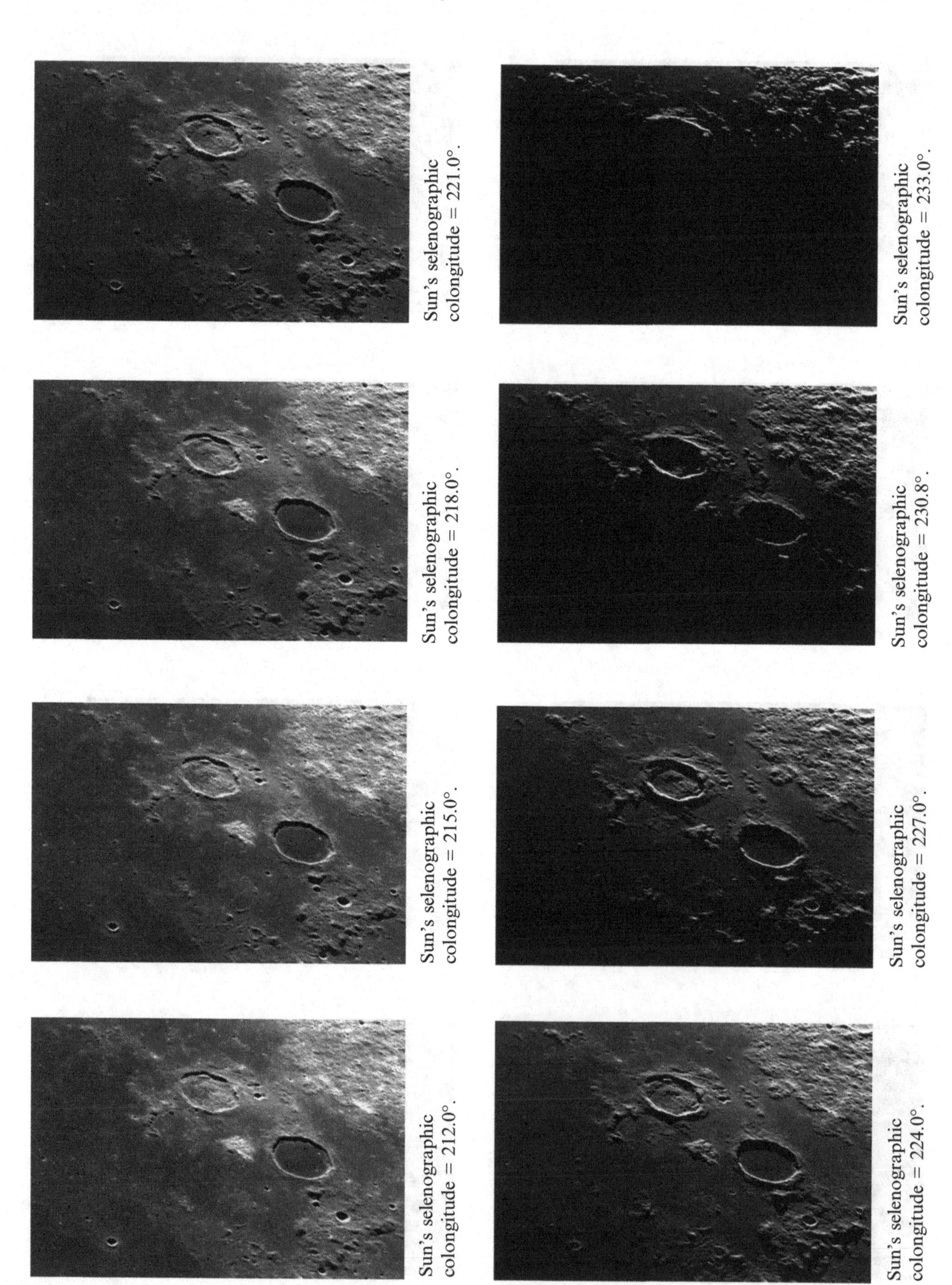

Sun's selenographic colongitude = 212.0°.

Sun's selenographic colongitude = 215.0°.

Sun's selenographic colongitude = 218.0°.

Sun's selenographic colongitude = 221.0°.

Sun's selenographic colongitude = 224.0°.

Sun's selenographic colongitude = 227.0°.

Sun's selenographic colongitude = 230.8°.

Sun's selenographic colongitude = 233.0°.

Selected Area 12a – Vitello (Sunrise)

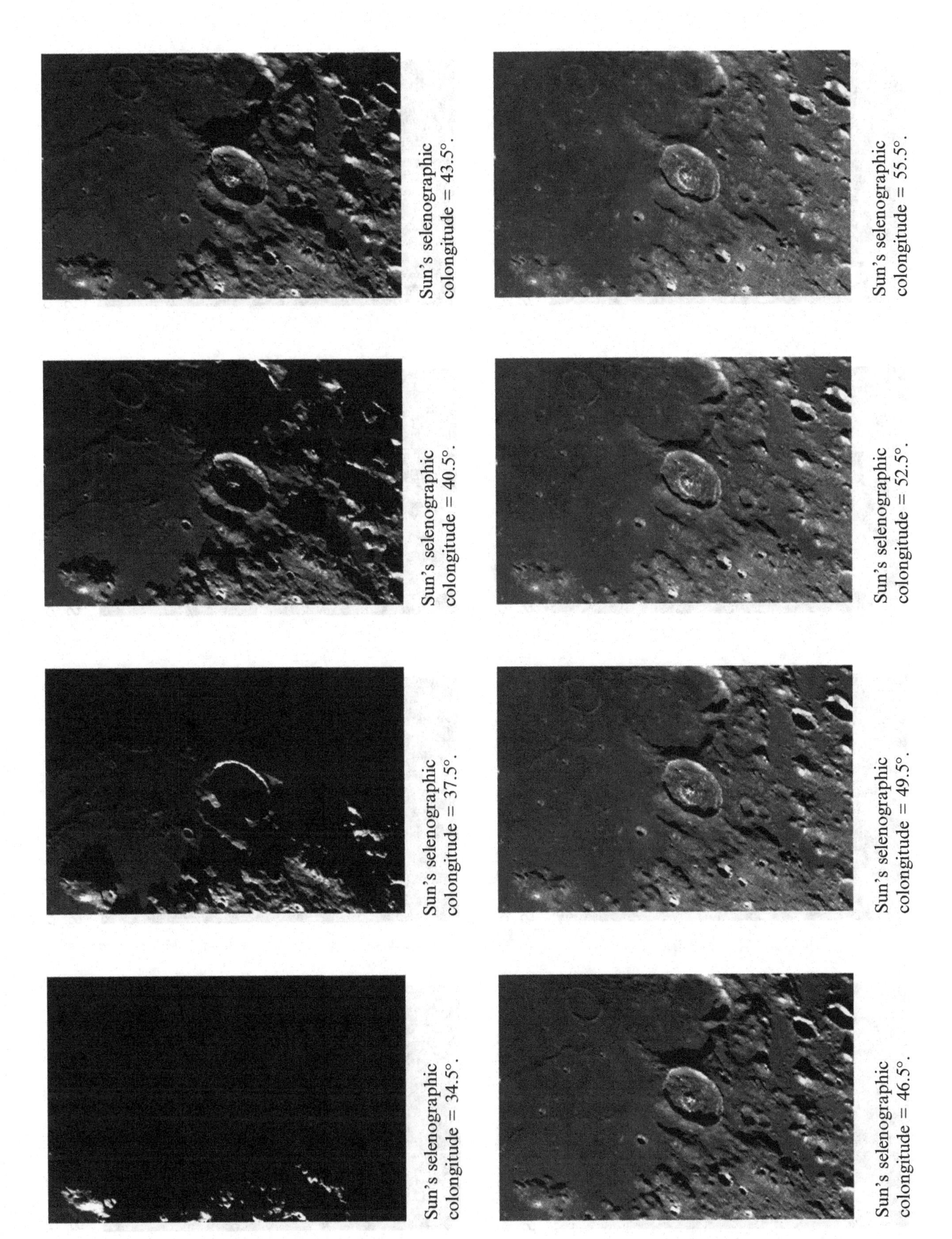

Sun's selenographic colongitude = 34.5°.

Sun's selenographic colongitude = 37.5°.

Sun's selenographic colongitude = 40.5°.

Sun's selenographic colongitude = 43.5°.

Sun's selenographic colongitude = 46.5°.

Sun's selenographic colongitude = 49.5°.

Sun's selenographic colongitude = 52.5°.

Sun's selenographic colongitude = 55.5°.

Selected Area 12a – Vitello (Sunset)

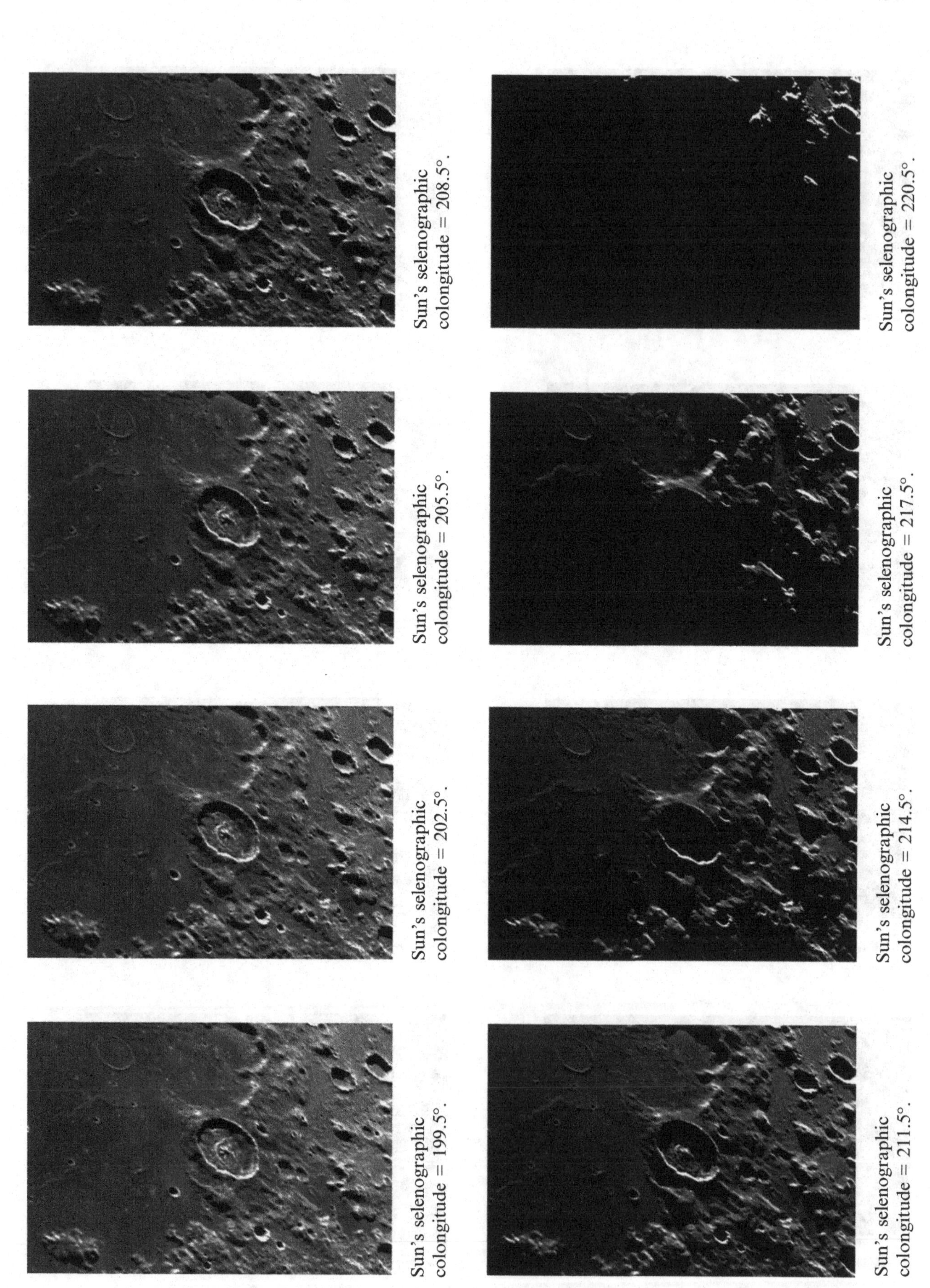

Sun's selenographic colongitude = 208.5°.

Sun's selenographic colongitude = 205.5°.

Sun's selenographic colongitude = 202.5°.

Sun's selenographic colongitude = 199.5°.

Sun's selenographic colongitude = 220.5°.

Sun's selenographic colongitude = 217.5°.

Sun's selenographic colongitude = 214.5°.

Sun's selenographic colongitude = 211.5°.

Selected Area 12b – Wargentin (Sunrise)

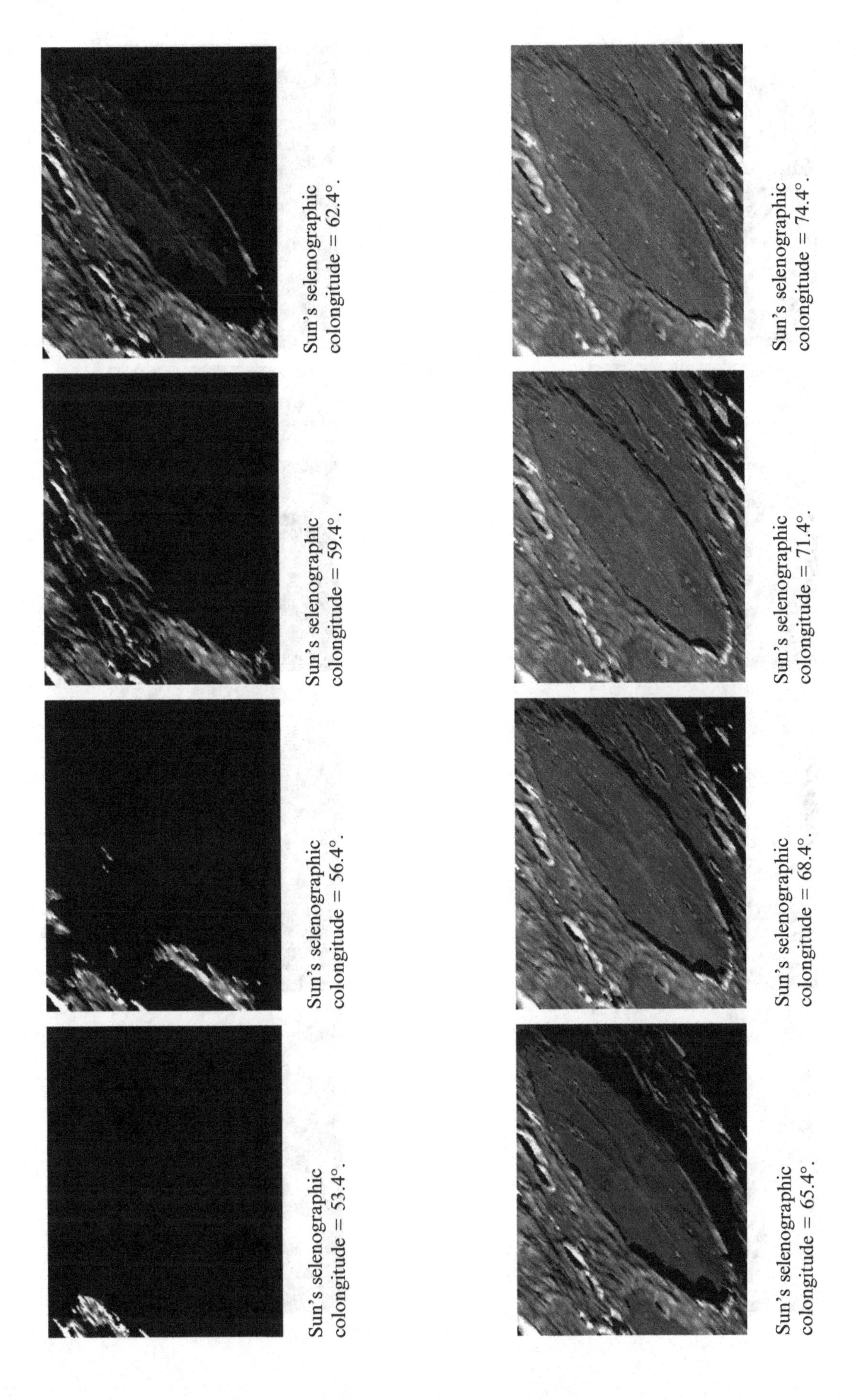

Sun's selenographic colongitude = 53.4°.

Sun's selenographic colongitude = 56.4°.

Sun's selenographic colongitude = 59.4°.

Sun's selenographic colongitude = 62.4°.

Sun's selenographic colongitude = 65.4°.

Sun's selenographic colongitude = 68.4°.

Sun's selenographic colongitude = 71.4°.

Sun's selenographic colongitude = 74.4°.

Selected Area 12b – Wargentin (Sunset)

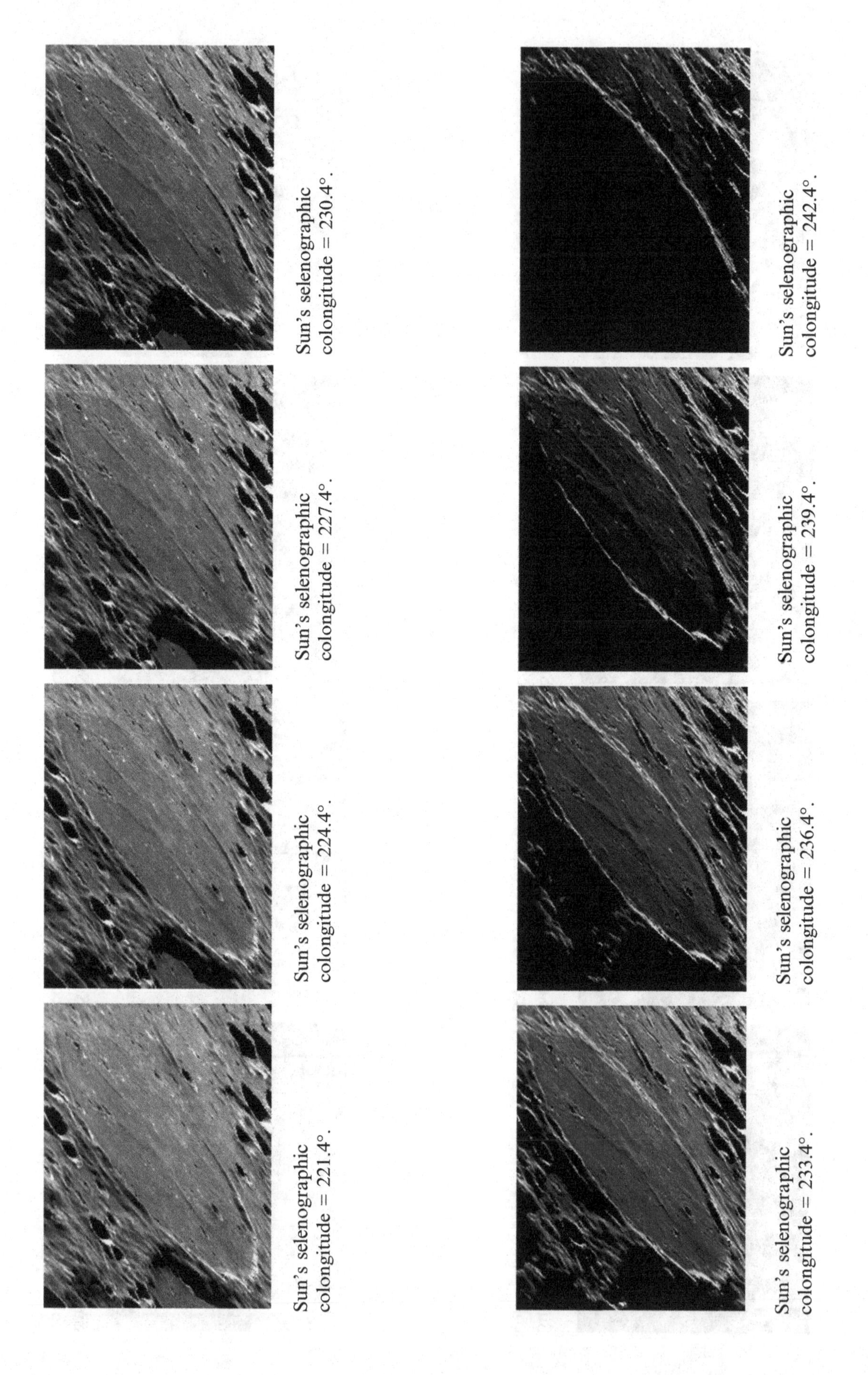

Sun's selenographic colongitude = 221.4°.

Sun's selenographic colongitude = 224.4°.

Sun's selenographic colongitude = 227.4°.

Sun's selenographic colongitude = 230.4°.

Sun's selenographic colongitude = 233.4°.

Sun's selenographic colongitude = 236.4°.

Sun's selenographic colongitude = 239.4°.

Sun's selenographic colongitude = 242.4°.

Selected Area 13a – Alphonsus (Sunrise)

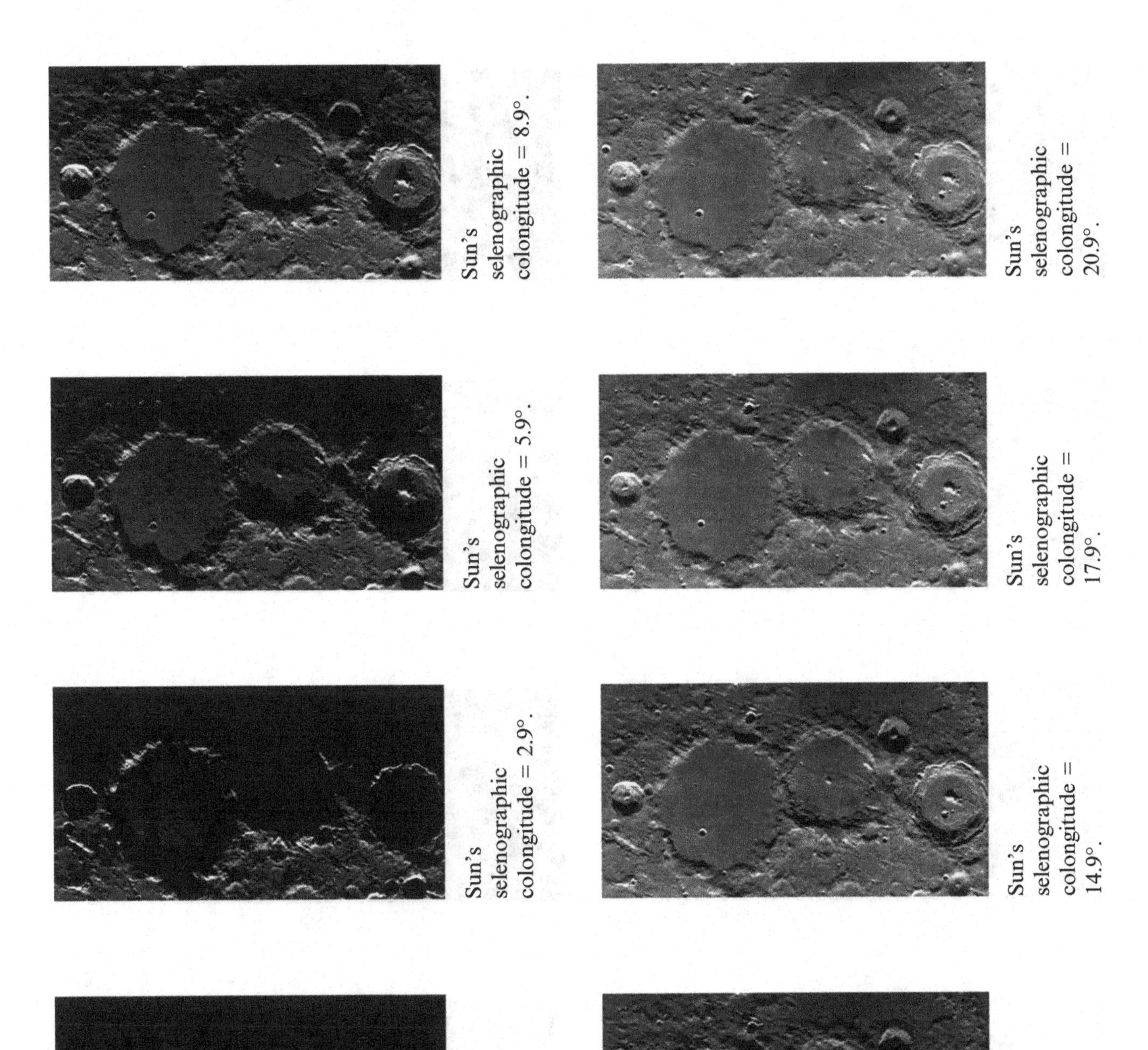

Sun's selenographic colongitude = 359.9°.

Sun's selenographic colongitude = 2.9°.

Sun's selenographic colongitude = 5.9°.

Sun's selenographic colongitude = 8.9°.

Sun's selenographic colongitude = 11.9°.

Sun's selenographic colongitude = 14.9°.

Sun's selenographic colongitude = 17.9°.

Sun's selenographic colongitude = 20.9°.

Selected Area 13a – Alphonsus (Sunset)

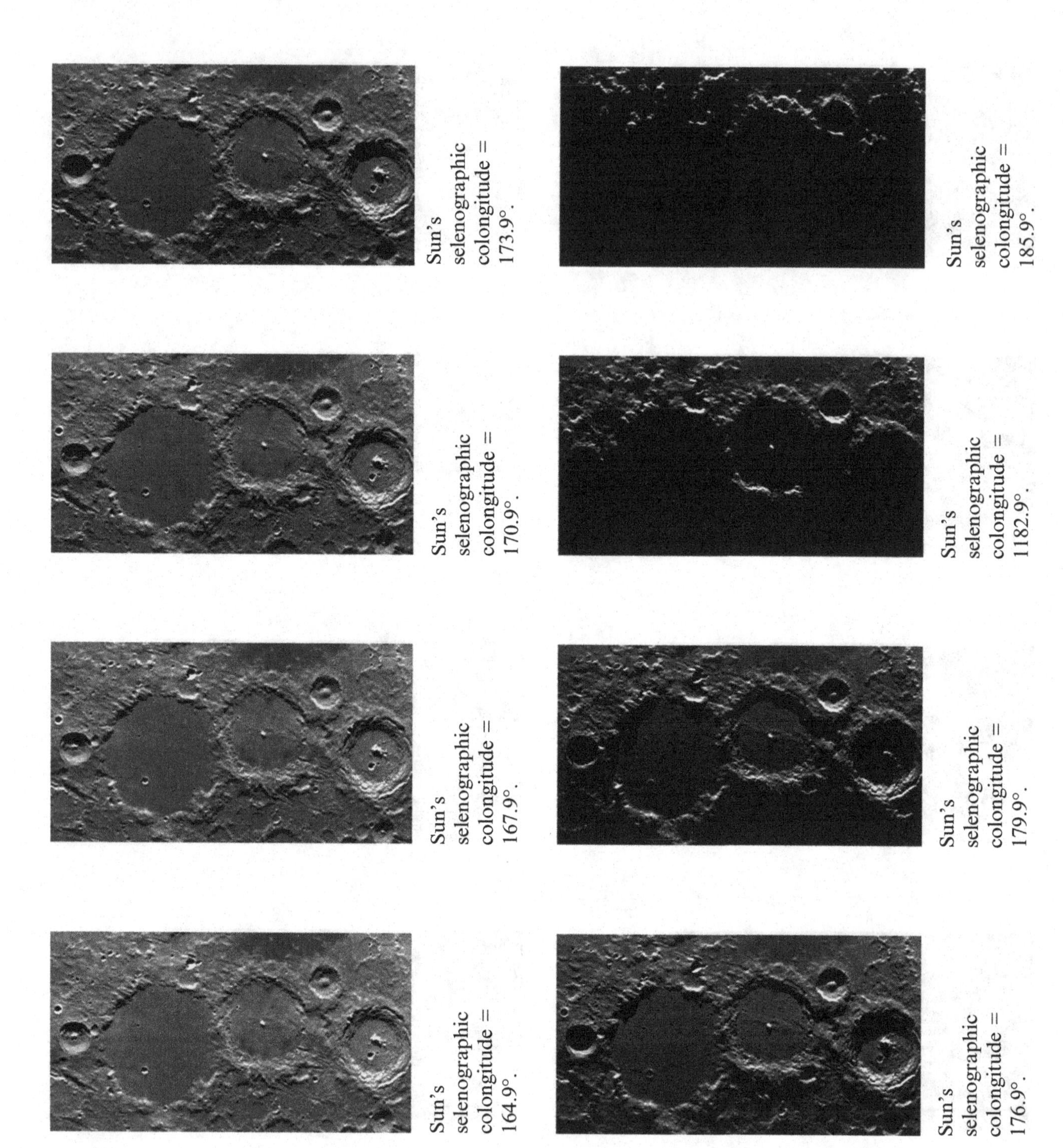

Sun's selenographic colongitude = 173.9°.

Sun's selenographic colongitude = 185.9°.

Sun's selenographic colongitude = 170.9°.

Sun's selenographic colongitude = 1182.9°.

Sun's selenographic colongitude = 167.9°.

Sun's selenographic colongitude = 179.9°.

Sun's selenographic colongitude = 164.9°.

Sun's selenographic colongitude = 176.9°.

Selected Area 13b – Rupes Recta (Sunrise)

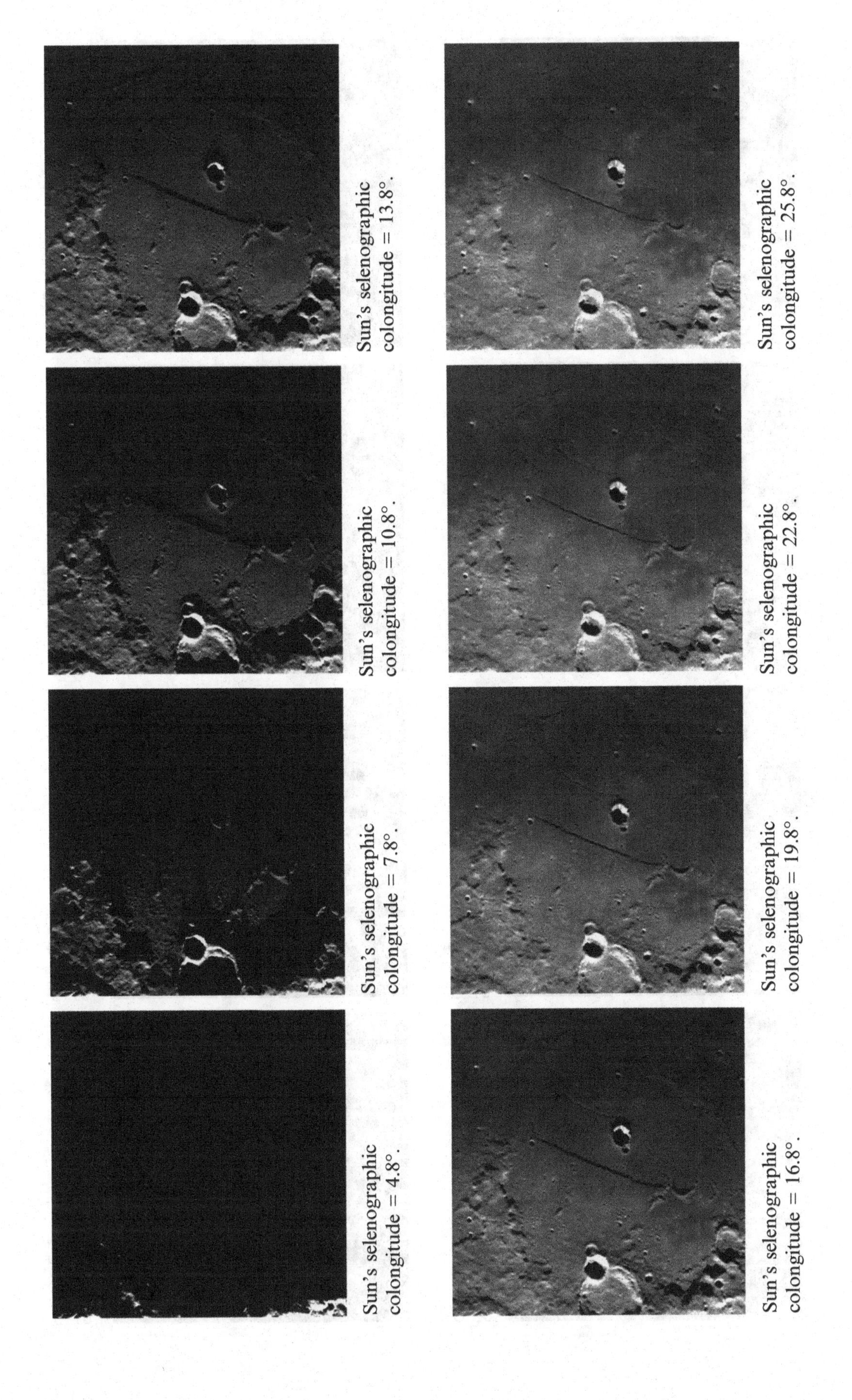

Sun's selenographic colongitude = 4.8°.

Sun's selenographic colongitude = 7.8°.

Sun's selenographic colongitude = 10.8°.

Sun's selenographic colongitude = 13.8°.

Sun's selenographic colongitude = 16.8°.

Sun's selenographic colongitude = 19.8°.

Sun's selenographic colongitude = 22.8°.

Sun's selenographic colongitude = 25.8°.

Selected Area 13b – Rupes Recta (Sunset)

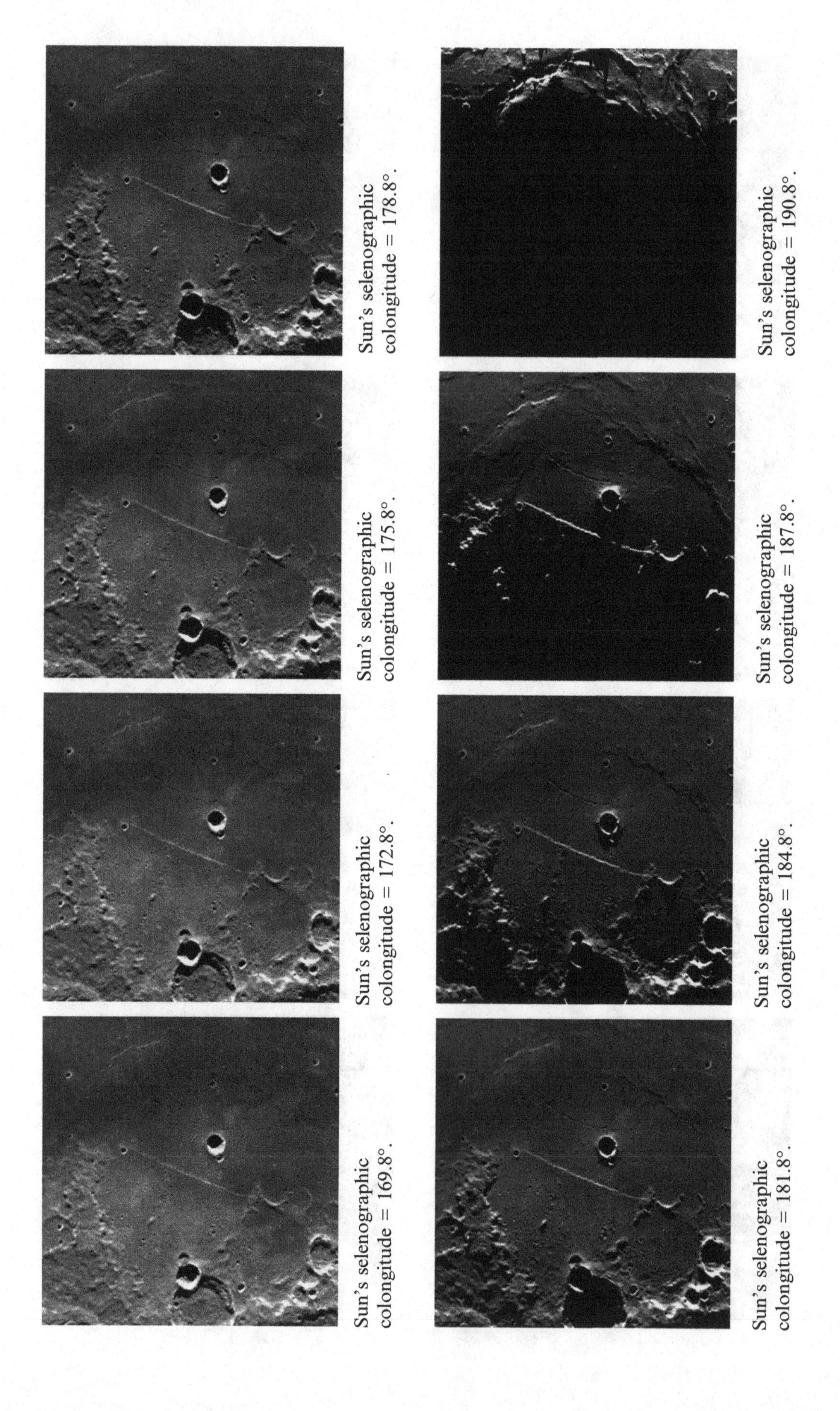

Sun's selenographic colongitude = 169.8°.

Sun's selenographic colongitude = 172.8°.

Sun's selenographic colongitude = 175.8°.

Sun's selenographic colongitude = 178.8°.

Sun's selenographic colongitude = 181.8°.

Sun's selenographic colongitude = 184.8°.

Sun's selenographic colongitude = 187.8°.

Sun's selenographic colongitude = 190.8°.

Selected Area 14a – Clavius (Sunrise)

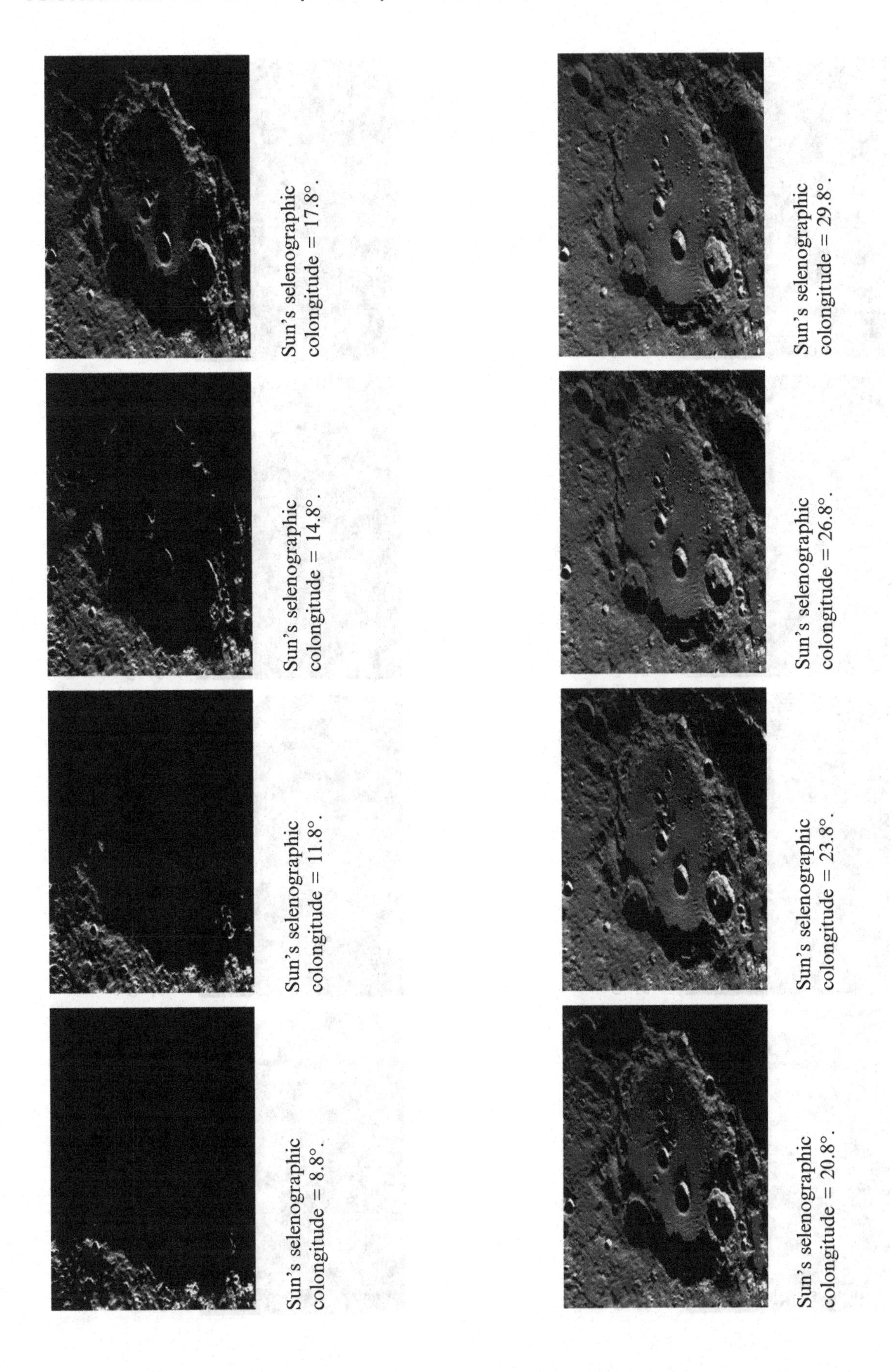

Sun's selenographic colongitude = 8.8°.

Sun's selenographic colongitude = 11.8°.

Sun's selenographic colongitude = 14.8°.

Sun's selenographic colongitude = 17.8°.

Sun's selenographic colongitude = 20.8°.

Sun's selenographic colongitude = 23.8°.

Sun's selenographic colongitude = 26.8°.

Sun's selenographic colongitude = 29.8°.

Selected Area 14a – Clavius (Sunset)

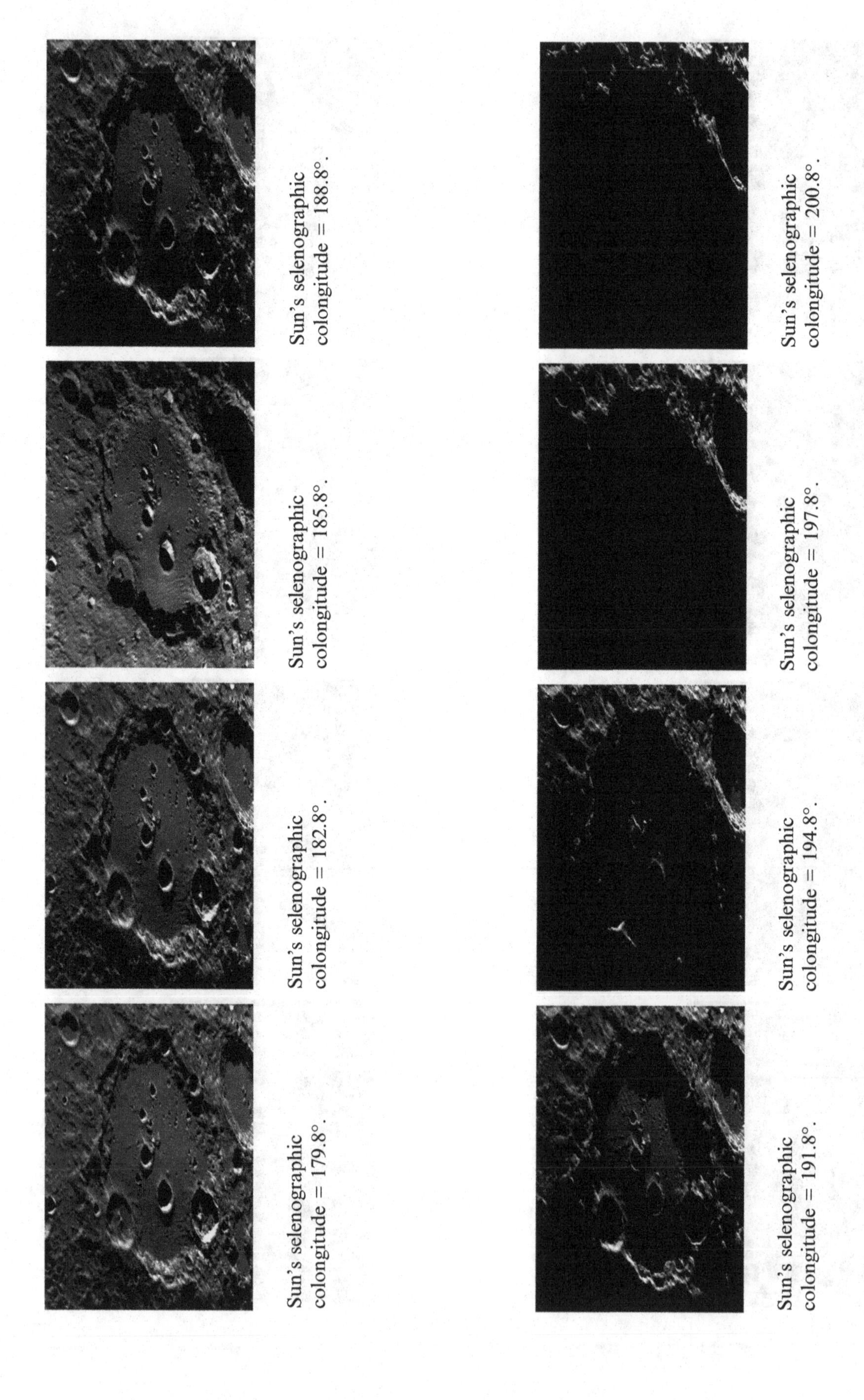

Sun's selenographic colongitude = 179.8°.

Sun's selenographic colongitude = 182.8°.

Sun's selenographic colongitude = 185.8°.

Sun's selenographic colongitude = 188.8°.

Sun's selenographic colongitude = 191.8°.

Sun's selenographic colongitude = 194.8°.

Sun's selenographic colongitude = 197.8°.

Sun's selenographic colongitude = 200.8°.

Selected Area 14b – Maurolycus (Sunrise)

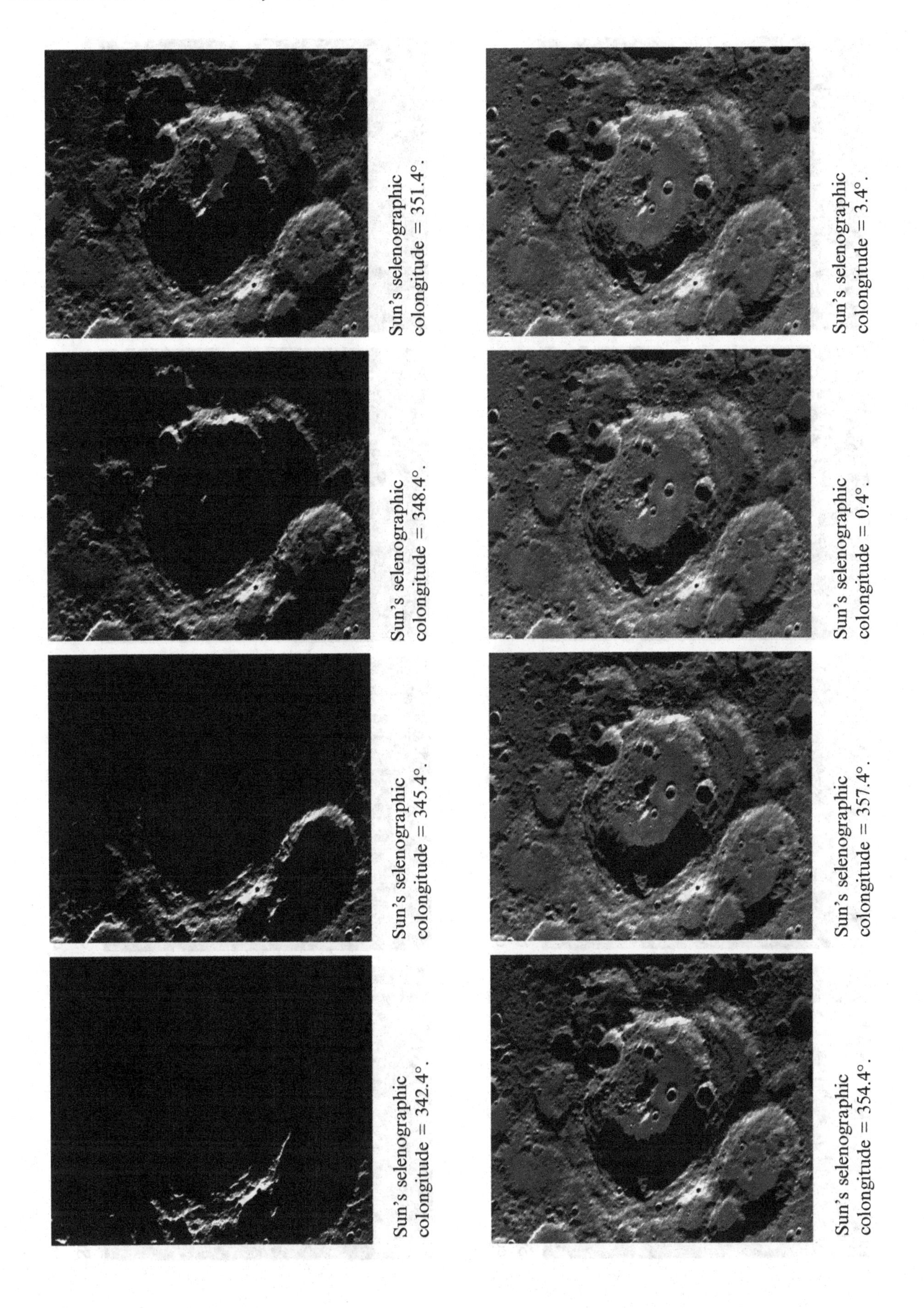

Sun's selenographic colongitude = 351.4°.

Sun's selenographic colongitude = 348.4°.

Sun's selenographic colongitude = 345.4°.

Sun's selenographic colongitude = 342.4°.

Sun's selenographic colongitude = 3.4°.

Sun's selenographic colongitude = 0.4°.

Sun's selenographic colongitude = 357.4°.

Sun's selenographic colongitude = 354.4°.

Selected Area 14b – Maurolycus (Sunset)

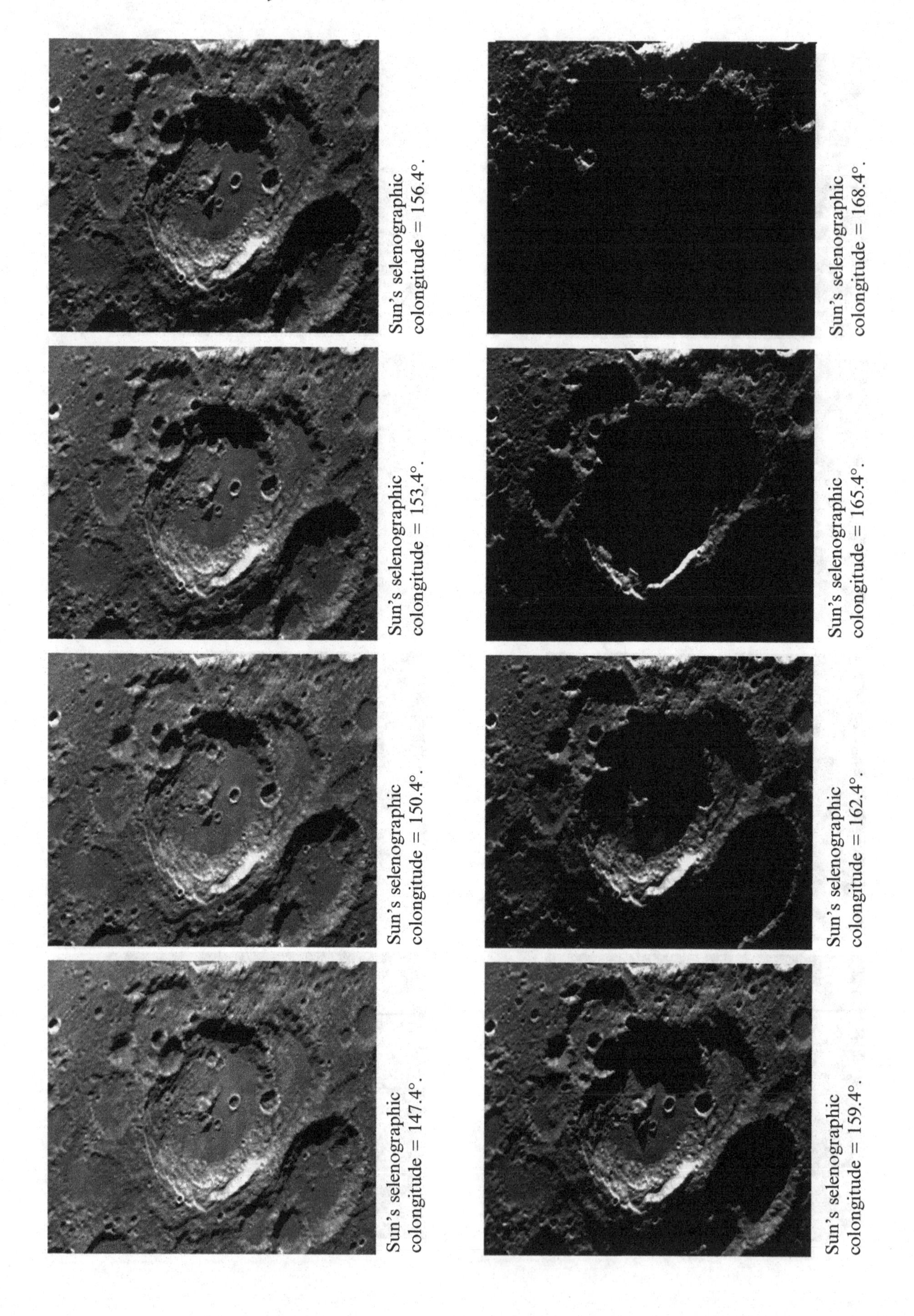

Sun's selenographic colongitude = 156.4°.

Sun's selenographic colongitude = 153.4°.

Sun's selenographic colongitude = 150.4°.

Sun's selenographic colongitude = 147.4°.

Sun's selenographic colongitude = 168.4°.

Sun's selenographic colongitude = 165.4°.

Sun's selenographic colongitude = 162.4°.

Sun's selenographic colongitude = 159.4°.

Selected Area 15a – Petavius (Sunrise)

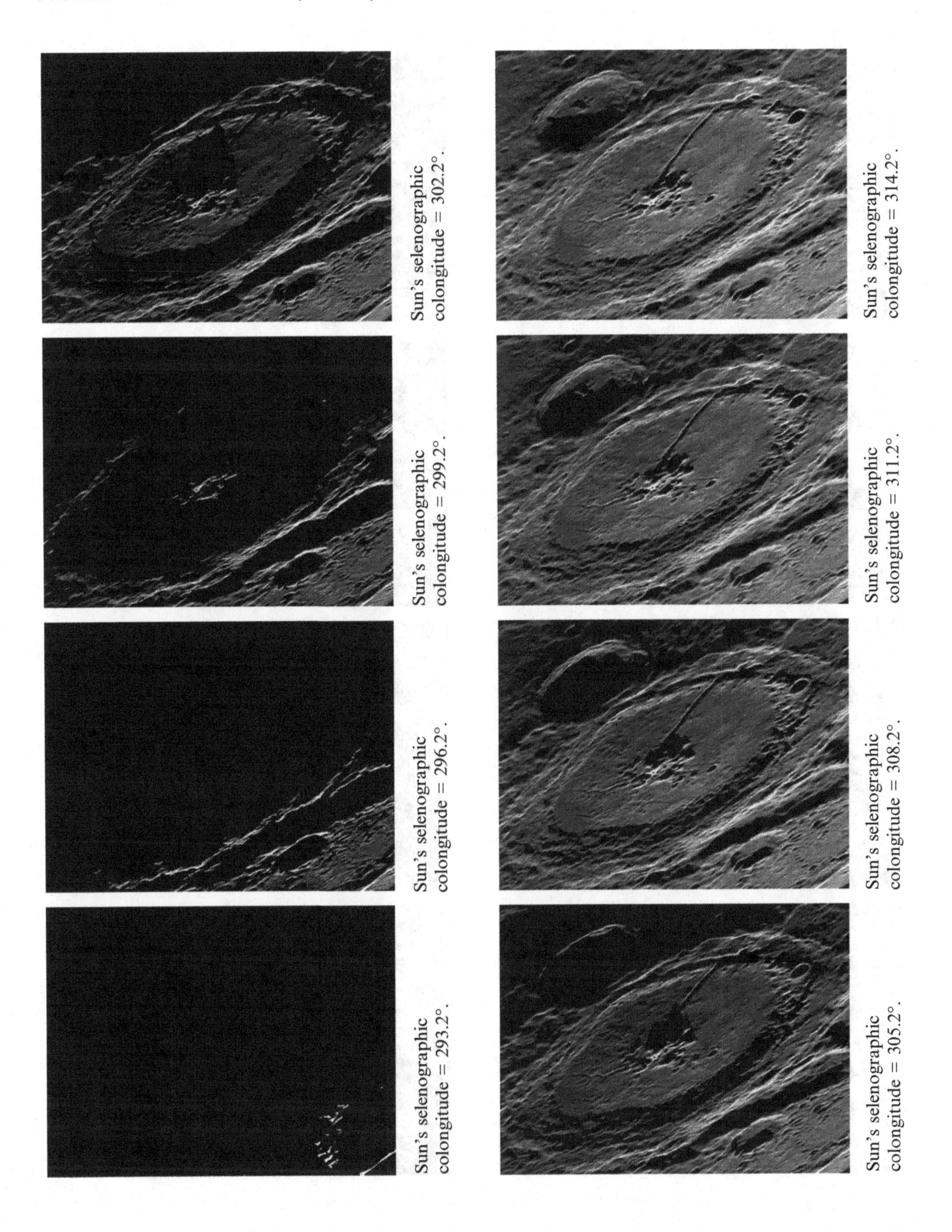

Sun's selenographic colongitude = 293.2°.

Sun's selenographic colongitude = 296.2°.

Sun's selenographic colongitude = 299.2°.

Sun's selenographic colongitude = 302.2°.

Sun's selenographic colongitude = 305.2°.

Sun's selenographic colongitude = 308.2°.

Sun's selenographic colongitude = 311.2°.

Sun's selenographic colongitude = 314.2°.

Selected Area 15b – Petavius (Sunset)

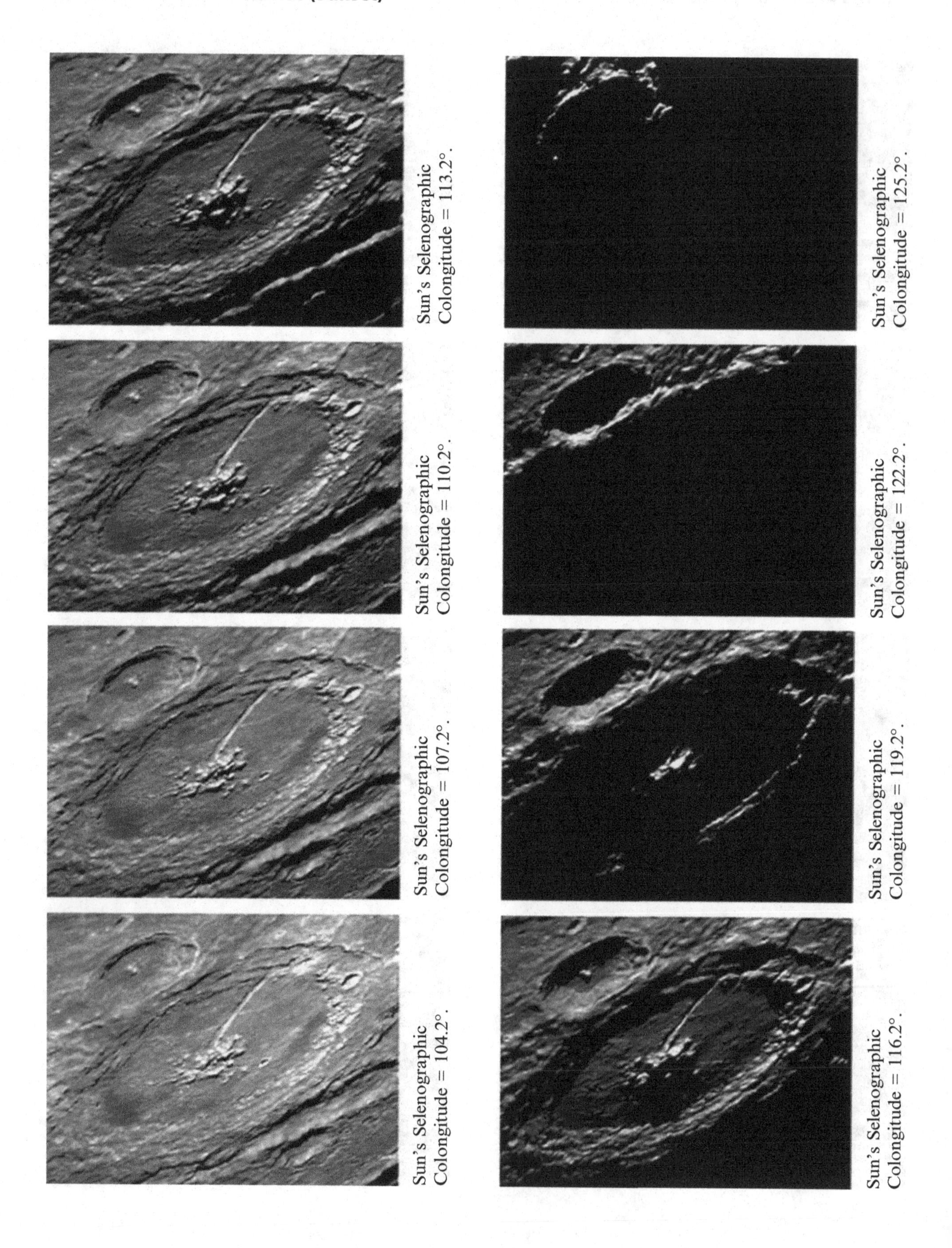

Selected Area 15b – Theophilus (Sunrise)

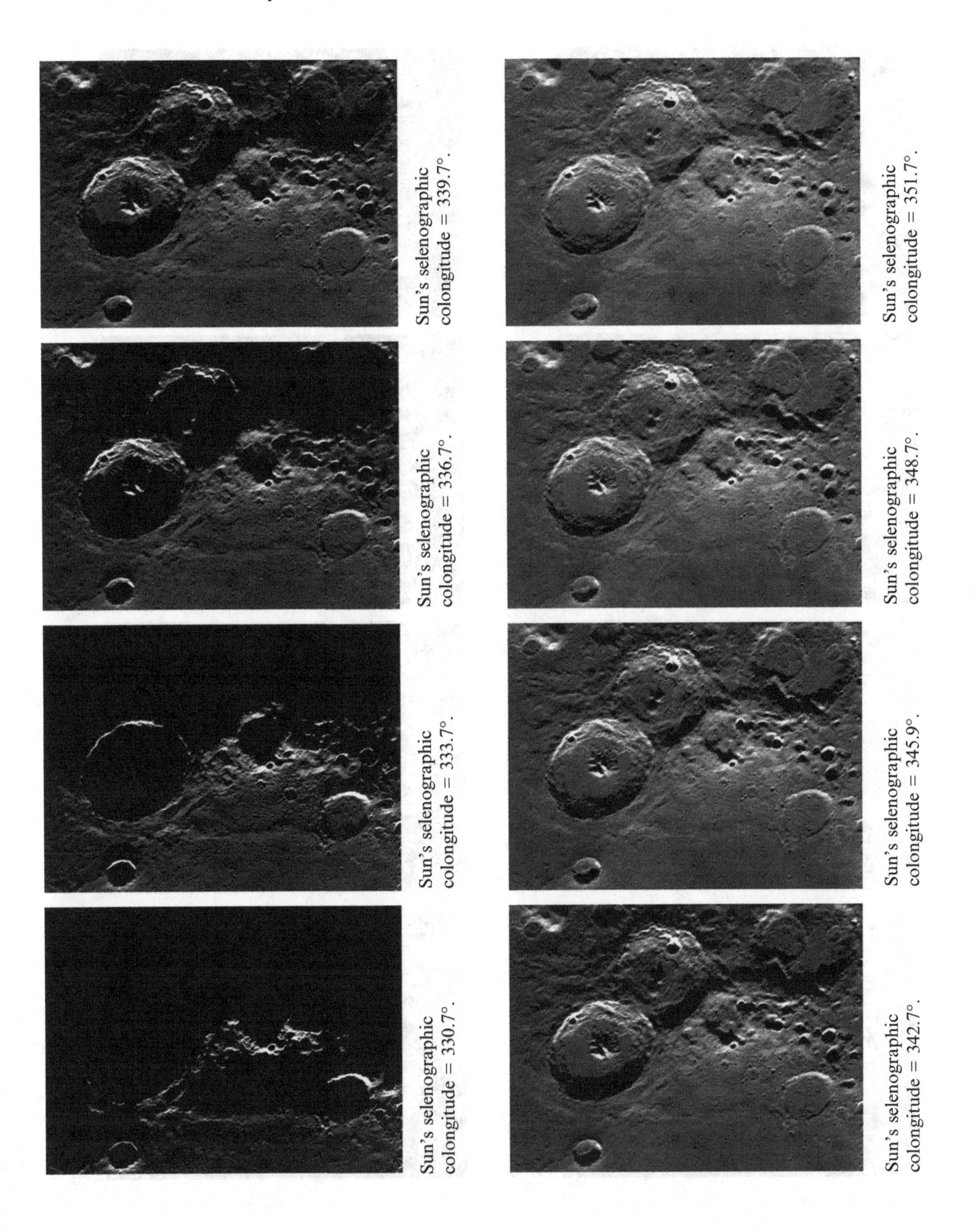

Sun's selenographic colongitude = 339.7°.

Sun's selenographic colongitude = 336.7°.

Sun's selenographic colongitude = 333.7°.

Sun's selenographic colongitude = 330.7°.

Sun's selenographic colongitude = 351.7°.

Sun's selenographic colongitude = 348.7°.

Sun's selenographic colongitude = 345.9°.

Sun's selenographic colongitude = 342.7°.

Selected Area 15b – Theophilus (Sunset)

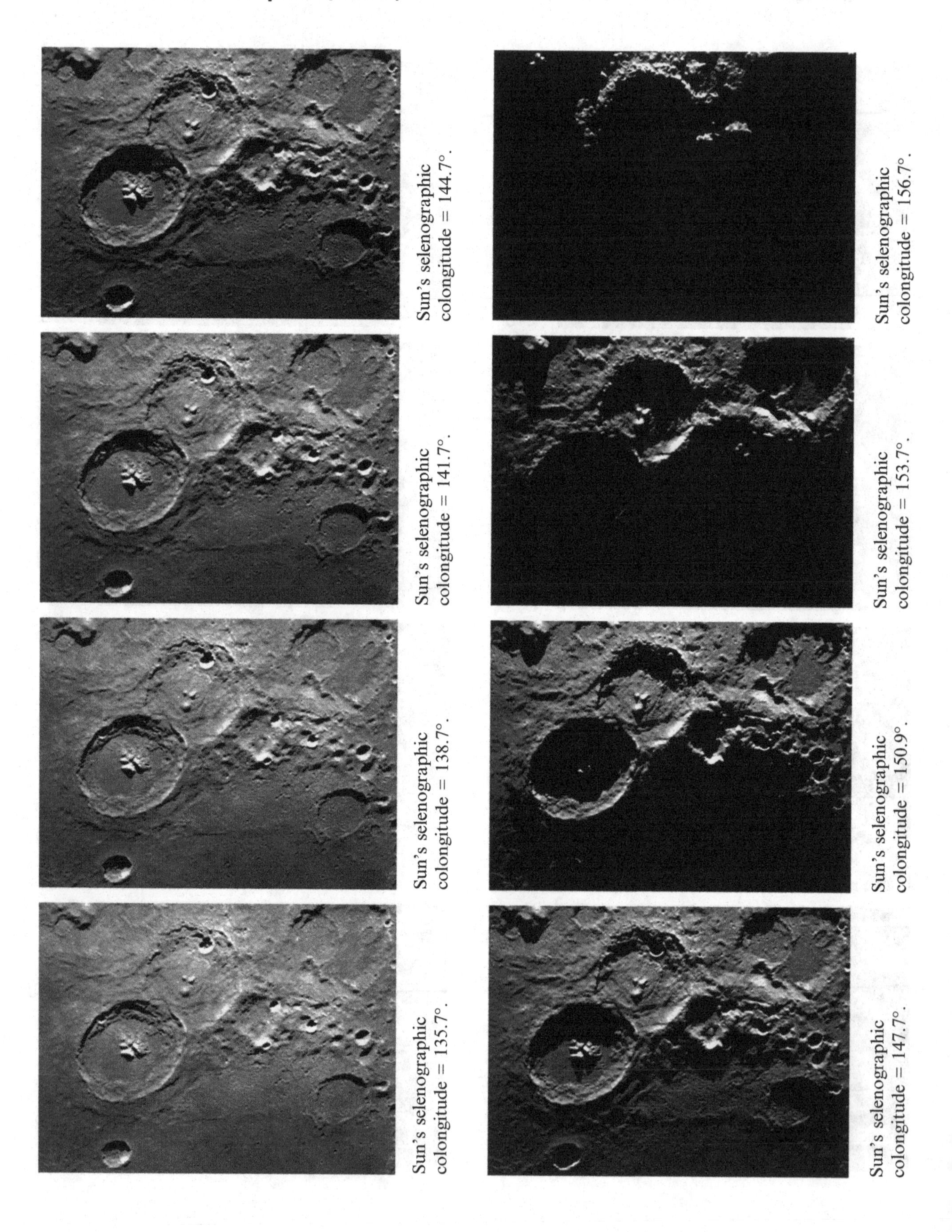

Sun's selenographic colongitude = 144.7°.

Sun's selenographic colongitude = 141.7°.

Sun's selenographic colongitude = 138.7°.

Sun's selenographic colongitude = 135.7°.

Sun's selenographic colongitude = 156.7°.

Sun's selenographic colongitude = 153.7°.

Sun's selenographic colongitude = 150.9°.

Sun's selenographic colongitude = 147.7°.

Selected Area 16a – Janssen (Sunrise)

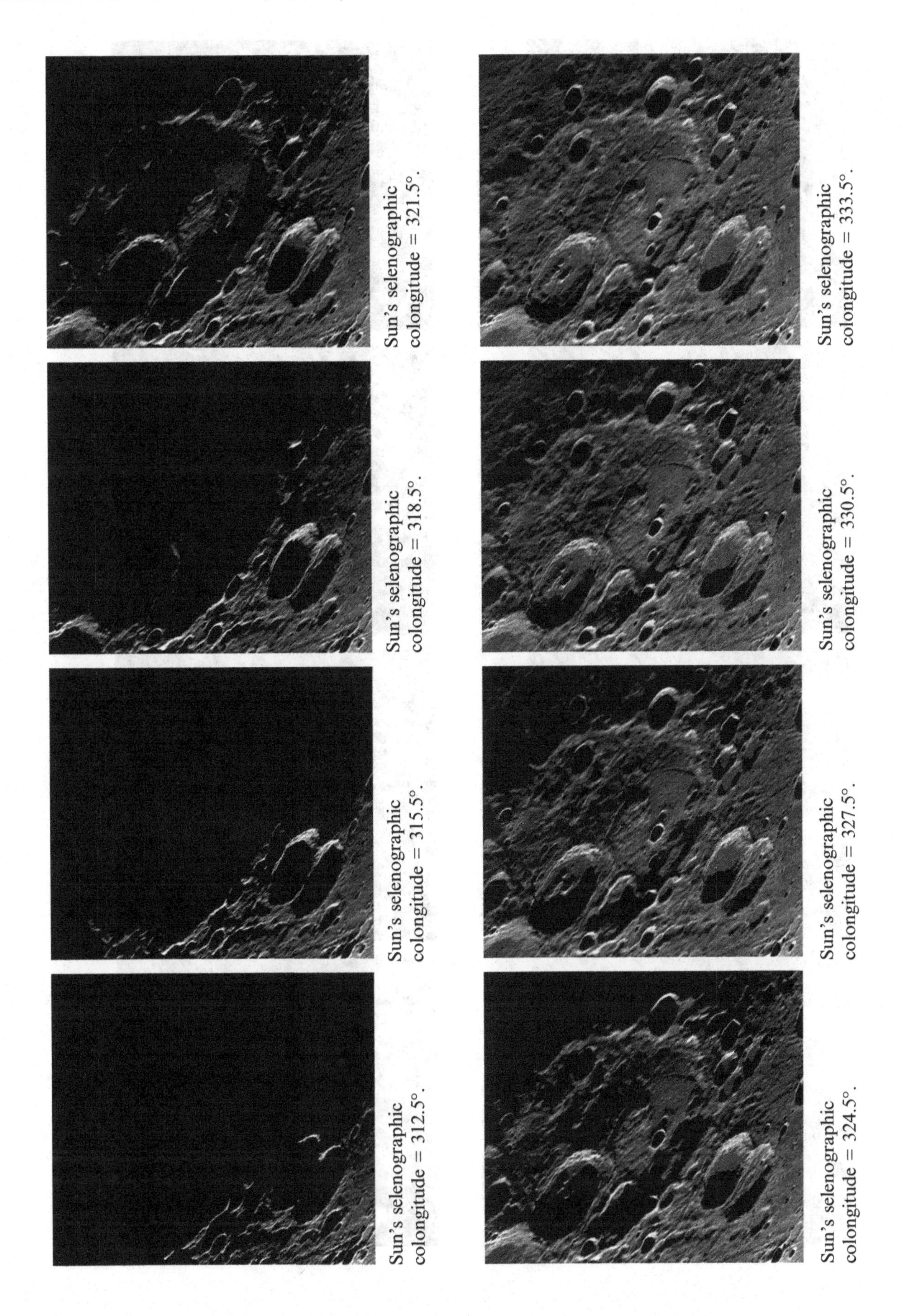

Sun's selenographic colongitude = 312.5°.

Sun's selenographic colongitude = 315.5°.

Sun's selenographic colongitude = 318.5°.

Sun's selenographic colongitude = 321.5°.

Sun's selenographic colongitude = 324.5°.

Sun's selenographic colongitude = 327.5°.

Sun's selenographic colongitude = 330.5°.

Sun's selenographic colongitude = 333.5°.

Selected Area 16a – Janssen (Sunset)

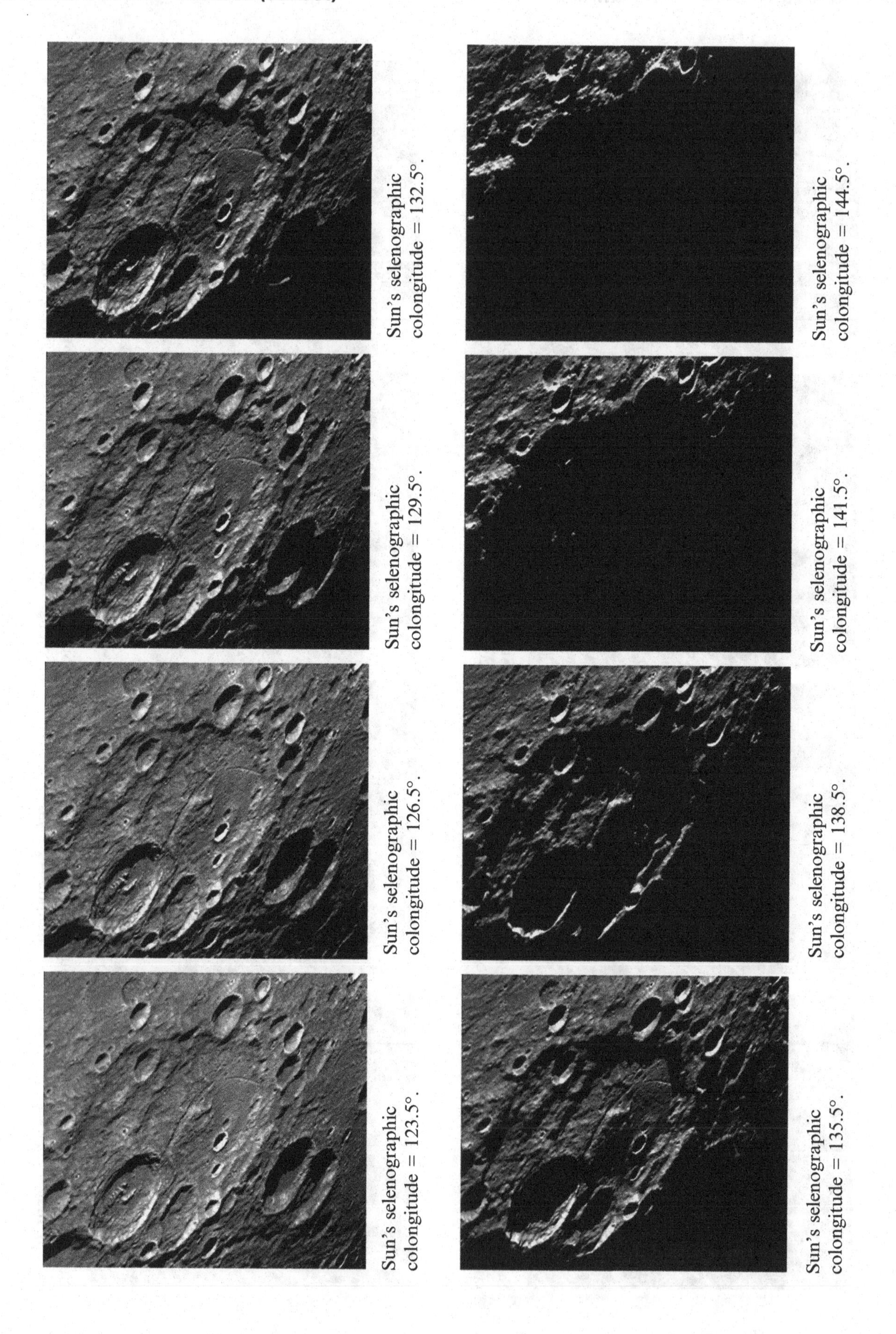

Sun's selenographic colongitude = 123.5°.

Sun's selenographic colongitude = 126.5°.

Sun's selenographic colongitude = 129.5°.

Sun's selenographic colongitude = 132.5°.

Sun's selenographic colongitude = 135.5°.

Sun's selenographic colongitude = 138.5°.

Sun's selenographic colongitude = 141.5°.

Sun's selenographic colongitude = 144.5°.

Selected Area 16b – Piccolomini (Sunrise)

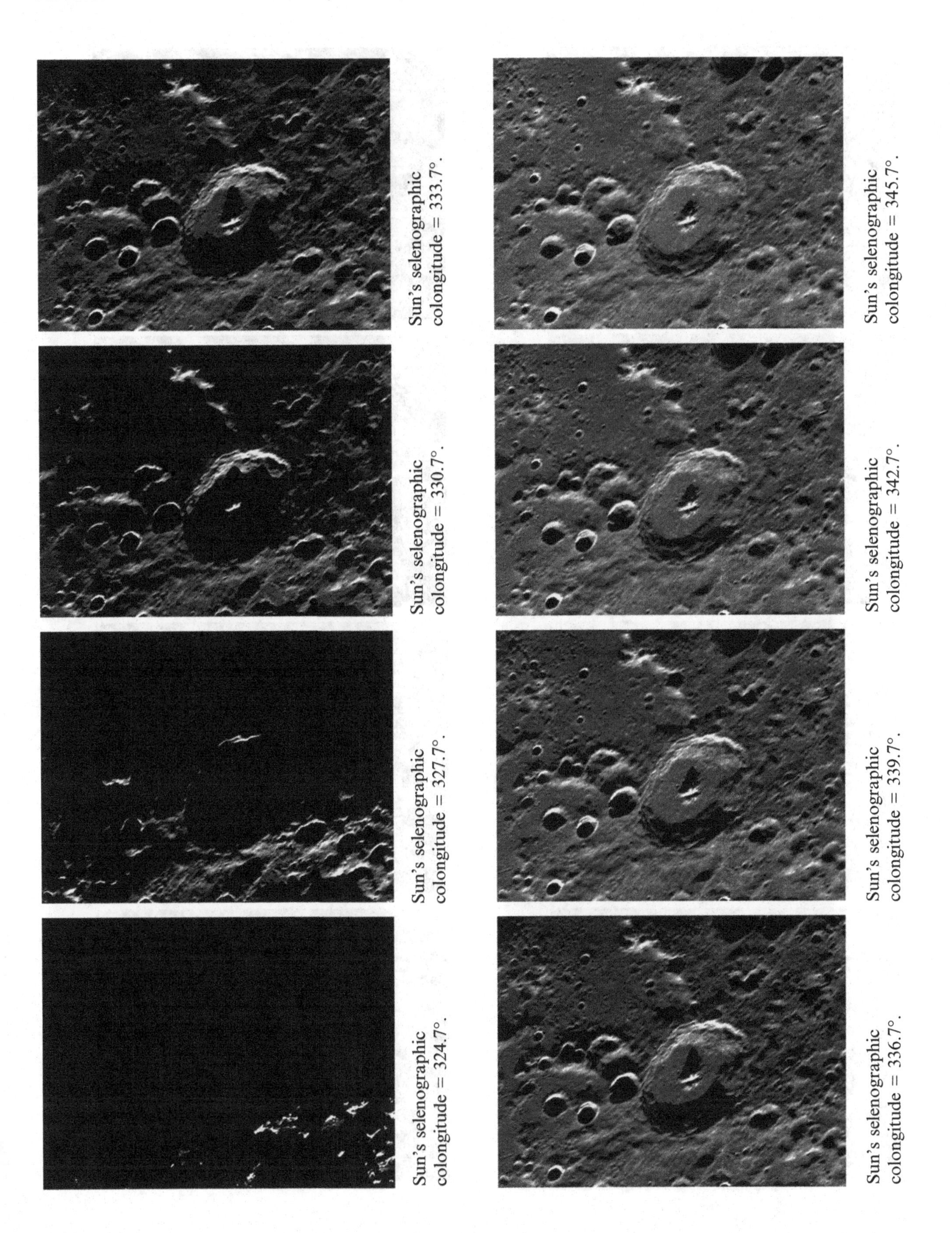

Sun's selenographic colongitude = 324.7°.

Sun's selenographic colongitude = 327.7°.

Sun's selenographic colongitude = 330.7°.

Sun's selenographic colongitude = 333.7°.

Sun's selenographic colongitude = 336.7°.

Sun's selenographic colongitude = 339.7°.

Sun's selenographic colongitude = 342.7°.

Sun's selenographic colongitude = 345.7°.

Selected Area 16b – Piccolomini (Sunset)

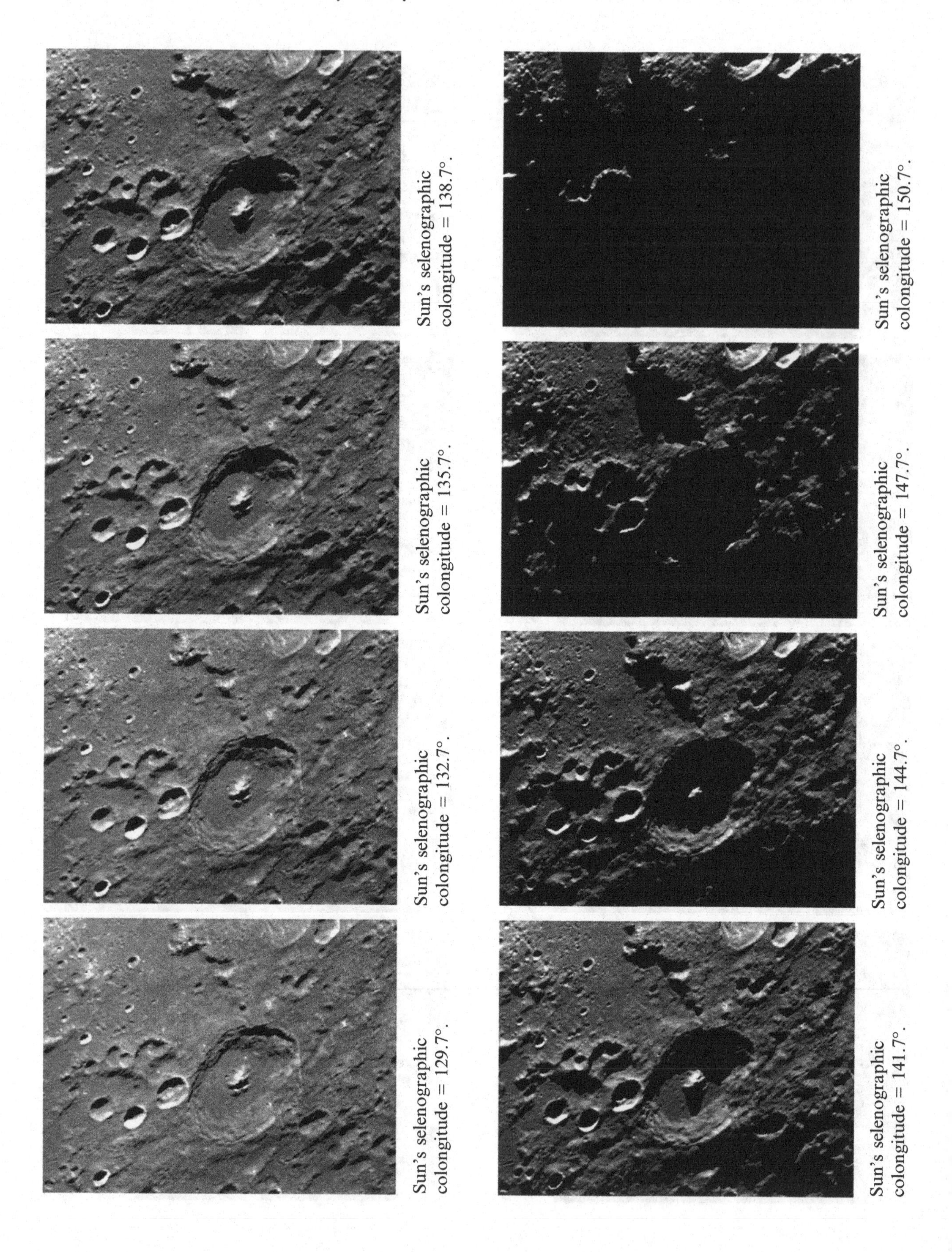

Appendices

Appendix 1

Details about Commander Hatfield's Photographic Plates

Date	UT	Sun's selenographic colongitude (degrees)	Moon's age (days)	Focal ratio	Plate
1965 Nov 08	21:32	91.3	15.3	f/29	7e/2
1966 Jan 08	22:28	113.6	17	f/24	3b, 15c
1966 Jan 31	19:47	32.1	10.1	f/24	5b
1966 Feb 02	19:44	56.3	12.1	f/24	7b, 8b
1966 Feb 06	23:26	106.7	16.3	f/24	14c, 16b
1966 Mar 05	22:59	75.1	13.5	f/41	10e/1
1966 Apr 27	2152	0.7	7.1	f/24	15a
1966 Apr 28	2046	12.3	7.9	f/30	1a, 13a
1966 May 23	2033	317.5	3.4	f/24	15e/2
1966 May 23	2034	317.5	3.4	f/24	3e/2, 4b
1966 May 23	2036	317.5	3.4	f/24	16e
1966 May 28	2122	19.0	8.4	f/30	14a
1966 May 29	2103	31.0	9.4	f/30	2c
1966 Jun 23	2109	336.6	5.0	f/24	16a
1966 Aug 06	02:11	144.4	18.9	f/30	2b
1966 Aug 06	02:15	144.5	18.9	f/30	1b, 13b
1966 Aug 09	03:15	181.6	21.9	f/30	5a, 6a
1966 Aug 09	03:17	181.6	22.0	f/30	9a, 9f/2, 11b, 12c/1
1966 Oct 06	05:16	170.5	21.4	f/30	9d, 10d, 10e/2
1966 Oct 28	22:19	87.4	14.7	f/30	7e/1, 8c
1966 Oct 28	22:29	87.5	14.7	f/30	11c, 12d
1966 Oct 29	22:48	99.8	15.7	f/30	4d
1966 Nov 04	06:09	164.3	21.0	f/30	10a
1966 Nov 22	18:14	29.6	10.2	f/30	2a
1966 Nov 23	21:42	43.6	11.3	f/30	6c
1966 Nov 28	21:25	104.2	16.3	f/30	15b
1966 Nov 28	21:27	104.2	16.3	f/30	4c
1966 Dec 01	23:36	141.7	19.4	f/30	14b
1966 Dec 07	06:55	206.2	24.6	f/30	7c, 8d
1966 Dec 23	22:32	48.9	11.8	f/30	6d
1966 Dec 23	22:36	48.9	11.8	f/30	9e
1966 Dec 25	20:36	72.2	13.8	f/30	7d,11d,11e/1,12b

(continued)

A.C. Cook, *The Hatfield SCT Lunar Atlas: Photographic Atlas for Meade, Celestron, and Other SCT Telescopes: A Digitally Re-Mastered Edition*, DOI 10.1007/978-1-4614-8639-8, © Springer Science+Business Media New York 2014

Date	UT	Sun's selenographic colongitude (degrees)	Moon's age (days)	Focal ratio	Plate
1966 Dec 25	20:40	72.2	13.8	f/30	6b, 8a
1966 Dec 25	21:35	72.7	13.8	f/30	5c, 9c
1966 Dec 25	21:37	72.7	13.8	f/30	1c
1967 Jan 21	17:58	39.2	11.0	f/30	9b, 10b
1967 Jan 24	23:10	78.2	14.2	f/30	7a, 11a
1967 Jan 24	23:14	78.2	14.2	f/30	13d
1967 Feb 16	17:57	355.4	7.3	f/30	3a, 4a
1967 Feb 19	18:01	31.9	10.3	f/30	5d
1967 Feb 23	21:04	82.0	14.5	f/30	10c, 12a, 12e/1
1967 Feb 24	22:59	95.1	15.6	f/30	16c, 16f
1967 Feb 24	23:00	95.1	15.6	f/30	15d, 15e/1
1967 Feb 24	23:53	95.6	15.6	f/30	3d
1967 Feb 26	23:34	119.6	17.6	f/30	13c, 16d
1967 Feb 26	23:36	119.7	17.6	f/30	3c
1967 Mar 18	18:35	0.9	7.6	f/30	1d, 2d
1967 Mar 19	19:25	13.5	8.6	f/60	2e/2
1967 Mar 19	19:28	13.6	8.6	f/60	9f/1
1967 Mar 19	19:59	13.8	8.6	f/30	13f, 14d
1967 Mar 20	20:29	26.3	9.7	f/60	6e/1
1967 Mar 20	20:41	26.4	9.7	f/60	5e/1
1967 Mar 20	20:46	26.4	9.7	f/60	14e/2
1967 Mar 20	20:50	26.4	9.7	f/60	13g/2
1967 Mar 20	20:56	26.5	9.7	f/60	9g/1
1967 Mar 21	19:08	37.7	10.6	f/60	5e/2
1967 Mar 21	19:23	37.9	10.6	f/60	6e/2
1967 Mar 21	19:32	37.9	10.6	f/60	9g/2
1967 Mar 23	19:38	62.3	12.6	f/60	12e/2
1967 Mar 23	19:44	62.4	12.6	f/60	8e/2
1967 Apr 17	19:50	7.3	7.9	f/52	13e/2
1967 Apr 17	19:51	7.3	7.9	f/52	13g/1
1967 Apr 20	21:06	44.5	11.0	f/52	11e/2
1967 Apr 21	19:52	56.0	11.9	f/52	8e/3b
1967 Apr 21	19:58	56.1	11.9	f/52	12c/2
1967 Apr 21	20:33	56.4	12.0	f/52	8e/1
1967 Apr 23	21:50	81.4	14.0	f/52	8e/3c
1967 May 13	20:02	324.7	4.2	f/52	3g/1
1967 May 13	20:32	325.0	4.3	f/7.25	17
1967 May 15	20:00	349.1	6.2	f/52	3f/2
1967 May 15	20:08	349.2	6.2	f/52	3g/2
1967 May 15	20:15	349.3	6.2	f/52	3f/1
1967 May 16	20:08	1.4	7.2	f/52	13e/1
1967 May 16	20:12	1.4	7.2	f/52	2e/1
1967 May 16	20:16	1.5	7.2	f/52	1e/1
1967 May 16	20:17	1.5	7.2	f/52	1e/2
1967 May 16	20:21	1.5	7.3	f/52	14e/1
1967 May 20	21:10	50.7	11.3	f/52	8e/3a
1967 Aug 20	23:04	95.8	14.9	f/30	3e/1, 4e

Appendix 2

TLP Identification Flow Charts

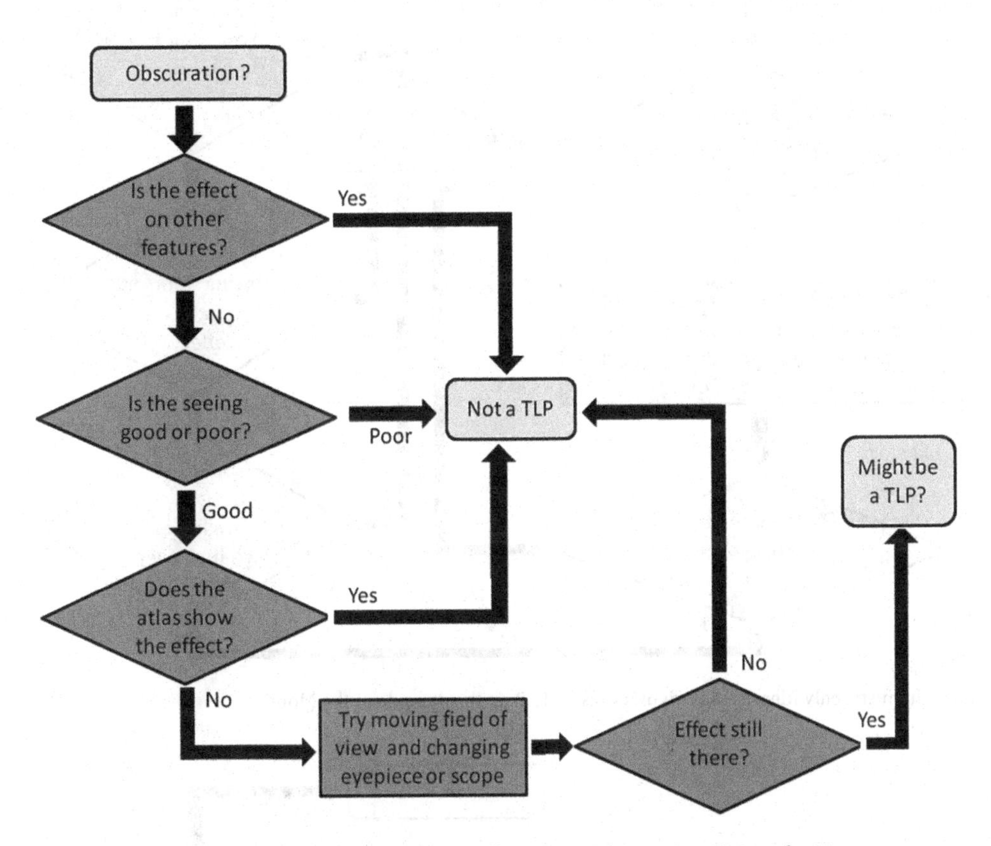

Fig. A.1 How to avoid mistakenly identifying atmospheric seeing effects for an obscuration TLP on the Moon

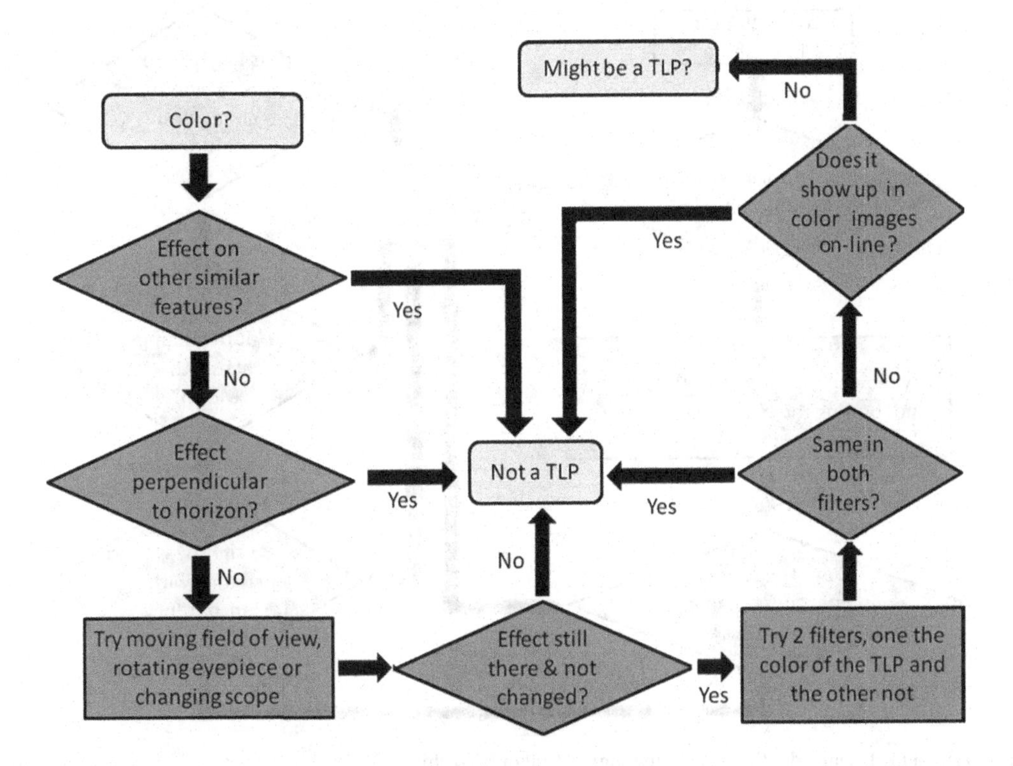

Fig. A.2 How to avoid mistakenly identifying color as TLP on the Moon

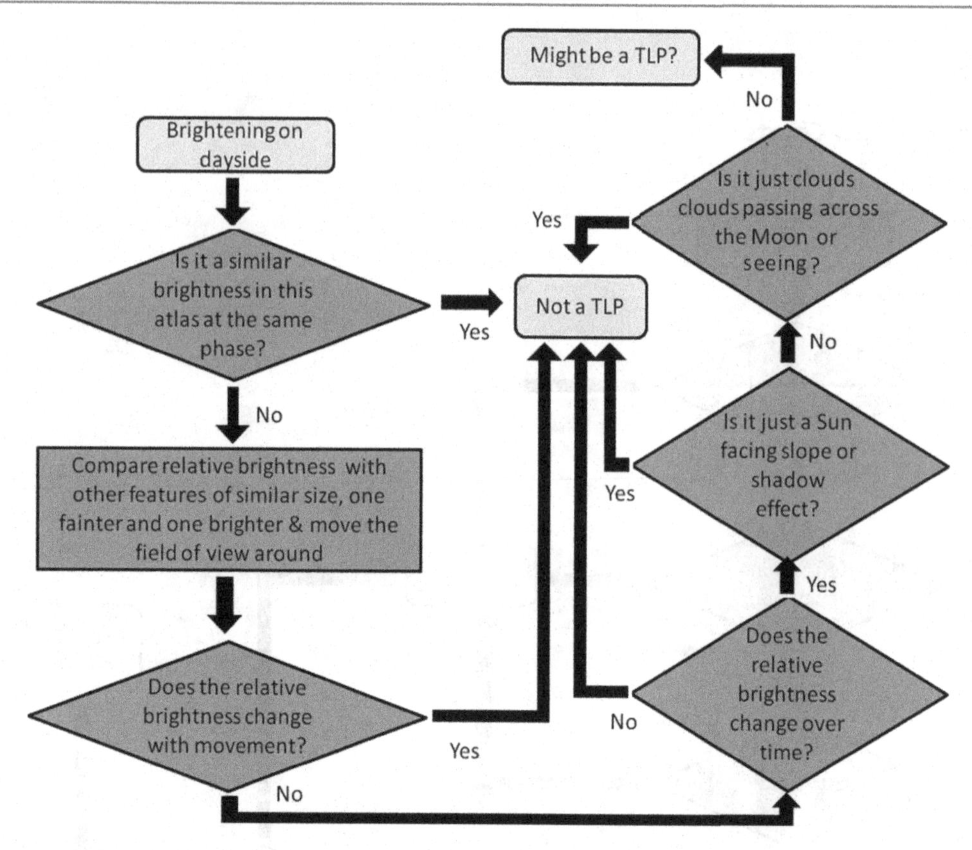

Fig. A.3 How to avoid mistakenly identifying brightenings as TLP on the day side of the Moon

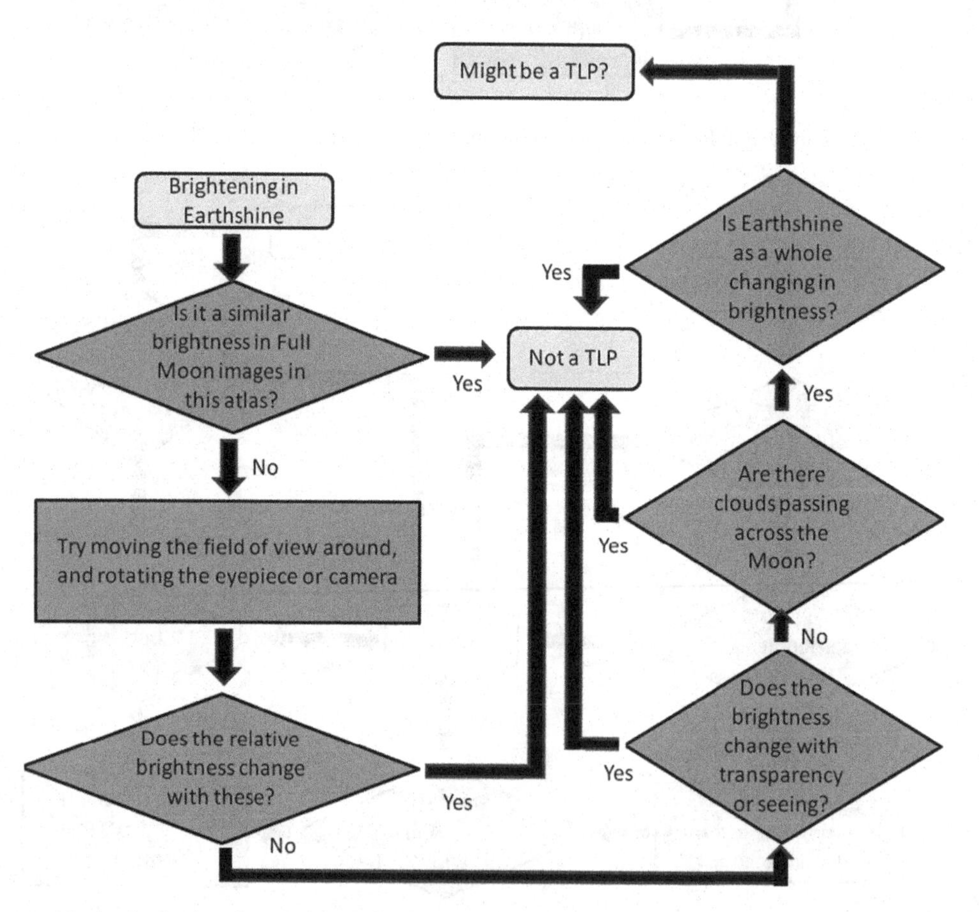

Fig. A.4 How to avoid mistakenly identifying brightenings as TLP in Earthshine

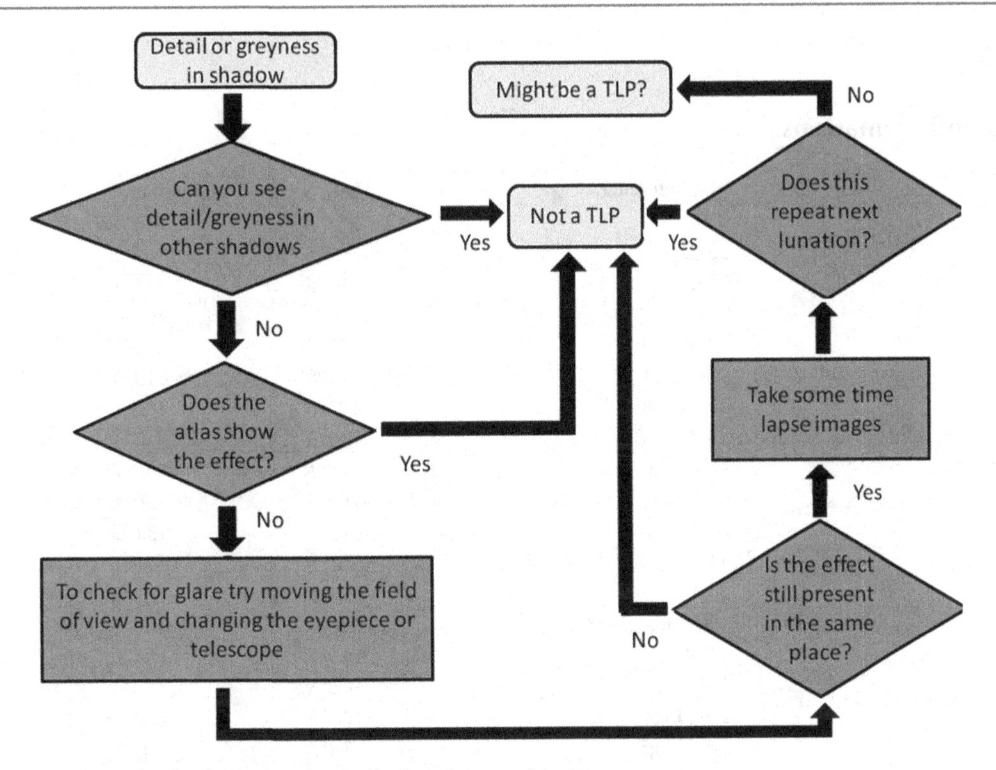

Fig. A.5 How to avoid mistakenly identifying TLP in shadowed areas of the Moon

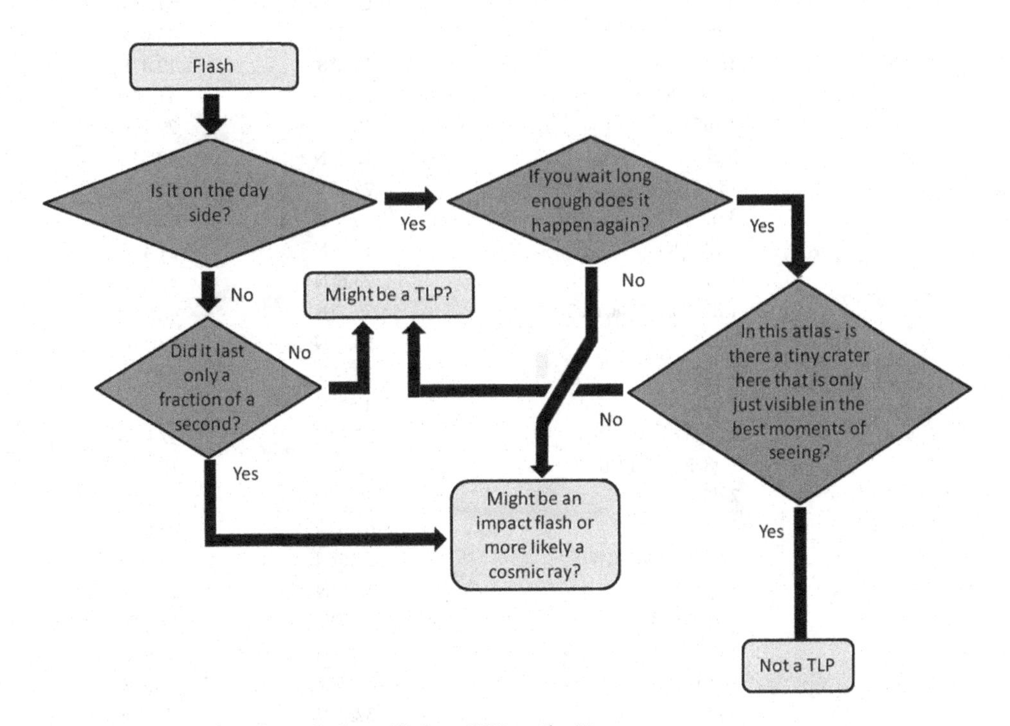

Fig. A.6 How to avoid mistakenly identifying flashes of light as TLP on the Moon

Appendix 3

Index of Named Formations

Feature	Map reference	Lat.	Lon.	Size (km)
Abbot	3 c2	5.5° N	54.7°E	10
Abenezra	13 d3	21.0°S	11.9°E	43
Abulfeda	13 d5	13.9°S	13.9°E	62
Acosta	15 c6	5.7° S	60.1°E	13
Adams	15 d1, 16 b7	31.9°S	68.3°E	63
Agatharchides	9f4, 11 3b	19.9°S	31.1°W	52
Agrippa	1e2, 13 d8	4.1° N	10.5°E	44
Airy	13 e4	18.2°S	5.6° E	39
Al-Bakri	1 c4, 3 h5	14.3°N	20.2°E	12
Albategnius	13 e5	11.2°S	4.0° E	131
Alexander	2 d3	40.2°N	13.6°E	93
Alfraganus	13 b6	5.4° S	19.0°E	21
Alhazen	3 b4	15.8°N	71.7°E	36
Aliacensis	10 a8, 13f1, 14e8	30.6°S	5.1° E	80
Almanon	13 c4	16.9°S	15.1°E	48
Al-Marrakushi	15 c6	10.5°S	55.7°E	9
Alpetragius	9 a4, 13g4	16.1°S	4.6° W	40
Alphonsus	9 a4, 13 g5	13.4°S	2.9° W	111
Alpine Valley*	2 f5	49.2°N	3.6° E	155
Ameghino	3 b1	3.3° N	57.0°E	9
Ammonius	13 g6	8.5° S	0.9° W	9
Amundsen	14 d1	84.4°S	83.1°E	103
Anaxagoras	2 g7, 6 b8	73.5°N	10.1°W	51
Anaximander	6 e7	67.0°N	51.4°W	69
Anaximenes	6 d8	72.5°N	45.0°W	81
Anděl	13 d5	10.4°S	12.4°E	33
Angström	5 h8, 6 h2, 8 c3	29.9°N	41.7°W	10
Ansgarius	15 a5	12.9°S	79.7°E	95
Anville	3 c2	1.9° N	49.5°E	10
Apianus	13 e2, 14 e8	27.1°S	7.8° E	63
Apollonius	3 b2	4.6° N	60.9°E	51
Arago	1 c2, 3 h3	6.2° N	21.4°E	26
Aratus	1 f6	23.6°N	4.5° E	10
Archimedes	1 h7, 2 g1, 5 b8, 6 a2	29.7°N	4.0° W	81
Archytas	2 f6	58.8°N	5.0° E	31
Argelander	13 e4	16.6°S	5.8° E	34
Ariadaeus	1 d2	4.5° N	17.3°E	10
Aristarchus	7 c8, 8 d2	23.7°N	47.5°W	40
Aristillus	1 g8, 2 f2, 5 a8, 6 a2	33.8°N	1.2° E	55
Aristoteles	2 d5	50.2°N	17.3°E	88
Arnold	2 c7	67.0°N	35.8°E	93
Artsimovich	5 h8, 6 g1, 7 b8, 8 c2	27.6°N	36.6°W	8
Aryabhata	3 e3	6.2° N	35.2°E	22
Arzachel	9 a3, 13 g4	18.3°S	2.0° W	97
Asada	3 c3	7.2° N	49.9°E	13
Asclepi	14 c4	55.2°S	25.5°E	41
Aston	8 g4	32.8°N	87.7°W	49
Atlas	2 a4,4 g6	46.7°N	44.3°E	87
Atwood	15 c8	5.9° S	57.7°E	29
Autolycus	1 g7, 2 f1, 5 a8	30.7°N	1.5° E	39

(continued)

Feature	Map reference	Lat.	Lon.	Size (km)
Auwers	1 d4	15.0°N	17.2°E	20
Auzout	3 b3	10.2°N	64.0°E	33
Azophi	13 d3	22.2°S	12.7°E	48
Baade	12 g6	44.8°S	82.0°W	58
Babbage	6 g6, 8 b7	59.6°N	57.4°W	147
Baco	14 c4	51.1°S	19.2°E	65
Baillaud	2 d8	74.6°N	37.2°E	89
Bailly	10 g2, 12 d2	66.5°S	69.5°W	301
Baily	2 b5, 4 h6	49.8°N	30.5°E	26
Balboa	7 g7	19.2°N	83.2°W	69
Ball	10 c7, 14 h7	35.9°S	8.5° W	40
Balmer	15 b3	20.3°S	70.2°E	136
Banachiewicz	3 a2	5.3° N	80.0°E	99
Bancroft	2 h1, 5 b7, 6 b1	28.1°N	6.4° W	13
Banting	1 d6	26.6°N	16.4°E	5
Barkla	15 b5	10.7°S	67.2°E	41
Barocius	14 d5	45.0°S	16.8°E	83
Barrow	2 f7	71.4°N	8.0° E	89
Bartels	7 g7, 8 h1	24.5°N	89.8°W	55
Bayer	10 g4, 12 c4	51.6°S	35.2°W	49
Beals	4 c4	37.1°N	86.5°E	53
Beaumont	13 a4, 15 g4	18.1°S	28.8°E	51
Beer	5 c7, 6 b1	27.1°N	9.1° W	9
Behaim	15 b4	16.6°S	79.4°E	56
Beketov	1 b4, 3 f5	16.2°N	29.2°E	8
Bellot	15 d5	12.5°S	48.2°E	18
Bernoulli	3 d8, 4 d4	34.8°N	60.6°E	48
Berosus	3 c8, 4 c4	33.4°N	69.9°E	75
Berzelius	4 e4	36.6°N	51.0°E	51
Bessarion	7 b6	14.8°N	37.3°W	10
Bessel	1 d5	21.7°N	17.9°E	16
Bettinus	10 g2, 12 c2	63.4°S	45.1°W	72
Bianchini	6 f5, 8 a6	48.8°N	34.4°W	38
Biela	16 f3	55.0°S	51.7°E	77
Bilharz	15 c6	5.8° S	56.3°E	45
Billy	11 e5	13.8°S	50.2°W	46
Biot	15 d3	22.7°S	51.0°E	13
Birmingham	2 g7, 6 b7	65.1°N	10.7°W	90
Birt	9 b3, 13 h3	22.4°S	8.7° W	16
Black	15 a5	9.2° S	80.4°E	19
Blagg	1 g1, 13 f8	1.2° N	1.4° E	5
Blancanus	10 e2, 12 a2	63.7°S	21.8°W	106
Blanchinus	13 f2	25.4°S	2.4° E	60
Bliss	2 h5, 6 c6	53.1°N	13.8°W	23
Bobillier	1 d5	19.6°N	15.4°E	6
Bode	1 h3, 5b3	6.7° N	2.5° W	18
Boguslawsky	14 c2	72.9°S	43.3°E	95
Bohnenberger	15 e5	16.3°S	40.0°E	32
Bohr	7 h5	12.7°N	86.5°W	70
Bombelli	3 c2	5.3° N	56.1°E	10
Bond G.	1 a7, 3 f8, 4 f3	32.4°N	36.3°E	19
Bond W.	2 f7, 6 a7	65.2°N	3.9° E	156
Bonpland	9 c6	8.4° S	17.3°W	59
Boole	6 g7, 8 b8	63.8°N	87.3°W	61

(continued)

Feature	Map reference	Lat.	Lon.	Size (km)
Borda	15 e3	25.2°S	46.5°E	47
Borel	1 b5, 3 g6, 4 g1	22.4°N	26.4°E	5
Born	15 b6	6.0° S	66.8°E	15
Boscovich	1 e3	9.7° N	11.0°E	42
Bouguer	6 f5, 8a6	52.3°N	35.8°W	22
Boussingault	14 b2, 16 h1	70.2°S	53.7°E	128
Bowen	1 e5	17.6°N	9.1° E	8
Brayley	5 h6, 7 b7, 8 c1	20.9°N	36.9°W	14
Breislak	14 c5	48.3°S	18.3°E	49
Brenner	16 f6	39.1°S	39.1°E	90
Brewster	1 a5, 3 f6, 4 f1	23.3°N	34.7°E	10
Brianchon	6 e8	74.7°N	88.3°W	137
Briggs	7 f8, 8 g2	26.4°N	69.2°W	37
Brisbane	16 d4	49.2°S	68.7°E	44
Brown	10 e5, 12 a4	46.5°S	18.1°W	34
Bruce	1 g1, 13 f8	1.2° N	0.4° E	6
Buch	14 c6	38.9°S	17.7°E	51
Bullialdus	9 e3	20.7°S	22.4°W	61
Bunsen	7 f5	41.3°N	85.4°W	55
Burckhardt	3 d8, 4 d3	31.1°N	56.4°E	56
Bürg	2 b4, 4 h5	45.0°N	28.2°E	41
Burnham	13 e5	13.9°S	7.2° E	24
Büsching	14 c7	38.0°S	19.9°E	53
Byrgius	11 g3	24.8°S	65.4°W	84
Cabeus	10 d1	85.3°S	41.8°W	101
Cajal	1 a3, 3 f4	12.6°N	31.1°E	9
Calippus	2 e3	38.9°N	10.7°E	30
Cameron	3 d3	6.2° N	45.9°E	11
Campanus	9 f2, 10 g8, 11 a1, 12 b8	28.1°S	27.9°W	46
Capella	15 f7	7.6° S	34.9°E	48
Capuanus	9 f1, 10 g7, 12 b7	34.1°S	26.8°W	60
Cardanus	7 g5	13.3°N	72.5°W	50
Carlini	6 e2	33.7°N	24.1°W	11
Carmichael	3 e5, 4 e1	19.5°N	40.4°E	20
Carpenter	6 e7	69.5°N	51.2°W	59
Carrel	1 b3, 3 g4	10.7°N	26.7°E	16
Carrington	4 e5	44.0°N	62.0°E	28
Cartan	3 b2	4.2° N	59.3°E	16
Casatus	10 e1, 12 a1	72.7°S	30.6°W	103
Cassini	2 f3	40.2°N	4.6° E	56
Catharina	13 b4, 15 h4	18.0°S	23.5°E	99
Cauchy	3 e3	9.6° N	38.6°E	12
Cavalerius	7 g4	5.1° N	66.9°W	59
Cavendish	11 e3	24.6°S	53.8°W	53
Caventou	5 f8, 6 f2, 8 a3	29.7°N	29.4°W	3
Cayley	1 d2, 13 c8	3.9° N	15.1°E	14
Celsius	14 c7	34.1°S	20.0°E	39
Censorinus	3 f1, 15 f8	0.4° S	32.7°E	4
Cepheus	4 f5	40.7°N	45.8°E	39
Chacornac	1 b7, 2a1, 3 f8, 4 g3	29.9°N	31.7°E	50
Challis	2 f8	79.6°N	9.1° E	53
C. Herschel	6 f3, 8 a4	34.5°N	31.3°W	14
Chevallier	4 f5	45.0°N	51.6°E	52
Chladni	1 g2	4.0° N	1.1° E	13

(continued)

Feature	Map reference	Lat.	Lon.	Size (km)
Cichus	9 e1, 10f7, 12 a7	33.3°S	21.2°W	39
Clairaut	14 d5	47.9°S	13.9°E	77
Clausius	9 h1, 12 e7	36.9°S	43.9°W	24
Clavius	10 d3, 12 a2, 14 g3	58.6°S	14.8°W	231
Cleomedes	3 c7, 4 d2	27.5°N	55.5°E	131
Cleostratus	6 h7, 8 c7	60.4°N	77.4°W	64
Clerke	1 b5, 3 f6, 4 g1	21.7°N	29.8°E	7
C. Mayer	2 e6	63.2°N	17.2°E	38
Cobra's Head*	7 d8, 8 e1	24.5°N	49.3°W	8
Colombo	15 d5	15.3°S	46.1°E	79
Condon	3 b1, 15 b8	1.8° N	60.4°E	34
Condorcet	3 b3	12.1°N	69.7°E	75
Conon	1 g6	21.7°N	2.0° E	21
Cook	15 d4	17.6°S	48.8°E	45
Copernicus	5 e4	9.6° N	20.1°W	96
Cremona	6 f7	67.2°N	90.9°W	85
Crile	3 d4	14.2°N	46.0°E	9
Crozier	15 d5	13.6°S	50.7°E	23
Crüger	11 g5	16.7°S	66.9°W	46
Curtius	10 b2, 14 e2	67.1°S	4.1° E	96
Cusanus	2 c8	71.8°N	69.2°E	61
Cuvier	10 a4, 14 e4	50.3°S	9.7° E	77
Cyrillus	13 a5, 15 h5	13.3°S	24.1°E	98
Cysatus	10 c2, 14 f2	66.2°S	6.4° W	48
Daguerre	15 f6	11.9°S	33.6°E	46
Dalton	7 g6	17.1°N	84.4°W	61
Daly	3 b2	5.8° N	59.5°E	15
Damoiseau	7 f2, 11 f7	4.9° S	61.2°W	37
Daniell	1 b8, 2 a2, 4 g4	35.4°N	31.1°E	28
Darney	9 e5	14.6°S	23.6°W	15
d'Arrest	1 d1, 13 c8	2.2° N	14.6°E	30
Darwin	11 g4	19.8°S	69.2°W	122
Daubrée	1 d4	15.7°N	14.8°E	15
da Vinci	3 d3	9.1° N	44.9°E	37
Davy	9 b5, 13 h5	11.9°S	8.2° W	34
Dawes	1 b4, 3 g5	17.2°N	26.3°E	18
Debes	3 d7, 4 d3	29.5°N	51.6°E	32
Dechen	8 d5	46.1°N	68.2°W	12
de Gasparis	11 e2	25.8°S	50.8°W	31
Delambre	13 c7	2.0° S	17.4°E	51
de La Rue	2 a6, 4 h7	59.2°N	52.9°E	135
Delaunay	13 f3	22.3°S	2.6° E	45
Delisle	5 g8, 6 g2, 8 b3	30.0°N	34.7°W	25
Delmotte	3 c7, 4c2	27.1°N	60.2°E	31
Deluc	10 c4, 14 f3	55.0°S	3.0° W	47
Dembowski	1 f2, 13 e8	2.9° N	7.3° E	26
Democritus	2 c6	62.3°N	34.9°E	38
Demonax	14 c1	78.2°S	59.1°E	122
De Morgan	1 d1, 13 c8	3.3° N	14.9°E	10
Desargues	6 f8	70.4°N	73.3°W	85
Descartes	13 c5	11.8°S	15.6°E	48
Deseilligny	1 c5, 3 h6, 4 h1	21.1°N	20.6°E	6
de Sitter	2 e8	79.8°N	38.4°E	64
Deslandres	9 b1, 10 c7, 13 h1, 14 g7	32.4°S	6.0° W	227

(continued)

Feature	Map reference	Lat.	Lon.	Size (km)
de Vico	11 f4	19.7°S	60.3°W	22
Dionysius	1 dl, 13 c8	2.8° N	17.3°E	17
Diophantus	5 g8, 6 g1, 7 a8, 8 b2	27.6°N	34.3°W	18
Dollond	13 c5	10.5°S	14.4°E	11
Donati	13 e3	20.8°S	5.0° E	36
Doppelmayer	9 h3, 11 c2, 12 d8	28.5°S	41.5°W	65
Dorsa Lister	3 g6, 4 g1	19.8°N	23.5°E	180
Dorsa Smirnov	3 g7, 4 g2	26.4°N	25.5°E	222
Dove	14 b5, 16 g5	46.9°S	31.4°E	30
Draper	5 e5	17.6°N	21.8°W	8
Drebbel	12 e6	40.9°S	49.1°W	30
Drygalski	10 e1	79.6°S	87.2°W	162
Dubyago	3 b2	4.4° N	69.9°E	48
Dunthorne	9 g2, 10 h8, 11 b1, 12 c8	30.1°S	31.8°W	15
Eddington	7 f7, 8g1	21.5°N	71.9°W	120
Egede	2 e4	48.7°N	10.6°E	34
Eichstadt	11 h3	22.7°S	78.4°W	50
Eimmart	3 b6, 4 c2	23.9°N	64.8°E	45
Einstein	7 h6	16.5°N	88.4°W	181
Elger	9 f1, 10 g7, 12 c7	35.4°S	29.9°W	22
Encke	7 c4, 11 b8	4.6° N	36.7°W	28
Endymion	4 g7	53.6°N	56.5°E	122
Epigenes	2 g7, 6 b7	67.5°N	4.5° W	55
Epimenides	10 g6, 12 b6	40.9°S	30.4°W	23
Eppinger*	9 e6,	9.4° S	25.7°W	6
Eratosthenes	5 c5	14.5°N	11.4°W	59
Esclangon	3 e6, 4 e1	21.5°N	42.1°E	15
Euclides	7 a1, 9 f7, 11 a6	7.4° S	29.6°W	12
Euctemon	2 e8	76.3°N	30.5°E	63
Eudoxus	2 d4	44.3°N	16.2°E	70
Euler	5 g7, 8 a2	23.3°N	29.2°W	26
Fabbroni	1 b4, 3f5	18.7°N	29.3°E	11
Fabricius	16 f5	42.8°S	41.8°E	79
Fahrenheit	3 b4	13.1°N	61.7°E	7
Faraday	10 a6, 14 e6	42.5°S	8.7° E	69
Fauth	5 e3	6.2° N	20.2°W	12
Faye	13 f3	21.4°S	3.7° E	36
Fermat	13 c3	22.7°S	19.8°E	38
Fernelius	10 a6, 14e6	38.3°S	4.8° E	68
Feuillée	2 h1, 5 c7, 6 b1	27.4°N	9.5° W	9
Firmicus	3 b3	7.3° N	63.3°E	53
Flammarion	5 b1, 13 g7	3.3° S	3.8° W	76
Flamsteed	7 d2, 9 h8, 11 d7	4.5° S	44.3°W	19
Fontana	11 f5	16.1°S	56.8°W	31
Fontenelle	2 h6, 6 c7	63.4°N	18.9°W	38
Foucault	6 f5, 8 a6	50.5°N	39.9°W	23
Fourier	11 e1, 12 f8	30.3°S	53.1°W	52
Fracastorius	15 g4	21.4°S	33.0°E	121
Fra Mauro	5 e1, 9 c6	6.1° S	17.0°W	97
Franck	1 a5, 3 f6, 4 f1	22.6°N	35.6°E	12
Franklin	4 e4	38.7°N	47.6°E	56
Franz	3 e5	16.6°N	40.2°E	25
Fraunhofer	16 d6	39.5°S	59.0°E	55
Fredholm	3 d5, 4 d1	18.3°N	46.5°E	13
Furnerius	15 d1, 16 c7	36.1°S	60.5°E	135
Galen	1 f6	21.9°N	5.0° E	9

(continued)

Feature	Map reference	Lat.	Lon.	Size (km)
Galilaei	7 f5	10.5°N	62.8°W	16
Galle	2 d6	55.9°N	22.3°E	21
Galvani	8 e6	49.6°N	84.5°W	77
Gambart	5 d2, 9c8	0.9° N	15.3°W	25
Gardner	1 a4, 3f5	17.7°N	33.8°E	18
Gärtner	2 c6	59.2°N	34.7°E	102
Gassendi	9 h5, 11 c4	17.5°S	40.0°W	110
Gaudibert	15 e6	10.9°S	37.8°E	33
Gauricus	9 c1, 10 d7	33.9°S	12.8°W	80
Gauss	3 c8, 4 c4	36.0°N	79.0°E	171
Gay-Lussac	5 e5	13.9°N	20.8°W	25
G. Bond	1 a7, 3 f8, 4 f3	32.4°N	36.3°E	19
Geber	13 d3	19.5°S	13.8°E	45
Geminus	3 d8, 4 d4	34.3°N	56.6°E	83
Gemma Frisius	14 d7	34.4°S	13.4°E	89
Gerard	8 e5	44.5°N	80.6°W	99
Gibbs	15 b3	18.3°S	84.2°E	79
Gilbert	15 a7	3.2° S	76.1°E	100
Gill	16 f2	63.8°S	75.9°E	64
Gioja	2 f8	83.3°N	1.7° E	42
Glaisher	3 c4	13.1°N	49.3°E	16
Glushko	7 h4	8.1° N	77.6°W	40
Goclenius	15 d6	10.1°S	45.1°E	73
Goddard	3 a4	15.0°N	89.1°E	93
Godin	1 e1, 13 d8	1.8° N	10.1°E	34
Goldschmidt	2 f7, 6 b8	73.0°N	3.8° W	115
Golgi	7 e8, 8 f2	27.8°N	59.9°W	6
Goodacre	13 d1, 14 d8	32.7°S	14.0°E	44
Gould	9 d4	19.3°S	17.3°W	33
Greaves	3 c4	13.2°N	52.8°E	13
Grimaldi	7 g1, 11 g7	5.4° S	68.4°W	173
Grove	1 b8, 2 a3, 4 g5	40.3°N	33.0°E	27
Gruemberger	10 c2, 14 g2	67.0°S	10.4°W	92
Gruithuisen	6 g2, 8 c3	32.9°N	39.8°W	16
Guericke	9 c5	11.6°S	14.2°W	61
Gutenberg	15 e6	8.7° S	41.2°E	69
Gyldén	13 f7	5.3° S	0.3° E	45
Hagecius	14 a3, 16 g3	60.0°S	46.7°E	80
Hahn	3 c7, 4 c3	31.2°N	73.6°E	87
Haidinger	10 f6, 12 b6	39.2°S	25.2°W	21
Hainzel	10 h6, 12 c6	41.2°S	33.6°W	71
Hale	14 b1, 16 h1	74.1°S	91.7°E	84
Hall	1 a7, 3 f8, 4 f4	33.8°N	36.7°E	32
Halley	13 e6	8.1° S	5.7° E	36
Hanno	16 e3	56.5°S	71.4°E	60
Hansen	3 b4	14.0°N	72.4°E	41
Hansteen	11 e5	11.5°S	52.0°W	45
Harding	8 e5	43.5°N	71.6°W	23
Hargreaves	15 b7	2.2° S	64.1°E	18
Harlan	16 c6	38.3°S	79.6°E	63
Harpalus	6 f5, 8a6	52.7°N	43.5°W	40
Hartwig	7 h1, 11 h7	6.3° S	80.4°W	78
Hase	15 d1, 16 c8	29.5°S	62.7°E	82
Hausen	10 h2, 12 e2	65.0°S	88.4°W	163
Hecataeus	15 b3	22.1°S	79.5°E	143
Hedin	7 h3	2.9° N	76.5°W	157

(continued)

Feature	Map reference	Lat.	Lon.	Size (km)
Heinrich	5 d7, 6 d1	24.8°N	15.4°W	7
Heinsius	10 e6, 12 a5	39.4°S	17.8°W	62
Heis	5 g8, 6 f2, 8 b3	32.4°N	32.0°W	14
Helicon	6 e4	40.4°N	23.1°W	24
Hell	9 b1, 10 c7, 13 h1, 14 h7	32.4°S	7.9° W	33
Helmholtz	14 b2, 16 g1	68.5°S	65.2°E	103
Henry	11 f3	24.0°S	57.0°W	39
Henry Fréres	11 f3	23.6°S	59.0°W	42
Heraclitus	10 a4, 14 e4	49.2°S	6.2° E	85
Hercules	2 a4, 4 g6	46.9°N	39.2°E	71
Herigonius	9 g6, 11 b5	13.4°S	34.0°W	15
Hermann	7 f2, 11e8	0.9° S	57.4°W	15
Herodotus	7 d8, 8 e1	23.2°N	49.8°W	34
Herschel	13 g6	5.7° S	2.1° W	39
Herschel C.	6 f3, 8 a4	34.5°N	31.3°W	14
Herschel J.	6 e7	62.2°N	41.9°W	163
Hesiodus	9 d2, 10e8	29.4°S	16.4°W	43
Hevelius	7 g3, 11 f8	2.1° N	67.5°W	117
Hill	3 e6, 4 e1	20.9°N	40.8°E	16
Hind	13 e6	7.9° S	7.3° E	28
Hippalus	9 f3, 11 b2	24.9°S	30.4°W	57
Hipparchus	13 e7	5.4° S	4.9° E	144
Holden	15 c4	19.1°S	62.5°E	48
Hommel	14 b4, 16 h3	54.7°S	32.9°E	114
Hooke	4 e5	41.1°N	54.8°E	33
Horrebow	6 e6	58.8°N	40.9°W	25
Horrocks	13 e7	4.0° S	5.8° E	30
Hortensius	5 g3, 7a4	6.5° N	28.0°W	14
Huggins	10 b6, 14 f6	41.1°S	1.6° W	66
Humason	8 e3	30.7°N	56.7°W	4
Humboldt	15 c1, 16 b8	26.9°S	80.8°E	199
Huxley	5 c6	20.2°N	4.6° W	3
Hyginus	1 f3	7.8° N	6.3° E	9
Hypatia	3 h1, 13 b7, 15 h7	4.3° S	22.6°E	36
Ibn Battuta	15 d6	7.0° S	50.4°E	12
Ibn-Rushd	13 b5, 15 h6	11.7°S	21.7°E	31
Ideler	14 c5	49.3°S	22.2°E	38
Inghirami	12 f5	47.5°S	69.0°W	95
Isidorus	15 f6	8.0° S	33.5°E	41
Jacobi	10 a3, 14 e3	56.9°S	11.3°E	66
Jansen	1 b3, 3 g4	13.6°N	28.6°E	23
Janssen	14 a5, 16 f5	45.0°S	41.1°E	201
Jenkins	3 a1, 15 a7	0.5° N	78.0°E	38
J. Herschel	6 e7	62.2°N	41.9°W	163
Joy	1 f6	25.0°N	6.6° E	5
Julius Caesar	1 d3	9.1° N	15.2°E	85
Kaiser	10 a7, 14 e7	36.5°S	6.5° E	53
Kane	2 d6	63.0°N	25.8°E	55
Kant	13 b5	10.6°S	20.2°E	31
Kapteyn	15 b5	10.8°S	70.5°E	50
Kästner	15 a6	6.8° S	78.9°E	113
Keldysh	2 a5, 4 g6	51.2°N	43.7°E	32
Kepler	7 c5	8.1° N	38.0°W	29
Kies	9 e2, 10 f8, 12 a8	26.3°S	22.7°W	45
Kinau	10 a3, 14 d3	60.8°S	14.9°E	42
Kirch	2 g3, 6 b3	39.3°N	5.6° W	12

(continued)

Feature	Map reference	Lat.	Lon.	Size (km)
Kircher	10 f2, 12 c2	67.0°S	45.5°W	71
Kirchhoff	3 e8, 4 f3	30.3°N	38.8°E	24
Klaproth	10 e2, 12 a1	69.8°S	26.7°W	110
Klein	13 f5	12.0°S	2.5° E	43
König	9 e3, 11 a2	24.2°S	24.7°W	23
Krafft	7 g6	16.6°N	72.7°W	51
Krieger	8 d3	29.0°N	45.6°W	23
Krogh	3 b3	9.4° N	65.6°E	20
Krusenstern	13 e2	26.4°S	5.8° E	46
Kuiper	9 e6	9.8° S	22.7°W	7
Kundt	9 b5	11.6°S	11.6°W	11
Kunowsky	5 h3, 7 b4, 11 a8	3.2° N	32.5°W	18
la Caille	13 f3	23.7°S	1.0° E	61
la Condamine	6 e6	53.5°N	28.2°W	38
Lacroix	12 f7	37.9°S	59.2°W	36
Lacus Aestatis	11 g5	14.8°S	68.6°W	86
Lacus Autumni	11 h5	11.8°S	83.2°W	196
Lacus Hiemalis	7 h2, 11 h8	15.0°N	14.0°E	48
Lacus Mortis	2 b4, 4h5	45.1°N	27.3°E	159
Lacus Somniorum	1 b8, 2 a3, 4 g4	37.6°N	30.8°E	425
Lacus Veris	11 h5	16.5°S	85.9°W	383
Lade	1 e1, 13 d7	1.4° S	10.0°E	58
Lagalla	10 f5, 12b5	44.6°S	23.1°W	65
Lagrange	11 g1, 12 g8	33.3°S	71.8°W	150
Lalande	5 c1, 9 a6, 13 h7	4.5° S	8.7° W	24
Lamarck	11 g3	23.2°S	69.9°W	100
Lambert	5 e7, 6 e1	25.8°N	21.0°W	30
Lamé	15 b4	14.7°S	64.5°E	84
Laméch	2 d3	42.8°N	13.1°E	13
Lamont	1 c2, 3 g3	5.1° N	23.4°E	74
Landsteiner	5 d8, 6 c2	31.3°N	14.8°W	6
Langley	8 d6	51.2°N	86.0°W	59
Langrenus	15 c6	8.9° S	61.0°E	132
Lansberg	5 g2, 7 a3, 9 e8	0.3° S	26.7°W	40
la Pérouse	15 b5	10.6°S	76.2°E	80
Lassell	9 b4, 13 h4	15.5°S	8.0° W	23
Lavoisier	8 f4	38.1°N	81.4°W	70
Lawrence	3 d3	7.4° N	43.3°E	24
Leakey	3 e1, 15 e7	3.2° S	37.4°E	12
Lee	9 h2, 11 c1, 12 d8	30.6°S	40.8°W	41
Legendre	15 c1, 16 b8	29.0°S	70.0°E	81
le Gentil	10 f1, 12 c1	74.6°S	76.1°W	125
Lehmann	12 f6	40.0°S	56.1°W	54
le Monnier	1 b6, 3 f7, 4 g2	26.6°N	30.4°E	60
Lepaute	9 g2, 10 h7, 12 c7	33.3°S	33.7°W	16
Letronne	7 d1, 9 h7, 11 c5	10.5°S	42.5°W	118
le Verrier	6 d4	40.3°N	20.6°W	21
Lexell	10 c7, 13 g1, 14 g7	35.8°S	4.4° W	62
Licetus	10 a5, 14 e5	47.2°S	6.5° E	75
Lichtenberg	8 f3	31.8°N	67.7°W	20
Lick	3 c4	12.3°N	52.8°E	33
Liebig	11 e3	24.3°S	48.3°W	38
Lilius	10 b4, 14 e4	54.6°S	6.1° E	61
Lindbergh	15 c7	5.4° S	52.9°E	13
Lindenau	13 b1, 14 b8, 16 h7	32.4°S	24.8°E	53
Lindsay	13 d6	7.0° S	13.0°E	32

(continued)

Feature	Map reference	Lat.	Lon.	Size (km)
Linné	1 e7, 2 d1	27.7°N	11.8°E	2
Lippershey	9 c2	26.0°S	10.5°W	7
Littrow	1 a5, 3 f6, 4 f1	21.5°N	31.4°E	30
Lockyer	14 a5, 16 g5	46.3°S	36.6°E	35
Loewy	9 g4, 11 b3	22.7°S	32.9°W	22
Lohrmann	7 g2, 11 g8	0.4° S	67.4°W	31
Lohse	15 c5	13.7°S	60.3°E	43
Longomontanus	10 e4, 12 a4	49.6°S	21.9°W	146
Louville	6 g4, 8 b5	44.1°N	46.0°W	35
Lubbock	15 e7	4.0° S	41.8°E	13
Lubiniezky	9 e4	17.9°S	23.9°W	43
Luther	1 c7, 2b2, 3 h8, 4 h3	33.2°N	24.1°E	9
Lyell	3 e4	13.6°N	40.5°E	31
Lyot	16 d4	50.7°S	84.6°E	138
Maclaurin	3 b1, 15 b7	1.9° S	68.0°E	54
Maclear	1 c3, 3 h4	10.5°N	20.1°E	20
MacMillan	5 c7, 6 b1	24.2°N	7.9° W	7
Macrobius	3 d6, 4 d1	21.2°N	46.0°E	63
Mädler	15 g6	11.0°S	29.7°E	27
Maestlin	7 c4	4.9° N	40.7°W	7
Magelhaens	15 e5	12.0°S	44.1°E	37
Maginus	10 c4, 14 g4	49.9°S	6.8° W	181
Main	2 f8	80.9°N	10.5°E	47
Mairan	6 g4, 8 c5	41.6°N	43.5°W	40
Malapert	10 c1, 14 f1	84.8°S	12.3°E	70
Mallet	16 e5	45.5°S	54.1°E	59
Manilius	1 e4	14.4°N	9.1° E	38
Manners	1 c2, 3 h3, 13 b8	4.6° N	20.0°E	15
Manzinus	14 d2	67.5°S	26.1°E	98
Maraldi	1 a4, 3 f6, 4 f1	19.3°N	34.8°E	40
Marco Polo	1 h4, 5 a5	15.5°N	2.0° W	30
Mare Anguis	3 b6, 4b1	22.4°N	67.6°E	146
Mare Australe	16 e3	40.4°S	94.5°E	612
Mare Cognitum	9 e6	10.5°S	22.3°W	350
Mare Crisium	3 c5, 4 c1	16.2°N	59.1°E	556
Mare Fecunditatis	3 c1, 15 c7	7.8° S	53.7°E	840
Mare Frigoris	2 e5, 6 c6	57.6°N	0.0° W	1446
Mare Humboldtianum	4 f7	56.9°N	81.5°E	231
Mare Humorum	9 g4, 11 c3	24.5°S	38.6°W	420
Mare Imbrium	2 h3, 5 e7, 6 d3, 8 a3	34.7°N	14.9°W	1146
Mare Insularum	9 d7	7.8° N	30.6°W	512
Mare Marginis	3 a4, 4 a1	12.7°N	86.5°E	358
Mare Nectaris	15 f5	15.2°S	34.6°E	339
Mare Nubium	9 d3	20.6°S	17.3°W	715
Mare Serenitatis	1 c6, 2 c1, 3 h7, 4 h2	27.3°N	18.4°E	674
Mare Smythii	15 a7	1.7° S	87.1°E	374
Mare Spumans	3 b1, 15 b8	1.3° N	65.3°E	143
Mare Tranquillitatis	1 b2, 3 f3, 13 a8, 15 g8	8.4° N	30.8°E	876
Mare Undarum	3 b2	7.5° N	68.7°E	245
Mare Vaporum	1 f4	13.6°N	3.7° E	242
Marinus	16 c6	39.4°S	76.5°E	57
Marius	7 e5	11.9°N	50.8°W	40
Markov	6 h6, 8 c6	53.4°N	62.8°W	41
Marth	9 f2, 10 g7, 11 b1, 12 b7	31.2°S	29.4°W	7
Maskelyne	1 a1, 3 f2	2.1° N	30.1°E	22
Mason	2 b4, 4 h5	42.7°N	30.5°E	33

(continued)

Feature	Map reference	Lat.	Lon.	Size (km)
Maupertuis	6 e5	49.7°N	27.3°W	44
Maurolycus	14 d6	41.9°S	13.9°E	115
Maury	1 a8, 4 f4	37.1°N	39.7°E	17
Mayer C.	2 e6	63.2°N	17.2°E	38
Mayer T.	5 g5, 7 a6	15.5°N	29.2°W	33
McClure	15 d5	15.4°S	50.2°E	24
McDonald	5 e8, 6 d2	30.4°N	20.9°W	7
Mee	10 g5, 12 c5	43.6°S	35.3°W	136
Menelaus	1 d4	16.2°N	15.9°E	27
Mercator	9 f2, 10 g8, 11 a1, 12 b8	29.3°S	26.2°W	46
Mercurius	4 e6	46.6°N	66.0°E	64
Mersenius	11 e3	21.5°S	49.3°W	84
Messala	4 d4	39.3°N	60.1°E	122
Messier	3 d1, 15 d8	1.9° S	47.6°E	14
Metius	16 e6	40.4°S	43.4°E	86
Meton	2 e7	73.4°N	19.2°E	125
Milichius	5 g4, 7 a5	10.0°N	30.3°W	12
Miller	10 b6, 14 f6	39.4°S	0.7° E	61
Mitchell	2 d5	49.8°N	20.2°E	32
Moigno	2 d7	66.2°N	28.9°E	33
Moltke	3 g1, 13 a7, 15 h8	0.6° S	24.2°E	6
Monge	15 d4	19.3°S	47.6°E	34
Mons Ampére	5 b6	19.3°N	3.7° W	30
Mons Argaeus	1 b5, 3 g6, 4 g1	19.3°N	29.0°E	61
Mons Bradley	1 g6	21.7°N	0.4° E	76
Mons Delisle	5 g8, 6 g2, 8 b3	29.4°N	35.8°W	32
Mons Gruithuisen Delta	6 g3, 8 b4	36.1°N	39.6°W	27
Mons Gruithuisen Gamma	6 g3, 8 b5	36.6°N	40.7°W	20
Mons Hadley	1 f6	26.7°N	4.2° E	26
Mons Hansteen	11 e5	12.2°S	50.2°W	31
Mons Herodotus	7 d8, 8 e2	27.5°N	52.9°W	7
Mons Huygens	5 b6	19.9°N	2.9° W	42
Mons La Hire	5 f8, 6 e1	27.7°N	25.5°W	22
Mons Maraldi	3 f6, 4 f1	20.4°N	35.5°E	16
Mons Pico	2 h4, 6 b4	45.8°N	8.9° W	24
Mons Piton	2 g3, 6 a4	40.7°N	0.9° W	23
Mons Rümker	8 d5	40.8°N	58.3°W	73
Mons Vinogradov	5 g7, 7 a7	22.3°N	32.5°W	29
Mons Wolff	5 b5	17.0°N	6.7° W	33
Montanari	10 e5, 12 a4	45.8°S	20.8°W	77
Mont Blanc	2 f4	45.4°N	0.5° E	22
Montes Agricola	8 e2	29.1°N	54.2°W	161
Montes Alpes	2 f4, 2 g5, 6 a5	48.2°N	0.6° W	334
Montes Apenninus	1 g5, 5 a6	19.3°N	0.6° W	600
Montes Archimedes	1 h7, 5 b7, 6 b1	25.4°N	5.4° W	147
Montes Carpatus	5 f5	14.6°N	23.6°W	334
Montes Caucasus	1 f8, 2 e2	37.3°N	10.0°E	444
Montes Cordillera	11 h5	4.2° S	96.5°W	956
Montes Haemus	1 c4, 3 h5	16.9°N	12.6°E	385
Montes Harbinger	6 h1, 7 b8, 8 c2	26.9°N	41.3°W	93
Montes Jura	6 e5, 8 a6	47.5°N	36.1°W	421
Montes Pyrenaeus	15 e5	14.6°S	41.0°E	167
Montes Recti	6 d5	48.3°N	19.7°W	83
Montes Riphaeus	7 a1, 9 f6, 11 a5	7.5° S	27.6°W	190
Montes Rook	11 h3	9.3° S	94.7°W	651
Montes Spitzbergen	1 h8, 2 g2, 6 b3	34.4°N	5.2° W	59

(continued)

Feature	Map reference	Lat.	Lon.	Size (km)
Montes Taurus	3 e7, 4 e2	27.1°N	40.3°E	166
Montes Teneriffe	2 h4, 6 c5	47.9°N	13.2°W	112
Moretus	10 c2, 14 f2	70.6°S	6.0° W	114
Morley	15 b7	2.8° S	64.6°E	14
Mösting	5 b1, 9 a7, 13 h8	0.7° S	5.9° W	25
Mouchez	2 g8, 6 c8	78.4°N	26.8°W	83
Müller	13 f6	7.7° S	2.0° E	23
Murchison	1 g2	5.1° N	0.2° W	58
Mutus	14 c3	63.6°S	29.9°E	76
Naonobu	15 c7	4.7° S	57.9°E	32
Nasireddin	10 b6, 14 f6	41.1°S	0.0° E	50
Nasmyth	12 e4	50.5°S	56.4°W	78
Natasha	5 g6, 8b1	20.0°N	31.2°W	11
Naumann	8 e4	35.4°N	62.0°W	10
Neander	15 f1, 16 e7	31.4°S	39.9°E	49
Nearch	14 b3, 16 h3	58.6°S	39.0°E	73
Neison	2 d7	68.2°N	25.0°E	51
Neper	3 a2	8.7° N	84.6°E	144
Neumayer	14 b2, 16 h1	71.2°S	70.8°E	80
Newcomb	3 e7, 4 e3	29.8°N	43.7°E	40
Newton	10 c1	76.7°S	17.2°W	75
Nicolai	14 b6, 16 h5	42.5°S	25.9°E	41
Nicollet	9 c3	22.0°S	12.6°W	15
Nielsen	8 d3	31.8°N	51.8°W	10
Nobili	3 a1, 15 a7	0.1° N	75.9°E	42
Nöggerath	12 d5	48.9°S	45.8°W	34
Nonius	10 a7, 13 f1, 14 f7	34.9°S	3.7° E	71
Norman	9 f6, 11 a5	11.8°S	30.4°W	10
Oceanus Procellarum	7 e4, 8 e2, 11 c8	20.7°N	56.7°W	2592
Oenopides	6 h6, 8 c7	57.2°N	64.2°W	73
Oersted	4 f5	43.1°N	47.2°E	41
Oken	16 c5	43.8°S	76.1°E	79
Olbers	7 h4	7.3° N	76.1°W	73
Opelt	9 d4	16.4°S	17.7°W	49
Oppolzer	1 h1, 13 f7	1.5° S	0.5° W	41
Orontius	10 c6, 14 g6	40.8°S	4.5° W	93
Palisa	9 a5, 13 h6	9.5° S	7.3° W	33
Palitzsch	15 c2, 16 b8	28.0°S	64.4°E	42
Pallas	1 h2, 5 a3	5.4° N	1.7° W	46
Palmieri	11 d2	28.6°S	47.8°W	39
Palus Epidemiarum	9 f2, 10 g7, 12 b7	32.0°S	27.5°W	300
Palus Putredinis	1 g7, 2 f1, 5 a7, 6 a1	27.4°N	0.0° E	180
Palus Somni	3 d4	13.7°N	44.7°E	163
Parrot	13 f5	14.7°S	3.3° E	68
Parry	9 c6	7.9° S	15.8°W	45
Pascal	6 e8	74.4°N	70.5°W	108
Peary	2 f8	88.6°N	25.7°E	85
Peirce	3 c5	18.2°N	53.3°E	18
Peirescius	16 d5	46.5°S	67.8°E	62
Pentland	10 a2, 14e2	64.6°S	11.3°E	56
Petavius	15 d2, 16 c8	25.4°S	60.8°E	184
Petermann	2 c8	74.4°N	67.7°E	73
Peters	2 d7	68.1°N	29.4°E	16
Petit	3 b2	2.3° N	63.5°E	5

(continued)

Feature	Map reference	Lat.	Lon.	Size (km)
Phillips	15 c2, 16 b8	26.6°S	75.6°E	104
Philolaus	6 c8	72.1°N	32.9°W	70
Phocylides	12 e4	52.8°S	57.3°W	115
Piazzi	11 f1, 12 g7	36.2°S	68.0°W	100
Piazzi Smyth	2 g3, 6 a4	41.9°N	3.3° W	13
Picard	3 c4	14.5°N	54.7°E	22
Piccolomini	13 a1, 14 a8, 15 g2, 16 g8	29.7°S	32.2°E	88
Pickering	13 e7	2.9° S	7.0° E	15
Pictet	10 c5, 14 g5	43.6°S	7.6° W	60
Pilâtre	12 e3	60.2°S	86.7°W	64
Pingré	12 e3	58.6°S	74.0°W	88
Pitatus	9 c1, 10 e8	29.9°S	13.6°W	101
Pitiscus	14 b4, 16 h4	50.7°S	30.6°E	82
Plana	2 b3, 4 h5	42.2°N	28.2°E	43
Plato	2 h5, 6 b5	51.6°N	9.4° W	101
Playfair	13 e3	23.6°S	8.4° E	46
Plinius	1 c4, 3 g5	15.3°N	23.7°E	43
Plutarch	3 b6, 4 b2	24.1°N	79.0°E	70
Poisson	13 e1, 14 d8	30.4°S	10.5°E	41
Polybius	13 b3, 15 h3	22.4°S	25.6°E	41
Poncelet	6 d8	75.9°N	54.5°W	68
Pons	13 b2, 16 h8	25.5°S	21.6°E	40
Pontanus	13 d2, 14 d8	28.5°S	14.3°E	56
Pontécoulant	16 f3	58.8°S	66.1°E	91
Porter	10 d3	56.1°S	10.3°W	51
Posidonius	1 b7, 2 a2, 3 g8, 4 g3	31.9°N	30.0°E	101
Prinz	7 c8, 8 d2	25.5°N	44.1°W	46
Proclus	3 d5	16.0°N	46.8°E	27
Proctor	10 c5, 14 g5	46.4°S	5.2° W	46
Promontorium Agarum	3 b4	14.0°N	65.9°E	62
Promontorium Agassiz	2 f3	42.4°N	1.7° E	19
Promontorium Archerusia	1 c4, 3 h5	16.8°N	22.0°E	11
Promontorium Deville	2 f4	43.3°N	1.1° E	17
Promontorium Fresnel	1 f7, 2 e1	28.6°N	4.8° E	20
Promontorium Heraclides	6 f4, 8 a5	40.6°N	34.1°W	50
Promontorium Kelvin	9 g3, 11 b2	26.9°S	33.5°W	45
Promontorium Laplace	6 e5	46.8°N	25.5°W	50
Promontorium Lavinium*	3 d4	15.0°N	49.0°E	2
Promontorium Olivium*	3 d4	15.0°N	49.0°E	2
Promontorium Taenarium	9 b3, 13 h4	18.7°S	7.4° W	70
Protagoras	2 e6	56.0°N	7.3° E	22
Ptolemaeus	13 g6	9.1° S	1.9° W	158
Puiseux	9 h3, 11 c2	27.8°S	39.2°W	24
Purbach	9 a2, 13 g2, 14 g8	25.6°S	2.0° W	119
Pythagoras	6 f7, 8 a8	63.7°N	62.9°W	145
Pytheas	5 e6	20.6°N	20.6°W	20
Rabbi Levi	14 b7, 16 h7	34.8°S	23.5°E	82
Raman	7 e8, 8 e2	27.0°N	55.1°W	8
Ramsden	9 g2, 10 h7, 12 c7	33.0°S	31.9°W	25
Rayleigh	3 b7	29.1°N	89.3°E	114
Réaumur	1 g1, 13 f7	2.4° S	0.8° E	51
Regiomontanus	9a, 10 b8, 13 g2, 14 f8	28.3°S	0.8° W	110
Regnault	8 d7	54.1°N	87.8°W	45
Reichenbach	15 e2, 16 d8	30.4°S	48.0°E	63

(continued)

Feature	Map reference	Lat.	Lon.	Size (km)
Reimarus	16 e4	47.8°S	60.4°E	48
Reiner	7 e4	6.9° N	55.0°W	29
Reiner Gamma	7 f4	7.4° N	59.0°W	73
Reinhold	5 f2, 9 d8	3.2° N	22.9°W	43
Repsold	8 d6	51.3°N	78.4°W	109
Rhaeticus	1 f1, 13 e8	0.1° N	4.9° E	45
Rheita	16 e6	37.1°S	47.1°E	71
Riccioli	7 h2, 11 g8	2.9° S	74.4°W	156
Riccius	14 b7, 16 h6	37.0°S	26.4°E	72
Riemann	4 c5	39.3°N	87.1°E	119
Rima Ariadaeus	1 d2	6.5° N	13.4°E	247
Rima Birt	9 b3	21.4°S	9.3° W	54
Rima Gay-Lussac	5 e5	13.1°N	22.3°W	40
Rima Hesiodus	9 e2	30.5°S	21.9°W	251
Rima Hyginus	1 f3	7.6° N	6.8° E	204
Rimae Repsold	8 d6	50.7°N	80.5°W	152
Rimae Sirsalis	11 f5	15.4°S	61.2°W	405
Rimae Triesnecker	1 f2	5.0° N	4.8° E	200
Ritchey	13 e5	11.1°S	8.5° E	24
Ritter	1 c1, 3 h2, 13 b8	1.9° N	19.2°E	30
Robinson	6 f6, 8 a7	59.0°N	46.0°W	24
Rocca	11 g5	12.9°S	72.8°W	84
Römer	1 a6, 3 f7, 4 f2	25.4°N	36.4°E	41
Rosenberger	14 a4, 16 g3	55.6°S	43.1°E	92
Ross	1 c3, 3 h4	11.7°N	21.7°E	24
Rosse	15 f4	18.0°S	35.0°E	11
Rost	10 f3, 12 c3	56.4°S	33.9°W	47
Rothmann	13 b1, 14 a8, 15 h2, 16 g7	30.8°S	27.7°E	43
Rupes Altai	13 b2, 14 a8, 15 h2, 16 g8	24.3°S	23.2°E	545
Rupes Liebig	11 d3	25.1°S	45.9°W	145
Rupes Recta	9 b3, 13 h3	21.7°S	7.8° W	116
Russell	7 f8, 8 g2	26.5°N	75.5°W	103
Rutherfurd	10 d3, 14 g3	61.1°S	12.4°W	47
Sabine	1 c1, 3 h2, 13 b8	1.4° N	20.0°E	30
Sacrobosco	13 c3	23.8°S	16.6°E	98
Santbech	15 e3	21.0°S	44.0°E	62
Santos-Dumont	1 f7, 2 f1	27.8°N	4.8° E	8
Sarabhai	1 c6, 3 h7, 4 h2	24.7°N	21.0°E	7
Sasserides	10 d6, 14 h6	39.3°S	9.4° W	82
Saunder	13 d7	4.3° S	8.7° E	44
Saussure	10 c5, 14 g5	43.4°S	4.0° W	55
Scheele	7c1, 9 g7, 11 c6	9.5° S	37.9°W	5
Scheiner	10 e3, 12 b2	60.3°S	28.0°W	110
Schiaparelli	7 e7, 8 f1	23.4°N	58.8°W	24
Schickard	12 e5	44.4°S	55.0°W	208
Schiller	10 g4, 12 c4	51.8°S	39.8°W	179
Schlüter	7 h1, 11 h7	5.9° S	83.4°W	88
Schmidt	1 d1, 13 d8	0.9° N	18.7°E	11
Schomberger	14 d1	76.6°S	24.6°E	88
Schröter	5 c2, 9 a8, 13 h8	2.8° N	7.0° W	37
Schröter's Valley*	7 d8, 8 e2	26.2°N	51.6°W	185
Schubert	3 a1, 15 a8	2.9° N	81.0°E	51
Schumacher	4 e5	42.4°N	60.8°E	61
Schwabe	2 c7	65.1°N	45.4°E	25
Scoresby	2 f8	77.7°N	14.1°E	54
Scott	14 d1	82.5°S	48.8°E	102

(continued)

Feature	Map reference	Lat.	Lon.	Size (km)
Secchi	3 d2	2.4° N	43.5°E	22
Seeliger	1 g1, 13 f7	2.2° S	3.0° E	9
Segner	10 g3, 12 d3	59.0°S	48.7°W	68
Seleucus	7 f7, 8 g1	21.1°N	66.6°W	45
Seneca	3 b6, 4 b2	26.6°N	79.8°E	51
Shapley	3 c3	9.3° N	56.8°E	25
Sharp	6 g4, 8 b6	45.8°N	40.3°W	37
Sheepshanks	2 e6	59.3°N	17.0°E	24
Short	10 c1, 14 f1	74.6°S	7.7° W	68
Shuckburgh	4 e5	42.6°N	52.7°E	37
Silberschlag	1 e2	6.2° N	12.5°E	13
Simpelius	10 b1, 14 e2	72.5°S	14.7°E	70
Sinas	1 a2, 3 f3	8.9° N	31.6°E	12
Sinus Aestuum	5 c4	12.2°N	6.6° W	317
Sinus Amoris	3 e6, 4 e1	20.0°N	37.3°E	189
Sinus Asperitatis	13 a6, 15 h7	5.4° S	27.5°E	219
Sinus Concordiae	3 d4	11.0°N	42.5°E	159
Sinus Iridum	6 e4, 8 a5	45.0°N	31.7°W	249
Sinus Lunicus	1 g8, 2 f1	32.3°N	2.1° W	119
Sinus Medii	1 g1, 13 f8	1.6° N	1.0° E	287
Sinus Roris	6 g5, 8 c6	50.3°N	50.9°W	195
Sinus Successus	3 c1, 15 c8	1.1° N	58.5°E	127
Sirsalis	11 f5	12.5°S	60.5°W	40
Smithson	3 c2	2.4° N	53.6°E	6
Snellius	15 d2, 16 c8	29.4°S	55.6°E	86
Somerville	15 b6	8.4° S	65.0°E	17
Sömmering	5 c2, 9 a7, 13 h8	0.2° N	7.5° W	28
Sosigenes	1 d3	8.7° N	17.6°E	17
South	6 f6, 8 b7	57.8°N	51.0°W	100
Spallanzani	14 b5, 16 h4	46.4°S	24.7°E	33
Spörer	13 g7	4.4° S	1.8° W	27
Spurr	1 h7, 2 g1, 5 a7, 6 a1	27.9°N	1.3° W	13
Stadius	5 d4	10.5°N	13.8°W	68
Stagg's Horn Mountains*	9 b2, 13 h3	26.0°S	7.0° W	20
Steinheil	16 f4	48.7°S	46.6°E	65
Stevinus	15 e1, 16 d7	32.5°S	54.1°E	72
Stiborius	14 a7, 15 h1, 16 g7	34.5°S	32.0°E	44
Stöfler	10 a6, 14 e6	41.3°S	5.8° E	130
Stokes	8 d6	52.4°N	88.1°W	54
Strabo	2 b7, 4 h8	62.0°N	54.4°E	55
Straight Wall*	9 b3, 13 h3	21.7°S	7.8° W	116
Street	10 d5, 14 h5	46.6°S	10.8°W	59
Struve	7 g7, 8 h1	23.4°N	76.6°W	164
Suess	7 d4	4.3° N	47.7°W	8
Sulpicius Gallus	1 e5	19.6°N	11.7°E	11
Swift	3 c5, 4 c1	19.3°N	53.4°E	10
Sylvester	6 c8	82.6°N	80.9°W	59
Tacitus	13 c4	16.2°S	19.0°E	40
Tacquet	1 c4, 3 h5	16.6°N	19.2°E	6
Tannerus	14 c3	56.5°S	21.9°E	28
Taruntius	3 d2	5.5° N	46.5°E	57
Taylor	13 c6	5.4° S	16.6°E	39
Tebbutt	3 c3	9.4° N	53.4°E	35
Tempel	1 e2, 13 d8	3.7° N	11.9°E	43
Thales	2 b7, 4 h8	61.7°N	50.3°E	31
Theaetetus	1 f8, 2 e2	37.0°N	6.1° E	26

(continued)

Feature	Map reference	Lat.	Lon.	Size (km)
Thebit	9 a2, 13 g3	22.0°S	4.1° W	55
Theon Junior	13 c7	2.4° S	15.8°E	17
Theon Senior	1 d1, 13 c7	0.8° S	15.4°E	18
Theophilus	13 a5, 15 g6	11.5°S	26.3°E	99
Theophrastus	3 e5	17.5°N	39.1°E	8
Timaeus	2 f6, 6 a7	62.9°N	0.6° W	31
Timocharis	5 d7, 6 c1	26.7°N	13.1°W	34
Tisserand	3 d6, 4 d1	21.4°N	48.1°E	35
T. Mayer	5 g5, 7 a6	15.5°N	29.2°W	33
Tolansky	9 c5	9.5° S	16.0°W	13
Torricelli	3 g1, 15 g7	4.8° S	28.4°E	31
Toscanelli	7 c8, 8 d2	28.0°N	47.6°W	7
Townley	3 b2	3.5° N	63.1°E	17
Tralles	3 d7, 4 d2	28.3°N	52.8°E	42
Triesnecker	1 g2	4.2° N	3.6° E	25
Trouvelot	2 f5	49.3°N	5.8° E	9
Turner	5 d1, 9 b7	1.4° S	13.3°W	11
Tycho	10 d5, 14 h5	43.3°S	11.4°W	86
Ukert	1 g3	7.7° N	1.4° E	22
Ulugh Beigh	8 g3	32.6°N	81.9°W	53
Väisälä	7 c8, 8 d2	25.9°N	47.9°W	8
Vallis Alpes	2f5	49.2°N	3.6° E	155
Vallis Baade	12 g5	45.6°S	77.3°W	207
Vallis Bouvard	12 h7	38.4°S	83.6°W	288
Vallis Capella	15 f6	7.4° S	35.1°E	106
Vallis Inghirami	12 g6	44.0°S	72.6°W	145
Vallis Palitzsch	15 c2	26.2°S	64.6°E	111
Vallis Rheita	16 e6	42.5°S	51.7°E	509
Vallis Schröteri	7 d8, 8 e2	26.2°N	51.6°W	185
van Albada	3 b3	9.3° N	64.3°E	23
Van Biesbroeck	8 d3	28.8°N	45.6°W	9
Van Vleck	15 a7	1.7° S	78.2°E	33
Vasco da Gama	7 h6	13.8°N	83.9°W	94
Vega	16 d5	45.4°S	63.2°E	74
Vendelinus	15 c4	16.5°S	61.5°E	132
Very	1 c6, 3 g7, 4 g2	25.6°N	25.3°E	5
Vieta	11 f2, 12 f8	29.3°S	56.5°W	89
Vitello	9 g2, 11 c1, 12 d8	30.4°S	37.6°W	43
Vitruvius	1 a4, 3 f5	17.7°N	31.3°E	29
Vlacq	14 a4, 16 g3	53.5°S	38.7°E	91
Vogel	13 e4	15.1°S	5.8° E	26
Volta	8 d7	54.1°N	84.4°W	128
von Behring	15 b6	7.7° S	71.7°E	38
von Braun	8 f5	41.0°N	78.1°W	60
Voskresenskiy	7 g8, 8 h2	27.9°N	88.1°W	49
Wallace	5 c6	20.2°N	8.8° W	27
Wallach	1 a1, 3 f2	4.9° N	32.3°E	6
Walther	10 b7, 13 g1, 14 f7	33.4°S	0.6° E	134
Wargentin	12 e5	49.5°S	60.4°W	86
Watt	16 f4	49.6°S	48.4°E	68
Watts	3 d3	8.8° N	46.3°E	16
W. Bond	2 f7, 6 a7	65.2°N	3.9° E	156
Webb	3 b1, 15 b8	1.0° S	60.0°E	22
Weierstrass	15 a7	1.2° S	77.1°E	32
Weigel	10 g3, 12 c3	58.4°S	39.4°W	36
Weinek	15 f2, 16 f8	27.6°S	37.1°E	32

(continued)

Feature	Map reference	Lat.	Lon.	Size (km)
Weiss	9 d1, 10 f7, 12 a7	31.8°S	19.6°W	67
Werner	10 a8, 13 f2, 14 f8	28.1°S	3.2° E	71
Whewell	1 e2, 13 c8	4.1° N	13.7°E	13
Wichmann	7 c1, 9 g7, 11 c6	7.5° S	38.2°W	10
Wildt	3 a3	9.0° N	75.8°E	12
Wilhelm	10 e5, 12 a5	43.3°S	20.9°W	105
Wilkins	13 c1, 14 c8	29.4°S	19.6°E	60
Williams	2 a3, 4 g5	42.0°N	37.3°E	36
Wilson	10 f2, 12 b1	69.3°S	42.8°W	67
Winthrop	7 d1, 11 d5	10.8°S	44.5°W	17
Wöhler	14 a7, 16 g6	38.3°S	31.3°E	26
Wolf	9 d3	22.8°S	16.7°W	26
Wollaston	8 d3	30.6°N	47.0°W	10
Wrottesley	15 d3	23.9°S	56.7°E	58
Wurzelbauer	9 d1, 10 e7	34.1°S	16.1°W	87
Xenophanes	8 c7	57.5°N	82.0°W	118
Yakovkin	12 f4	54.5°S	78.9°W	36
Yangel'	1 f5	16.9°N	4.7° E	8
Yerkes	3 c4	14.6°N	51.7°E	36
Young	16 e6	41.5°S	51.0°E	71
Zach	10 b3, 14 e3	61.0°S	5.4° E	70
Zagut	13 c1, 14 b8, 16 h7	31.9°S	21.9°E	79
Zähringer	3 e2	5.5° N	40.2°E	11
Zeno	4 d6	45.1°N	72.9°E	65
Zinner	7 e8, 8 f2	26.6°N	58.8°W	5
Zöllner	13 b6	8.0° S	18.7°E	46
Zucchius	10 g2, 12 d2	61.4°S	50.6°W	63
Zupus	11 e4	17.2°S	52.4°W	35

Older names, no longer used officially are marked with an asterisk (*), whilst on the map they are shown enclosed in brackets

Printed in the United States
By Bookmasters

Printed in the United States
By Bookmasters